THE BIOLOGY OF RACIAL INTEGRATION

THE BIOLOGY OF RACIAL INTEGRATION

K. F. DYER

BRISTOL: SCIENTECHNICA (Publishers) LTD.
1974

ISBN 0 85608 007 1

PRINTED IN GREAT BRITAIN AT THE PITMAN PRESS, BATH

For Maureen, Lesley, Geraldine, and Joanne

PREFACE

THE purpose of this book is to provide a reasonably comprehensive and objective presentation of the biological aspects of race mixing in man, comprehensible to and useful to a wide range of students of biology, anthropology, and sociology. It is commonly and truthfully said that man's racial problems are among the most urgent facing him today. As our world becomes more crowded and its inhabitants become more restless it is imperative that they learn to live in harmony with one another. Living in harmony almost inevitably leads to interracial mating and marriage, and it is the consequences of such events, both long-term and short-term, that we are here examining.

Few people would, I think, quarrel with the aims of comprehensiveness or objectivity as stated in the first sentence. It is probably also true that few people have stopped to consider just what we might mean by these terms, particularly in the context in which they are used here.

Objectivity is often claimed as one of the hallmarks of the scientific approach and of scientific method—and so it is. On the other hand it is rarely associated with the choice of problem in the first place. At this stage of the proceedings a scientist's personal preferences, insight, and prejudices are equally, if not more important, than his objective appraisal of the available evidence. Furthermore, no matter how objective one might be during the experiments or analysis of the particular situation in hand, there comes a stage when some decision has to be made on the basis of the evidence available. Should one, for example, accept a 0·05 or a 0·01 level of probability as significant or is neither secure enough? Clearly the answer to such a question will lie in the nature of the problem. If it is a potentially important research area and some association is suggested, one will continue on the basis of possible further results. On the other hand if one were testing some new drug for possible use in pregnant women, say, one would require safety margins considerably greater than these.

Consider the present study. The very fact that it appears to me necessary to write a book on racial integration probably means that I believe some useful principles will emerge from such a study, and that I believe that some of the many generalizations made about racial mixing by experts and non-experts can be substantiated or disproved. Therefore however objectively I have researched the facts and analysed them, and however hard I have tried to remain objective in the interpretation and presentation of my findings, it is probable that my pre-existing biases, prejudices, and opinions have influenced me quite significantly in what follows. Note, too, that at many points subjective opinions are required by the facts and freely given; this or that is probably significant, possibly harmful, could be overcome, is of minor importance, and so on. Others, by a different choice of words, would subtly convey a different impression, unintentionally perhaps, but quite definitely.

The ideal of comprehensiveness is no less elusive. In the present context, for instance, just how far back in man's history need we go to understand his present racial distribution. Among the many restraints on going too far back, such as limitations of time and space and one man's abilities, there are others, again bound up with one's pre-existing beliefs. Races, one might believe, are of recent or of very ancient origin and one would treat the history of the species *Homo sapiens* and its predecessors accordingly. Furthermore how comprehensive should one be in one's coverage of different examples of racial mixing? A small number of possibly representative examples might suffice to illustrate one viewpoint; exhaustive coverage might be needed to do justice to another. The object and likely readership of the book will also influence one's decision here.

I would anticipate in this respect that a wide variety of people interested in the biology of man would find at least parts of the book useful and interesting. It seems clear to me that race mixing in man has been such an inevitable accompaniment of his biological and social evolution that one could subtitle the present work *A Historical Biology of Man*. There is therefore a good deal of material presented here which would be part of any orthodox course in genetics, human biology, and anthropology at the tertiary level. Graduates, teachers, and students of these disciplines should find the book useful and, I hope, thought provoking. The level of treatment is such that it should be understandable to a wide variety of biologists without thereby sacrificing accuracy. To facilitate this understanding I have included a number of appendices which contain information on some more specialized aspects of genetics, psychology, and anatomy, and also a fairly comprehensive glossary.

I have tried to provide a balanced and fairly thorough review of the literature of this controversial subject. I have assumed a knowledge of elementary Mendelian genetics and also various aspects of formal human genetics. The starting points for my treatment are such books as Stern's comprehensive *Principles of Human Genetics* and Cavalli-Sforza and Bodmer's more recent masterly work on *The Genetics of Human Populations*. Any work in this field must also lean heavily on J. C. Trevor's scholarly monograph *Race Crossing in Man* and on the thorough research report of Morton, Chung, and Mi on the *Genetics of Interracial Crosses in Hawaii*. I am very grateful to these authors and to their respective publishers W. H. Freeman, Cambridge University Press, and S. Karger for permission to reproduce a number of figures and tables from each work.

I have included material from a wide variety of research papers old and new to illustrate our current state of knowledge on racial biology in man. And rather than run the risk of being thought to select particular cases I have probably alienated and bored readers with a mass of factual examples. I have tried to say something useful about every biological system in man which might show racial differences and which might therefore be affected in some way by racial crossing. I have nothing to say directly in this book on such matters as equality, prejudice, social behaviour, or political implications, although I certainly believe that some of the facts presented in this book, and many more besides, should be part of the intellectual currency of those who do express opinions on such matters. I shall feel satisfied if this book helps to introduce a few more facts into arguments which have been distinguished in the past more by opinion and emotion than by relevant facts and orthodox logic.

It would be impossible for one person to claim expertise in all the areas of knowledge discussed in this book and there are many areas no doubt where the treatment is perhaps superficial and inaccurate. One can only claim as justification that it is of value for there to be the unity of treatment and outlook which one author brings.

I am grateful to many people for permission to reproduce diagrams and tabular material from the sources indicated. I also wish to thank Mrs. M. Pagonis, Mrs. J. Muller, Mrs. E. Azzopardi, and Miss J. Snell for typing various drafts of the manuscript; Miss M. Kaighin for drawing some of the figures; and Miss L. Horney for very considerable assistance in the final stages. I have also pestered many people here at Monash and elsewhere with requests for information, advice, translations, and other assistance and I have been very lucky and am very grateful that such help was invariably forthcoming.

Monash University
Clayton, Victoria
April, 1972

K. F. D.

CONTENTS

CHAPTER 1

THE BIOLOGY OF RACE

1.*1*. The Problem of Racial Definition

THERE are almost as many definitions of the word 'race' as there are biologists who have thought about the problem. But the term is generally used in Biology to indicate the differentiation of a species into distinct populations which are defined in terms of physical or genetic differences and which often, though not always, inhabit different geographical areas of the species range. The existence of two or more races in a species sometimes, but not always, indicates the incipient formation of new species; but races in different species cannot be regarded as having equivalent taxonomic or biological status, and mere enumeration and definition of the races in any species is not itself an adequate biological or genetical description of that species.

'Race', in fact, is a term which is very difficult to define biologically, and the fact that there is no general agreement on what is or is not a race in different species has led to a great deal of controversy over the interpretation of the data which have been collected on races and racial differentiation, particularly in man. In line with most modern taxonomic thinking, 'race' is a term descriptive of populations not individuals. It implies that a population, or group of populations, is sufficiently different from all others in the species to be separately recognized. Use of the term does not imply that there is one racial type from which all other individuals in the race deviate, nor does it imply that there is no overlap in physical or genetic terms between any races so defined, it simply means that the population itself differs sufficiently to be distinguished from others.

The controversies of racial biology have been intensified because the data on racial differences were often collected in the past by people who adopted this typological approach. That is to say, they did regard each and every race as being typified by one ideal individual and all members of that race as merely showing irrelevant deviations from this type. Often, therefore, the data that have been collected in the past are singularly uninformative and irrelevant in the light of today's understanding of racial and species variation. Nowhere is this problem more acute than in the case of man himself. Here, of course, in addition to the biological problems, there are the biases and passions engendered by racial prejudices, all serving to obscure the true nature of human races and the differences between them.

How, in fact, does the term 'race' fit into modern schemes of classification in Biology? The fundamental unit of biological classification is the species: a species being defined as a group of organisms which are reproductively isolated from all other organisms under natural conditions. Although there is usually a complete lack of interbreeding with all other species, hybrids with other species may at times be formed, but when they occur they are almost invariably sterile.

Within any one species there are often many morphologically distinct populations which inhabit different geographic regions, and the unity of the species is preserved by the fact that there are usually a series of intermediate types linking these populations and allowing the exchange of genes between all the constituent populations. The degree of difference between different populations of a species can range from almost complete identity to a distinctiveness almost of species level. A species divided into these different populations is said to be 'polytypic'. Differences certainly exist between one individual and another in all species but these are merely examples of polymorphism (*see* Chapter 2). To emphasize the point, it is the genetic and physical differences existing between populations as a whole, not the individuals in the populations, which define the existence of subspecies or races.

A large number of different formal definitions of what constitutes a race in man or other animals have been proposed. Thus, for instance, W. C. Boyd in 1950 suggested 'a race is a breeding population or interbreeding unit, which differs from all others in the frequency of one or more genes'. Th. Dobzhansky in 1962 proposed that 'races in sexually reproducing organisms are Mendelian populations which differ in the incidence of some genetic variants in their gene pools'. Curt Stern, in 1960, with specific reference to man, wrote that a race is 'a geographically or culturally more or less isolated division of mankind whose corporate genic content (gene pool) differs from that of all other similar isolates'. There is a rather different emphasis to the definition of Coon, Garn, and Birdsell (1950) that 'a race is a population which differs phenotypically from all others with which it has been compared'. Brace (1969) adopts a similar definition, a race being defined as 'a group of mankind, members of which can be identified by the possession of distinctive physical characteristics'. The first UNESCO statement on race published on 18 July, 1950, under the title 'The UNESCO statement by Experts on Race Problems' approached the problem as follows:

'A race, from the biological standpoint, may therefore be defined as one of the group of populations constituting the species *Homo sapiens*. These populations are capable of interbreeding with one another but, by virtue of the isolating barriers which in the past kept them more or less isolated, exhibit certain physical differences as a result of their somewhat different biological histories. These represent variations, as it were, on a common theme.

'In short the term "race" designates a group or population characterized by some concentrations, relative as to frequency and distribution, of hereditary particles (genes) or physical characters, which appear, fluctuate and often disappear in the course of time by reason of geographic and/or cultural isolation.'

The substance of this rather lengthy definition is conveniently summarized by Barnett (1961) who says 'A race may be defined as a group which shares in common a certain set of genes, and which become distinct from other groups as a result of geographical isolation'. The emphasis on geographical isolation in several of these definitions is one which we shall bear in mind throughout this book.

None of these definitions mentions either the degree of genetic difference or the magnitude of the morphological differences. It would indeed be impossible to include a reference to any magnitude of difference in any racial definition, since if it is correct that species do arise from races by a process of gradual divergence, then there will be every gradation from the most minor genetic differences to forms that are almost too distinct to be regarded as races of one species but not distinct enough quite to be counted as separate species. How then can we decide which of these definitions, if any, is most appropriate to man?

The first task is to ascertain whether the group under investigation is in fact an interbreeding Mendelian population, the members of which are more likely to breed with themselves than with members of a different population. In the case of most animals and plants, with the notable exception of those domesticated by man and most parasitic species, residence in the same neighbourhood may fairly be taken as a diagnostic criteria. Man himself, however, is divided into different linguistic, religious, economic, and other groups, people from which may be as unlikely to marry as those living thousands of miles apart. Conversely, throughout historical times, and no doubt before, man has taken part in many and virtually continuous large-scale invasions, migrations, and transportations across the whole face of the earth and therefore a truly interbreeding Mendelian population in man is probably rather rarely established and then for only fairly short periods.

The second necessity is to determine the nature and extent of the genetic difference between two Mendelian populations once they have been shown not to interbreed to any great extent. In the case of a number of animals and plants the genetic basis of many of the observed phenotypic differences can be ascertained by careful experiment and observation. In man himself, of course, it is not possible to experiment and it is exceedingly unfortunate for race studies that many of the phenotypic features which have been closely observed and used in the past by anthropologists and human biologists to distinguish human races, have a rather complex genetic basis. Many are also rather profoundly influenced by the very different environments in which man lives. Even many genetic characters, which are not at first sight directly correlated with climatic or other environmental variables, change gradually and insensibly in frequency from one population to another in a regular and unbroken fashion, and therefore hardly seem appropriate characters on which to base the division of man into supposedly isolated and distinct races. It is clear, then, that neither of these conditions for objective racial classification is adequately met in man. This in itself immediately throws doubt upon rigid racial classifications and indicates that analysis of racial hybridization needs particular care.

Although it has been stated that races may be incipient species in many organisms, this is not the case in man. Nor are any of the races in man usually regarded as true subspecies. According to the definition of Ernst Mayr (1963): 'A subspecies is an aggregate of local populations of a species and differing taxonomically from other populations of the species'. In fact the various races of man are less different from each other than are the subspecies of most polytypic species of animals (*see* Chapter 2, Section 5), further emphasizing their particular nature. Some authorities have argued, however, on the basis of the 75 per cent rule, that the major geographical races are in fact subspecies. This 75 per cent rule states that if 75 per cent or more of one population can be distinguished from the individuals of another, then the two represent separate subspecies (Amadon, 1949). On this criteria, fairly widely but perhaps uncritically used in Zoology, the races of man certainly are distinct. On the other hand such a rule attaches no weight or relative importance to the criteria by which the individuals are distinguished, which is the reason it is now received with less favour by most biologists.

Nevertheless the basis for race formation is the same in all sexually reproducing organisms and consists in the fact, as we shall see, that no two individuals are identical nor are any two local populations. As our later analysis will show, it is virtually a contradiction in terms to speak of 'pure races', because races are founded on genetic variability and what differs from race to race is the degree of variability and the particular sample of that variability. Further, as we have just mentioned, man's history also makes it impossible to impose a rigid racial classification on, or to find pure races among, his present population. The concept of race in man, as in other organisms, is a dynamic one. Races are always transient, destined either to become subspecies and then new species or, as in man, to merge, to die out, or to evolve; and any racial classification must recognize this. A discussion of racial hybridization or integration is merely a concentration upon a small part, and an oft-repeated part, of a continually fluctuating and recurring biological process which can be observed in all sexually reproducing species; a process as fundamental and inevitable as evolution itself.

The whole of the above discussion clearly implies great difficulty in adequately defining the term 'race' as it applies to man and also shows the imprecision which surrounds those definitions most commonly used. Indeed, some people believe that races of the kind that are known in other animals do not exist in man at all; a point of view which has been cogently argued by many geneticists and anthropologists (*see*, particularly, Montague, 1964, 1969). These scientists assert that races were assumed to exist in man simply because they existed in so many other animal species; having been assumed to exist they were soon recognized among the variability of the human species; recognition led to a formal description and finally they were fairly rigidly classified. Those who deny the existence of race in man argue that, although it is possible to divide a group of related organisms into distinct units, namely species, it is impossible to divide a single species into groups larger than the panmictic population. As F. B. Livingstone (1969c) puts it, 'If two characters are used it may still be possible [to define distinct human races] but there will be some problem populations which, if you are an anthropologist, will be labelled composite or mixed. As the number of characters increases it becomes more nearly impossible to determine what the actual races really are'. This is a powerful criticism of the use of the word 'race' in view of Dobzhansky's (1962) words that '. . . how many races should be recognized is a matter of convenience and hence of judgement', or Boyd's (1950) statement that 'whatever races we choose to distinguish will be almost entirely arbitrary, and their distribution will depend on the particular characteristic on which we choose to base them', or S. L. Washburn's (1963) opinion that 'since races are open systems which are intergrading, the number of races will depend on the purpose of the classification'.

Such uncertainty obviously makes it difficult to discuss racial integration in a rigorous or comprehensive manner, since what one author would regard as racial integration would not be recognized as such by another. However, it is in 'the purpose of the classification' that the key to this difficult problem can be found. For there are three different and mutually exclusive reasons for defining human races and only by fully understanding each one can we arrive at a meaning of the word 'race' adequate for the purpose of this book.

The anthropological concept of the word 'race' is the oldest. As we have seen, it treats races as fixed and immutable entities comprising individuals each of whom possesses a certain aggregate of characters which, individually and collectively, serve to distinguish them from the individuals in all other groups. The prime aim of such anthropologists is simply to classify mankind and there is little to be done with the races that are described other than to enumerate them. Although it is admitted (Garn, 1961) that 'certain human differences transcend race and are more meaningfully distributed with respect to climate or disease', the anthropologist's belief in the prime influence of race in human evolution is shown by continuing the quotation from Garn: 'A geographical race is a collection of populations whose similarities are due to a long-continued confinement which set geographical limits.' Such a definition and general approach, it can be argued, really takes for granted the one thing which remains to be proved—namely, that the races of man correspond with a reality which can be measured as *units* of evolution rather than as the incomplete and transitory results of evolutionary processes still occurring.

With the rise of population genetics and the greater understanding of population structure, our notions of the extent of genetic variability in populations have changed. These changes have clearly necessitated a different approach to the definition of race in biology. For instance, the differences in gene frequency from one population to another can now be explained in terms of mutation, genetic drift, gene flow, and the relative fitness of genes in different environments. Furthermore, the variability in the frequency of many genes can be plotted from place to place, as can the incidence of many physical characteristics of whose genetic control we are as yet still ignorant. It is found that the distribution of most genes is decidedly non-random and sometimes rather discontinuous. High frequencies of some genes and low of others tend to be associated together in certain populations and in a systematic way with many physical traits. These associations form the basis of the racial classifications of Boyd (1950) and Dobzhansky (1962) and others who fully accept, and indeed have heavily contributed to, current concepts of population genetics. In other words, racial classifications along these lines are attempting to describe something of the ongoing evolution of the human species and the genetic changes which are accompanying it.

Finally there is the sociological aspect and definition of the term 'race'. However much we may deny that races of man exist as biological entities, we can hardly deny that they are important realities in sociological or cultural terms. In South Africa and some of the states of the U.S.A., for instance, the existence and definition of different races is written into the law. According to social scientists a race is defined as being 'a group of mankind, members of which can be identified by the possession of distinctive physical characteristics'. The reality of this definition in socio-political terms impinges upon the human biologist and anthropologist in two ways. Firstly it presents him with a definition of race from which he finds it exceedingly difficult to escape and which determines much of the research he does in this field. For instance, in the United States all hybrids between Europeans and Negroes are classified socially, and in many areas legally, as Negroes, whatever the proportions of their contributing genes. Similarly hybrids between Indians and Europeans are regarded as Indians. Such a situation has a profound influence upon the genetic processes taking place in these races and also upon the statistics of racial integration and composition which are collected in the U.S.A. In South America the position is normally reversed, i.e., the term 'Negro' is reserved only for those expressing the complete range of Negro characteristics and there are a number of officially recognized intermediate or hybrid classes, Mullattoes, Mestizos, etc., in addition to the three major racial types Europeans, Negroes, and Indians. The sociological and biological consequences are very different in North and South America and yet the three races concerned, Europeans, Amerindians, and Negroes, are the same in each case.

The second interaction between sociology and biology is due to the fact that matings within human breeding populations are determined by the dictates of culture as much as, if not more than, by specifically physical features. The pressures on people of the same religion to marry, for instance, is one aspect of this, as is the likelihood of people of similar economic circumstances to marry. The complex marriage rules of the Indian caste system or those among the primitive Indian tribes of South America are but two examples where social rules have had a profound influence on the genetic structure of the peoples concerned.

These latter interactions of sociology and biology directly determine the subject-matter of this book. Racial integration is normally thought of primarily in social terms and the sociological and biological consequences of integration are usually discussed without there normally being any very clear identification of the biological differences between the races concerned. It is the purpose of this book to remedy these deficiencies. The full biological principles upon which racial differentiation is founded will be discussed and a classification of races will be used which recognizes these principles and is based on them. We shall also take into account isolation between populations of man for social or cultural reasons rather than strictly racial, if the subsequent breakdown of isolation is, or is likely to be, of significance either scientifically or sociologically. In this way we can undertake a comprehensive review of the true biological effects of racial integration, where the definition of race is of direct sociological relevance. Thus we can, as biologists, meaningfully comment on one of the most important social problems of today.

1.2. Some Racial Classifications

If the races of mankind were defined and described with a view to indicating their evolutionary or phylogenetic relationships, as is usually the case with higher orders of biological classification, we would now present the evolutionary evidence and proceed to deduce an appropriate classification. The evidence is just not available to enable us to do this, however, and in any case this is not the philosophy behind the majority of the racial classifications which have been devised hitherto. The attempt by Carleton Coon (1963) to investigate the origin of races in this way, using largely fossil and archaeological evidence, has not generally been accepted as successful (*see*, for example, the review by Birdsell, 1963). From our present viewpoint, too, we are assuming the existence of many races which have a very uncertain biological status. As mentioned in the preceding section, any population which has been labelled a human race or racial group, whether on biological or social grounds, must be so accepted by us. Our analysis must be defined in terms of genetic and biological isolation and we must therefore, for the purpose of this book, proceed backwards from what would be the normal method. In other words, having defined our races or interracial crosses we can then examine their history and their genetic differences. This is not meant necessarily to give approval for one classification rather than another, merely to indicate how many so-called races we have to describe, not define, in genetic terms.

The first UNESCO statement on race published in 1950 has this to say about racial classification:

> 'Human races can be and have been differently classified by different anthropologists, but at the present time most anthropologists agree in classifying the greater part of present-day mankind into three major divisions, as follows:
>
> The Mongoloid Division
> The Negroid Division
> The Caucasoid Division
>
> The biological processes which the classifier has here embalmed, as it were, are dynamic not static. These divisions were not the same in the past as they are at present and there is every reason to believe they will change in the future. Many sub-groups or ethnic groups within these divisions have been described. There is no general agreement upon their number, and in any event most ethnic groups have not yet been either studied or described by the physical anthropologists.'

1.2.a. *Genetic Racial Classifications*

In attempting to arrive at a satisfactory racial classification for use in the present book, most of the early classifications which do not make use of genetical information can be ignored. The earliest classification of Linnaeus, for instance, in which four subspecies were recognized—*H.s. europaeus*, *H.s. asiaticus*, *H.s. afer*, and *H.s. americanus*—is now no more than of historical interest. Nevertheless, when W. C. Boyd (1950) proposed a sixfold racial classification based almost entirely on blood group studies, it was little different from that of the earlier anthropologists, with the exception that he postulated the existence of a hypothetical Early European race which, he suggested, is represented today by their descendants the Basques.

His classification was as follows:

1. Early European
2. European (Caucasoid)
3. African (Negroid)
4. Asiatic (Mongoloid)
5. Amerindian Group
6. Australoid

This early genetic classification was based principally on the ABO, MN, and Rhesus blood groups, the PTC locus, and the secretor locus. With more extensive serological and biochemical information at his disposal, Boyd in 1964 produced a somewhat finer classification of thirteen races divided into African, European, American, and Pacific groups.

African Group.
1. Africans
2. Asians
3. Indo-Dravidians

European Group.
4. Early European
5. Lapps
6. North-west European
7. Eastern and Central European
8. Mediterranean

American Group.
9. Amerindians, including the Eskimos

Pacific Group.
10. Indonesian
11. Melanesian
12. Polynesian
13. Australian Aborigines

This scheme, while owing a great deal to modern genetics, clearly rests on geographical associations. It is interesting that Boyd's classification recognizes the Amerindians as a separate race, since these are usually recognized as being ultimately of Mongoloid origin. These later classifications, of which Boyd's was a pioneer, have therefore taken an important step forward in recognizing the transitory nature of many races' existence. The recognition by Boyd of a separate Pacific group of races is also interesting here, since these, and the Australoids in particular, present peculiar problems for racial classifiers. A. S. Weiner, for instance, in 1948 divided his Mongoloid group into an Asiatic subgroup and a Pacific Island and Australian subgroup. Others, as we shall see, have regarded the Pacific Islanders as belonging to the Negroid group and the Australian as belonging to the Caucasoid group. The other important area in which many people would disagree with Boyd is his putting Asians and Africans, and for that matter the Indo-Dravidians, into one major group. The relationship between these races appears on many counts to be rather distant.

The position we have arrived at and the problems which confront us raise the same fundamental questions regarding the equivalence of the various races defined by different authors. To return to the modern evolutionary/genetical/geographical definitions of race, we can recognize that, if they are defined in terms of population groups on the one hand or small local populations on the other, we will end up with small or large numbers of defined races respectively. And this is perfectly valid since, as we have mentioned, the magnitude of genetic or morphological difference is largely immaterial to racial definition. Perhaps the most appropriate way of dealing with this situation and recognizing the enormous diversity in the human population is to define different types of human race.

Rensch in 1929 suggested the term '*geographical race*' to describe the broad geographically delimited groups of populations, and the term '*local race*' to refer to the race-populations themselves. Dobzhansky (1962) suggested the term '*microgeographical race*' to describe further the very small statistically distinct populations which cannot be delimited as circumscribed breeding populations. As Garn (1961) however suggests, it is more appropriate to use the term '*micro-race*' for this latter group, partly as being more convenient, but more importantly because it avoids any possible confusion with *geographical race*.

1.2.*a*.i. *Geographical Races*

Probably the most authoritative recent classification of major human races is that of S. M. Garn (1971). He recognizes nine major geographical races whose distributions are shown in *Fig.* 1.1 and whose major genetic and morphological characteristics are given below.

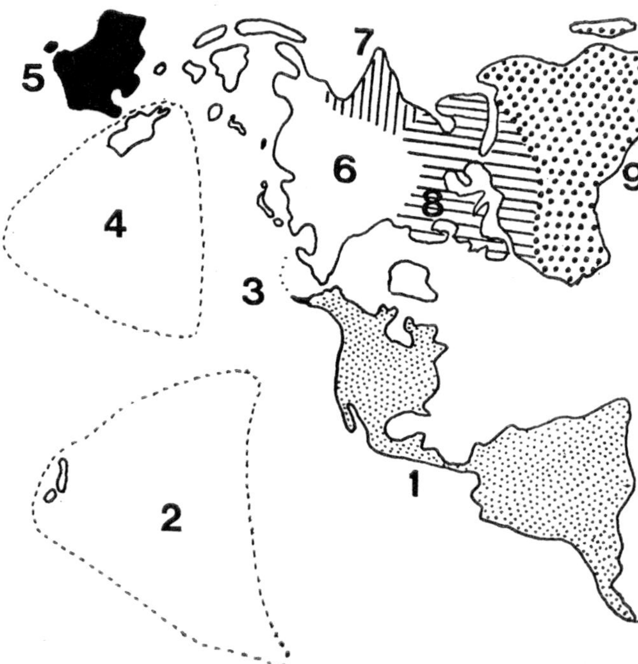

Fig. 1.1.—Polar projection map of the world showing the approximate limits of the geographical races described in the text. The races are: (1) Amerindian; (2) Polynesian; (3) Micronesian; (4) Melanesian; (5) Australian; (6) Asiatic; (7) Indian; (8) European; (9) African. (*From* Garn, 1971.)

1. *The Amerindian Geographical Race*

This comprises the pre-Columbian population of the Americas. Although undoubtedly of Mongoloid origin, they are distinguished from modern Asiatics by very low levels of the N gene in the MNS blood group system, by the total lack of the B blood group, and by the complete absence of the Rhesus-negative gene. They are similar to Asiatics in possessing β—AIB high excretors, the Diego-positive gene (Di^a), and at least one rare haemoglobin variant. They share with many Asiatics a high frequency of shovel-shaped incisor teeth, a low incidence of Carabelli's molar cusp, generally straight hair and rather sparse body hair, and general similarity of facial morphology.

2. *The Polynesian Geographical Race*

This race occupies a large part of the Pacific Ocean extending from New Zealand to Hawaii and Easter Island. They are low in blood group B, high in N, and only moderately high in the Duffy factor (Fy^a) which in other parts of the Pacific reaches 100 per cent. They have no Rhesus-negative and no A_2 alleles, thus differentiating them from other races, although as a race they are quite variable or polytypic.

3. *The Micronesian Geographical Race*

A large number of small islands in the central Pacific are the home of this race. Although similar physically to the Polynesians they differ in having rather higher frequency of blood group B and are almost completely Duffy positive (Fy^a) and Diego negative. They are also quite distinct from Asian populations.

4. *The Melanesian–Papuan Geographical Race*

Inhabiting New Guinea, New Britain, the Solomon Islands, and neighbouring groups, this geographical race contains hundreds of isolated linguistic and cultural groups. Although, particularly in the south where they are close, they share many elements of similarity with the Australian Aborigine, they are quite distinct genetically in possessing blood group B, S (in the MNS series), the thalassæmia gene, and in lower areas the G6PD deficiency. A common origin with the Australians is likely, with subsequent marked divergence due to selection and genetic drift in the very varied environments they now inhabit.

5. *The Australian Geographical Race*

This race is typically Pacific for the MNS blood group, having a considerable excess of N over M. Similarly, the Duffy (Fy^a) gene is very common, but they differ from other Pacific peoples in the virtual absence of the B gene, S in the MNS series, and having no G6PD deficiency and few of the atypical haptoglobins. Together these serological and biochemical peculiarities distinguish them from Melanesians and Polynesians as well as precluding any recent contact with Africa or mainland Asia. The fact that many have brown or even red hair and that many of the men show typical male pattern balding and leg (calf) balding are sometimes taken as indicating Caucasoid relationships, albeit rather distantly.

6. *The Asiatic Geographical Race*

A very variable race inhabiting Tibet, Mongolia, Northern Asia, and the islands of Japan, Taiwan, Phillipines, Borneo, and Indonesia. Serologically the race is characterized by rather high levels of the B gene, up to 40 per cent in some areas, but little, if any, of the gene A_2 or Rhesus-negative. The race possesses a number of characteristic morphological features such as 'inner' eye-folds, coarse hair, broad molars, shallow orbits, multiple suture bones, and a high frequency of shovelling of the incisors. The whole race is very variable, both in its genetics and morphology, since it inhabits a very wide range of varied environments.

7. *The Indian Geographical Race*

This race is characterized serologically by high frequencies of B, up to 35 per cent, but rather low frequencies of Rhesus-negative. M tends to be more frequent than N. Each of these characteristics differentiates them from Europe, although it is generally thought that the European and Indian races had a common origin. It is possible that particular diseases such as small-pox may account for the high frequency of blood group B in some areas, and the rather different climatic and nutritional selection of régimes on the Indian subcontinent have contributed to the observed divergence between the two races.

8. *The European Geographical Race*

The Europeans are characterized by a relatively high frequency of the Rhesus-negative gene and also of A_2 in the ABO system. The race occupies North Africa and the Middle East as well as Europe, and throughout displays hirsuitism, male pattern balding, and delayed tooth eruption. Pigmentation is very variable, as indeed are many other characters in this rapidly expanding race.

9. *The African Geographical Race*

A collection of peoples indigenous to Africa south of the Sahara who are characterized serologically by having up to 70 per cent of the R_0 Rhesus subtype, the Duffy variant *Fy*, and a number of distinctive haptoglobin and gamma-globulin variants (*see* Chapter 3). A variety of adaptations to malaria are known including the G6PD deficiency and abnormal haemoglobins (*see* Chapter 2 and Chapter 5 section 5.*1*). Other characters such as pigmentation, helical hair, and lip eversion are very variable, but in parts of Africa reach their maximum expression among men.

We have briefly considered Garn's major geographical races and their distinguishing characteristics. To each of them we could add the corollary that in many parts of their domain they are experiencing an influx of genes from other races. The Europeans have undergone an extraordinarily rapid population expansion since about 1500 and have spread around the world. Wherever they have gone they have mated with and married the native inhabitants. In Africa large numbers of Asians, as well as Europeans, have settled both in the recent and not so recent past and in America of course, either north or south, there are three major races present in significant numbers. In many parts of the Pacific also, Hawaii for instance, members of each of these major races are contributing together with the indigenous races to the gene pool of future generations.

Over the whole globe the major 'geographical races' are becoming more mixed up and will, in time no doubt, eventually lose or at least change their identity (*see* Chapter 8 for some predictions as to when this might come about).

1.2.*a*.ii. *Local Races*

If such processes are occurring for the major divisions of mankind we can be sure that they are happening at the level of the 'local races', and indeed the classifications of Coon and others (1950), Garn (1961, 1971), or Dobzhansky (1962) explicitly recognize this in some of the races they propose. We shall list here the thirty-four races named by Dobzhansky, merely as one example of several possible classifications. He himself constructed this system by amalgamating the racial classifications proposed by Coon and others (1950) and Garn (1961) and the nature of just what is a local race emerges fairly clearly from this list. They are grouped here (although not by Dobzhansky) into broadly related stocks but it should be mentioned that not everyone would agree with this grouping.

Caucasoid Races	
1. North-west European	Northern and western Europe
2. North-east European	Eastern Europe and Siberia
3. Alpine	Central Europe eastwards to the Black Sea
4. Mediterranean	All around the Mediterranean
5. Hindu	Indian subcontinent
6. Turkic	Turkestan, western China
7. Lapp	Extreme northern Europe
8. Ainu	Aboriginal population of northern Japan
Mongoloid Races	
9. Tibetan	Tibet
10. North Chinese	Northern and central China
11. Eskimo	Extreme North America
12. Southeast Asiatic	South China, Indo-China, Malaya, Indonesia
13. Classic Mongoloid	Siberia, Mongolia, Korea, Japan
14. North American Indian	Indigenous peoples of Canada and the U.S.A.
15. Central American Indian	Southern U.S.A. through to Bolivia
16. South American Indian	Indigenous peoples of South America
17. Fuegian	Southern South America

Negroid Peoples	
18. East African	Northern East Africa
19. Sudanese	Sudan
20. Forest Negro	West Africa and Congo
21. Bantu	South Africa and part of East Africa
22. Bushman and Hottentot	Aboriginal inhabitants of South Africa
23. African Pygmy	Equatorial Africa
Pacific Peoples	
24. Dravidian	Aboriginal populations of southern India and Ceylon
25. Negrito	Small statured people of the western Pacific Islands
26. Melanesian-Papuan	New Guinea to Fiji
27. Murrayian	Aboriginal population of southeastern Australia
28. Carpentarian	Aboriginal population of northern and central Australia
29. Micronesian	Islands of the western Pacific
30. Polynesian	Islands of the central and eastern Pacific
Emerging Peoples	
31. Neo-Hawaiian	The mixed population of Hawaii
32. Ladino	The mixture of Mongoloid, Caucasoid and Negroid in Central and South America
33. North American Coloured	The Negroid/Caucasoid hybrids of U.S.A.
34. South African Coloured	The Negroid/Caucasoid/Asian population of South Africa

Obviously within each of these five groupings there are areas of controversy. Have the emerging peoples really attained the status of races, for instance, or can we really divide densely populated Europe up into four different races? It seems unlikely that such questions could be decided merely by the collection of more data. In the last resort they reflect the opinion of the investigator. On the other hand in the case of other suggestions by Dobzhansky, the majority of research workers in the field would today disagree with him. For instance few can now see any justification for recognizing two different races of Australian Aborigines, and the racial status of the Pacific Negritos is also being increasingly questioned.

1.2.*a*.iii. *Micro-races*

The number of 'micro-races' in the world must run into many thousands and because of their very nature we would not expect any agreement between different authorities as to what constitutes a race and what does not. The level at which we are now looking is that of the old city-state of the Romans or the Athenians perhaps, or the American Indian Tribe or one of the famous three races of Gaul which Caesar recognized. It represents probably the tribal structure which man had during his hunting and early agricultural days. True the genetic insularity of such small groups of people has long vanished, even supposing it ever really existed, and yet sufficient genetic isolation exists between these micro-races for us still to be able to recognize their reality.

Some of the most important work in human genetics at present under way is being done on micro-races in various parts of the world. Prominent among this work is the very large project on South American Indian tribes being led by J. V. Neel and co-workers (*see*, for example, Neel and Salzano, 1966; Fitch and Neel, 1969) and the studies on African Pygmies being undertaken by Cavalli-Sforza, Bodmer, and their co-workers (*see*, for example, Bodmer and Bodmer, 1970). Other important studies include those on some Japanese islands (Nei and Imaizumi, 1966a, b) and on Aboriginal tribes (Robson and Parsons, 1967). The example we shall choose to illustrate the idea of micro-races is that of the indigenous population of Wales, which, because of its extreme western position and the nature of its terrain, is one of the few regions of Europe where the descendants of early pre-Roman stocks remain. Wales differs from all other countries of Western Europe in that about 95 per cent of present-day Welsh surnames are derived from baptismal names which came into general use in the late eighteenth and early nineteenth centuries. There are very few local names, trade names, or nicknames, and less than 50 names altogether are in common use in Wales. Ashley and Davies (1966) have shown that it is in fact possible to use the surname as a genetic marker in this situation. *Fig.* 1.2 from Watkin (1956, 1965) shows the gene frequency maps for the A and B genes in Wales and its neighbouring counties, and it is clear that there is considerable local differentiation.

We might indicate the following significant features: (1) The B gene is significantly higher in Western than in Eastern Wales. It is highest in isolated moorland areas, usually well over 10 per cent, and on the Black Mountain of Carmarthenshire it approaches 25 per cent. This high frequency, it is suggested, corresponds to the survival of early human settlers in this part of Wales. (2) There is a marked difference between North and South Wales in O gene frequency and this change is accompanied by differences in A gene frequency between North and South. Watkin (1956) points out that the North Welsh type of ABO frequencies are found in islands or coastal areas from Iceland and Ireland, via coastal France to the Mediterranean and White North Africa, as far east as the Aegean and Asia Minor. Watkin speculates whether these represent remnants of a related seafaring non-Celtic race. Berry (1969b) also has some observations on these peoples based on minor skeletal differences which we discuss in Chapter 2. (3) The extreme south of Pembrokeshire differs from other parts of Wales, particularly in a high A value (up to 50 per cent in the South East of the county) and Watkin (1960), by a number of detailed local comparisons, provides confirmatory evidence for the Viking settlement that historians have long suspected took place here. The fact that parts of

Norway, Sweden, and Denmark have similar A blood group frequencies to this part of the country is a strong pointer to the correctness of this conclusion.

We see then, in this rather small area, evidence for the existence of three micro-races, each of which has become somewhat modified by contact with the remainder of the rather heterogeneous Welsh population. What of this remaining Welsh population? Is it too a micro-race? Can it be distinguished from

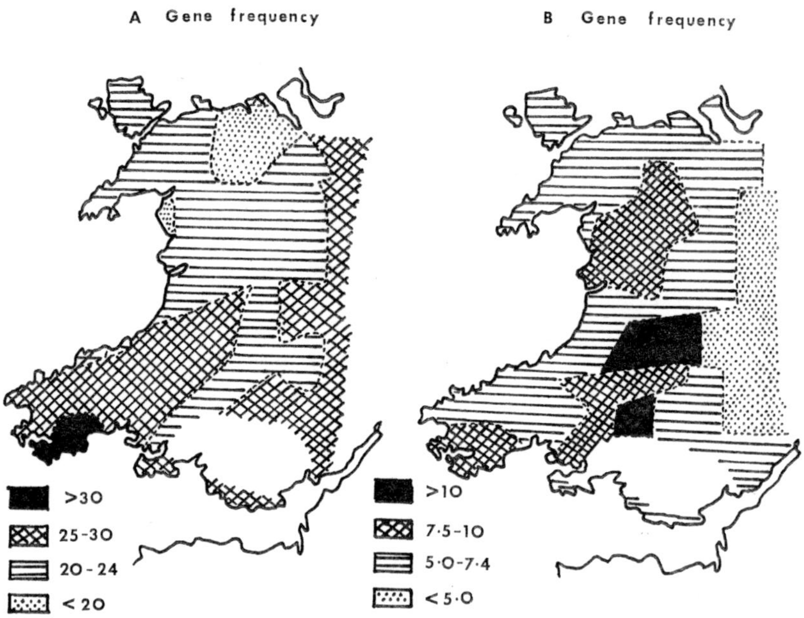

Fig. 1.2.—The frequencies of the genes for A and for B blood types among the inhabitants of different parts of Wales. (*From* Watkin, 1956.)

non-Welsh immigrants and neighbours? On this last point D. J. B. Ashley has conducted a number of surveys (Ashley 1966, 1967a, b, c, 1969; Ashley and Davies, 1966) which shows conclusively that for a number of characters, coronary artery disease, diabetes, vascular disease of the central nervous system, and gastric cancer, the incidence within the borders of Wales is different in Welsh and non-Welsh men and women. The fact that there is quite a high degree of assortative mating among the Welsh, coupled with this other evidence, suggests that the genetic isolation of the Welsh in Wales is quite marked; in other words they achieve the status of a micro-race. In this one small country of less than three million people it appears we can recognize perhaps four micro-races which are differentiated from one another to a greater or lesser extent. No doubt every local race in the world, let alone every major geographical race, is a similar patchwork and mosaic of minor micro-races. The idea of racial purity anywhere in the world thus dissolves into a meaningless and impossible abstraction.

1.2.b. *Non-genetic Racial Classifications*

By considering racial classification at these three levels we achieve a clearer idea of what is the modern, population genetics concept of race. We also make use of a classification of races which would be agreed to by a majority of anthropologists. For the sake of completeness, however, we will present a classification

1. *Negroid*
   - African Negroes
     - *a.* True Negro
     - *b.* Half Hamites
     - *c.* Forest Negroes
     - *d.* Bantu-speaking Negroids
     - *e.* Nilotic Negroes
     - *f.* Bushmen
     - *g.* Hottentot
   - Oceanic Negroes
     - *a.* Papuans
     - *b.* Melanesians
   - Asiatic Pygmies
     - *a.* Andamenese
     - *b.* Semang
     - *c.* Philippine Negritos
   - Oceanic Pygmies
     - *a.* New Guinea Pygmies

2. *Caucasoid*
   - *a.* Basic Mediterranean
   - *b.* Atlanto-Mediterranean
   - *c.* Irano-Afghan Mediterranean
   - *d.* Nordic
   - *e.* East Baltic
   - *f.* Lapps
   - *g.* Alpine
   - *h.* Dinaric
   - *i.* Armenoid
   - *j.* Hamites
   - *k.* Indo-Dravidian
   - *l.* Polynesians
   - Sub-Division Australoid or Archaic Caucasoid
     - *a.* Australian
     - *b.* Veddah
     - *c.* Pre-Dravidian
     - *d.* Ainu

3. *Mongoloid*
   - Classical Mongoloids
     - *a.* An undetermined number of ethnic groups in the older populations of Tibet, Mongolia, China, Korea, Japan, and Siberia
   - Arctic Mongoloids
     - *a.* Eskimo
     - *b.* Evenki or true Tungos
     - *c.* Kamchadales
     - *d.* Samoyedes
   - Palaeoasiatics
   - Neo-asiatics
   - Amer-Indians
   - Indo-Malay
     - *a.* Indonesian
     - *b.* Malay

presented by Montague (1960) which is probably fundamentally rooted in pregenetic classifications even though Montague himself fully accepts the full implications of modern genetic research in physical anthropology. Secondly, we will discuss the classification devised by Coon (1963, 1966) derived from and illustrative of his unique and probably not generally accepted ideas on racial origins. Montague's classification is set out on the previous pages.

The major features of Montague's classification are clearly not at variance to any major extent with others that we have discussed, and in many respects it is merely a more detailed division into fifty or more races. The classification which C. S. Coon adopts, however, in his book *The Origin of Races*, is rather different from any of these. He constructs a fivefold classification on basically

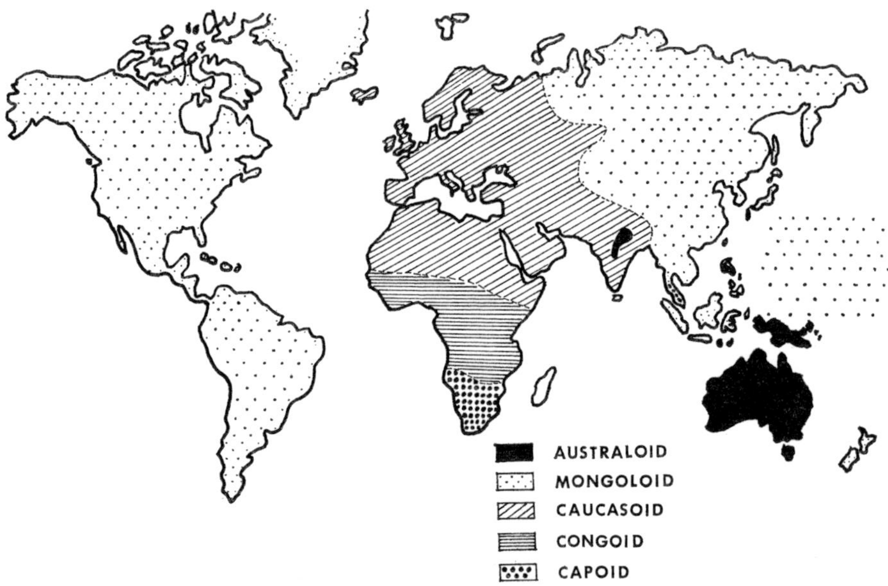

Fig. 1.3.—The five subspecies of *Homo sapiens* as they existed in 1492 according to Carleton S. Coon. (*From* Coon, 1963.)

geographical lines. His five races, which he formally recognizes as subspecies, are Caucasoid, Mongoloid, Australoid, Congoid, and Capoid. The novel features of this classification are the recognition of the Australoids, not restricted just to Australia incidentally, as a group equal in status to the Caucasoids and Mongoloids, and his splitting of the race, which most people recognize as Negroid, into two races which, he claims, are no more closely related than are the others. *Fig.* 1.3 shows what Coon believes to have been the distribution of his five subspecies prior to A.D. 1500 when large-scale recent migrations began. Among the Caucasoids Coon includes the Middle Eastern Whites from Morocco to West Pakistan, most of the inhabitants of India, and the Ainu of

Japan. Within the Mongoloids he includes most of the East Asiatics, Indonesians, Polynesians, Micronesians, Amerindians, and Eskimos. The Australoids include, as well as the Australian Aborigines, Melanesians, Papuans, some tribal Indians, and the various Negritos of South Asia and Oceania. The Congoids comprise the Negroes and Pygmies of Africa, and the Capoids include the Bushmen, Hottentots, and other relics of African tribes like the Sandawe of Tanganyika.

It is not the prime purpose of this book to assess the validity or otherwise of proposed racial classifications. Nevertheless the recognition of the racial status of individuals involved in interpopulation crosses must be one of our main concerns, and to this end we will in Chapter 3, after considering the nature and amount of genetic variation present within and between human races, return to consider racial origins somewhat more closely, not so much to devise a formal classification but to determine, as closely as we are able, the degree of genetic relationship between as large a number of the major races as possible. This is absolutely essential if we are to assess in any meaningful way the genetic outcome of crosses between them.

CHAPTER 2

THE NATURE AND DISTRIBUTION OF HUMAN GENETIC VARIATION

2.1. The Universality of Genetic Variation

We have seen that the fundamental criteria for the biological definition of races are essentially genetic. We must therefore examine the nature and the amount of genetic variation which is commonly present in human populations fully to understand the differences which exist between the races of man. Although this variation can, for convenience, be considered to be of three different kinds—genic or simple discontinuous variation, biometric or continuous variation, and epigenetic or quasi-continuous—each is really just a different biological expression of similar underlying genetic phenomena.

It is one of the cornerstones of the modern theory of evolution that genetic variation is essential to the future existence and evolution of any species. Indeed, it is an essential axiom of modern biology in general that in order to exist any organism must be adapted to its environment, i.e., that the individuals in the populations of which the species is composed must possess a variety of genotypes permitting them to survive under the changing variety of existing conditions. It is one of the greatest achievements of Charles Darwin and Alfred Wallace, in their joint discovery and presentation of the theory of evolution by natural selection, that they could appreciate the significance of intraspecific variability in evolutionary processes, rather than ignore it or dismiss it as most biologists of their time normally did (Dyer, 1970b).

This genetic variation exists and is preserved in many different ways in different organisms. Micro-organisms, for instance, have enormously large populations and very rapid rates of reproduction. Therefore, even though mutation rates are normally very low, they probably have enough newly arisen mutations at any given time to adapt to any new situation through the rapid reproduction of those most suited to the new environment. Multicellular sexually reproducing organisms normally have smaller populations with lower reproductive rates, and their genetic flexibility is usually achieved through the recombination of stored or concealed genetic variations already present in the population but unexpressed in the phenotypes of most of the individuals. Of course a casual observation of any human population shows that there is also a great deal of expressed variation, hair colour, eye colour, skin colour, height, weight, and so on, even within any of the races of man as normally defined. Nor is this situation unique to man, as routine observation of nearly all known species will easily demonstrate.

Our task in this section is to survey this genetic variability, both concealed and expressed, to discover how much genetic variation is usually present in the human species, to review the processes normally at work originating and maintaining this variation, and to describe the nature of that variation which particularly characterizes the different races of man. Each of these is an essential prerequisite to understanding the genetical and biological processes of racial integration.

2.1.a. *Genic Variation and Polymorphism*

We cannot carry out experiments on man, but we are able to collect an enormous amount of genetic and other data because of his very large population size and because of the historical records available to us. By so doing we are going some way to overcoming the disadvantage of man as a genetical subject. The Roman Catholic Church's insistence that all cousin marriages must be granted special permission, for instance, means that records of all such marriages exist in catholic countries and their reproductive outcome can be compared with that of marriages between non-relatives. Similarly, the large-scale work of the Atomic Bomb Casualty Commission in Japan during the post-war years resulted in an enormous amount of fundamental genetic information being gathered. And with the increasing emphasis throughout the world on social and preventive medicine, the rate of accumulation of knowledge of human genetics is increasing all the time (for an idea of just how rapidly *see* Dyer, 1968c; Lush, 1970; Cavalli-Sforza and Bodmer, 1971).

The study of marriages between relatives, consanguineous marriages, is particularly important in discovering hidden genetic variation. Closely related individuals have a higher chance of carrying the same alleles at any genetic locus than less closely related individuals, because they may be inheriting them from the same common ancestor. The offspring from consanguineous marriages are therefore more frequently homozygous for, i.e., carry two doses of, various alleles than are the offspring from non-consanguineous marriages.

Table 2.1.—MORTALITIES AMONG OFFSPRING OF MARRIAGES BETWEEN UNRELATED PERSONS AND BETWEEN FIRST COUSINS

DEATHS	PERIOD	UNRELATED		FIRST COUSINS	
		Number	Mortality (per cent)	Number	Mortality (per cent)
United States					
Young children	Before 1858	837	16·0	2778	22·9
Before age 20 years	18th–19th centuries	3,184	11·6	672	16·8
Children 0–10 years	1920–1956	167	2·4	209	8·1
*France**					
Stillbirth neonatal	1919–*c.* 1950	2,745	3·9	743	9·3
Infantile, juvenile, and later	1919–*c.* 1950	515	9·6	674	14·3
*Japan**					
Stillbirth	1948–1954	63,145	1·5	2798	1·6
Infants to 1 month	1948–1954	63,145	1·8	2798	2·8
Infants during first 9 months	1948–1954	17,331	4·7	822	6·6
Between 1 and 8 years	1948–1954	544	1·5	326	4·6

* Deaths of children without visible major malformations.

Table 2.2.—FREQUENCIES OF DISEASES AND OF PHYSICAL AND MENTAL DEFECTS AMONG OFFSPRING OF UNRELATED AND OF FIRST COUSIN MARRIAGES

POPULATION	UNRELATED		FIRST COUSINS	
	Number	Affected (per cent)	Number	Affected (per cent)
France	833	3·50	144	12·80
Japan	63,796	1·02	2846	1·69
Sweden	165	4·00	218	16·00
U.S.A.	163	9·82	192	16·15

An estimate of the number of recessive lethal, detrimental, or visible alleles carried in any population can therefore be ascertained by comparing the reproductive performance of these two sorts of marriages. Some of the results obtained using this approach are shown in *Tables* 2.1 and 2.2. Now in the data from France shown in *Table* 2.1 there is about a 10 per cent excess of total mortality (13·5 per cent as against 23·6 per cent) among the progeny of cousin marriages over that of the outbred offspring, and we know that this is due to the fact that $\frac{1}{16}$ of the genes that normally would be heterozygous become homozygous in the offspring of cousin marriages. If all the offspring were completely homozygous, i.e., if individuals could be formed by simply doubling each gamete rather than the uniting of two unlike cells, there would obviously be approximately 16 times as much mortality. There would, therefore, be an increase in mortality approaching 200 per cent (10·1 $\times$ 16). This in turn leads to the estimate that the average gamete in this French population carries about 2 lethal equivalents, where a lethal equivalent may be a single lethal gene or more than one detrimental gene. Since two gametes are needed to produce one zygote, it appears that, on average, each individual carries up to 4 lethal equivalents which are present in the heterozygous condition. Data from the other populations lead to results which are broadly similar to this.

There are, of course, further lethal equivalents causing early abortions and, as *Table* 2.2 shows, there are many non-lethal equivalents per person which cause anatomical and physiological abnormalities. All in all it seems from studies such as these that every person carries many more than 10 lethals or detrimentals distributed throughout his genome. Another interesting facet of the universality of human genetic variation demonstrated by this approach is that many of the individual genes uncovered vary somewhat in frequency from one population to another. Although the reason for variation such as this is largely unknown, it is an interesting observation on racial differentiation and has certain important consequences in subsequent racial amalgamations.

Although this sort of investigation shows that every individual carries a great deal of concealed variability, and therefore the populations which they comprise a great deal more, it makes no distinction as to the overall nature of genetic variation. That is to say, it gives no indication as to what proportion of the variants is being maintained in the population at a relatively low level by recurrent mutation or what proportion is in the process of being eliminated by natural selection, having been at a fairly high frequency at earlier generations, or finally what proportion is being selectively maintained as a permanent feature of the gene pool of man. It is an exceedingly difficult task to distinguish between the latter two possibilities, but the fact that the Rhesus and ABO blood groups are known in many of the primates, and the ability to taste phenylthiourea (otherwise phenylthiocarbamide or PTC) has been found in the apes (Fisher, Ford, and Huxley, 1939) and in rats and mice (Klein and DeFries, 1969) argue for the stability of variability at these loci and probably many more.

Permanent discontinuous genetic variation such as is presumably represented by these latter examples is termed *polymorphism*. E. B. Ford in 1940 defined this as 'the occurrence together in the same habitat of two or more discontinuous forms of a species in such proportions that the rarest of them cannot be maintained merely by recurrent mutation'.

Polymorphism in these terms has been demonstrated in man in a wide variety of ways in a wide variety of systems:—

1. Physiological, e.g., colour blindness, PTC tasting ability, sticky or dry ear wax.
2. Morphological, e.g., hair colour, eye colour, (although these are not completely discontinuous characters), hair form, dermatoglyphics.
3. Skeletal, e.g., fused carpal bones, brachymesophalangia and dental differences such as shovelling, extra cusps, etc.
4. Immunological, e.g., the numerous blood groups, the gamma globulin groups (Gm), the alpha globulins (Gc), and beta-lipoproteins (Ag) of human serum, the histocompatibility loci.
5. Enzymic, e.g., differences in erythrocyte glucose-6-phosphate dehydrogenase, acid phosphatases, phosphoglucomutase, etc.
6. Structural, e.g., differences in haemoglobin and serum proteins such as haptoglobins, which are due to amino-acid sequence differences.
7. Metabolic, e.g., differences in the rate of acetylation of isoniazid, a drug used in the treatment of TB, or deficiency in ability to metabolize β-amino-isobutyric aciduria.
8. Chromosomal, e.g., the presence or absence of satellites on some chromosomes and different sized Y chromosomes.

In all, more than 30 different human enzymes and proteins have now been shown to exhibit polymorphism in one or more of the major races. There are more than 30 blood groups known to control immunological differences of the red cells and there are in addition many histocompatibility loci of the white blood-cells polymorphic in all investigated populations. If we include in this enumeration all the physiological and morphological characters which are polymorphic, we arrive at a total of probably rather more than 100 loci known to be polymorphic in man. When we recall that our total knowledge of confirmed genetic loci in man is of the order of 500, we put into perspective these estimates of the extent of polymorphism in man. For recent reviews of human polymorphism consult Harris (1970), Price (1967), or Cavalli-Sforza and Bodmer (1971).

Amongst this wealth of genic variation, more of which is discovered each year, the prime problem is to assess overall levels of genic heterozygosity and the frequency of polymorphism; and it is only since the mid 1960s that any accurate estimates of these two important parameters of population structure have been possible for man, or for any other animals for that matter (Hubby

Table 2.3.—THE LEVELS OF POLYMORPHISM FOUND IN DIFFERENT ORGANISMS

ORGANISM	DEGREE OF POLYMORPHISM (per cent)	AVERAGE DEGREE OF HETEROZYGOSITY (per cent)	NO. OF LOCI STUDIED	REFERENCE
Drosophila melanogaster	54	23	10	O'Brien and MacIntyre, 1969
Drosophila pseudoobscura (isolated population)	25	4·4	24 (both *D. pseudoobscura* populations)	Prakash, Lewontin, and Hubby, 1969
Drosophila pseudoobscura (ecologically central)	46	12·3		
Drosophila persimilis	25	10·5	24	Prakash, 1969
Drosophila paulistorum	55–67	21	17	Richmond, 1972
Drosophila willistoni	58	18·4	28	Ayala, Powell, Tracy, Mournão, and Perez-Solas, 1972
Drosophila ananassae	50	13	4	Stone, Kojima, and Johnson, 1968
Drosophila obscura	58	11·1	33	Lakovaara and Saura, 1971
Drosophila nasuta	100	33	3	Stone, Kojima, and Johnson, 1968
Cepaea nemoralis (land snails)	63		27	Manwell and Baker, 1968
Cepaea hortensis (land snails)	52		27	Manwell and Baker, 1968
Acris crepitans (cricket frog)	62		21	Dessauer and Novo, 1969
Limulus polyphemus (horseshoe crab)	25	5·7	25	Selander, Yang, Lewontin, and Johnson, 1970
Phasianus colchicus (pheasant)	43		44	Baker, Manwell, Labisky, and Harper, 1966
Coturnix coturnix (quail)	54–58		24	Baker and Manwell, 1967
Mus musculus (Texas wild populations)	30	11	40	Selander and Yang, 1969
Mus musculus (two Danish subspecies)	22–30	7·8	41	Selander, Hunt, and Yang, 1969
Mus musculus (N. American feral populations)	29	10·3	17	Ruddle, Roderick, Shows, Weigh, Chipman, and Anderson, 1969
Peromyscus polionotus (wood mouse)	40	10	41	Selander and Yang, 1969
Homo sapiens (Europeans)	48 (both *H. sapiens* groups)	16·8	26 (both *H. sapiens* groups)	Hopkinson and Harris, 1971
Homo sapiens (Negroes)		16·2		

and Lewontin, 1966; Lewontin and Hubby, 1966). Of the estimates of average heterozygosity in man that have been made, R. C. Lewontin in 1967, on the basis of the degree of polymorphism shown by the various erythrocyte blood groups discovered at that time and their rate of discovery, estimated average heterozygosity in man at 12 per cent. H. Harris and his co-workers (Harris, 1966, 1969, 1970) based their estimate on surveys of randomly chosen blood enzymes and they estimated polymorphism at 30 per cent, with average heterozygosity at 16 per cent. *Table* 2.3 shows that these estimates are fully in line with those from other organisms.

From the particular point of view of the present book it is also very important indeed to note that this level of genetic variation appears to be similar in the different races of man. Subjective impressions based on blood group or enzyme frequencies are apt to be misleading, but Harris and his colleagues carried out an electrophoretic survey of twenty arbitrarily chosen enzymes in European and Negro populations and obtained the results shown in *Table* 2.4. The actual

Table 2.4.—ESTIMATED AVERAGE HETEROZYGOSITY PER LOCUS FROM A SURVEY OF 20 ARBITRARY CHOSEN ENZYMES IN EUROPEANS AND NEGROES (*Only loci where the incidence of heterozygotes was* >0·01 *are shown.*) (*From* Harris, 1970.)

ENZYME	INCIDENCE OF HETEROZYGOTES	
	Europeans	Negroes
Red-cell acid phosphatase	0·51	0·28
Phosphoglucomutase		
locus PGM_1	0·35	0·33
locus PGM_3	0·38	0·47
Adenylate kinase	0·09	—
Peptidase A	—	0·16
Peptidase D (prolidase)	0·02	0·10
Adenosine deaminase	0·11	0·06
Average heterozygosity per locus assuming 27 loci were screened (i.e., those detected electrophoretically)	0·054	0·052
Average heterozygosity per alleles determining all structural protein and enzyme variants	0·162	0·156

genes which are polymorphic vary between the races, but this is of course to be expected. The important point is that this similarity between the races in genetic variability suggests that few problems are likely to arise among possible hybrids from sudden influxes of new variability.

What exactly do these figures of polymorphism levels and average degrees of heterozygosity mean in terms of genetic variation? Most geneticists agree that even a relatively simple organism like *Drosophila* has 10,000 or more genetic loci, and man therefore has certainly no less and probably rather more, say a minimum of 50,000 loci. A conservative estimate of the number of polymorphic loci in *Drosophila*, from the data of *Table* 2.3, is considered to be 2,000 and the corresponding figure for man therefore perhaps 10,000. Now

Table 2.5.—THE RELATIONS BETWEEN THE NUMBER OF GENE LOCI, THE NUMBER OF ALLELES PER LOCUS AND THE NUMBER OF POSSIBLE GENOTYPES. (*From* Wallace, 1968.)

No. of Alleles	No. of Loci 1	2	3	10...	n
1	1	1	1	1...	1
2	3	9	27	59000...	3^n
3	6	36	216	6000000000...	6^n
.	.	.	.	.	.
.	.	.	.	.	.
.	.	.	.	.	.
.	.	.	.	.	.
.	.	.	.	.	.
K	$\frac{K(K+1)}{2}$	$\left[\frac{K(K+1)}{2}\right]^2$	$\left[\frac{K(K+1)}{2}\right]^3$..	$\left[\frac{K(K+1)}{2}\right]^{10}$..	$\left[\frac{K(K+1)}{2}\right]^n$

Table 2.5 shows the relationship between the number of gene loci, the number of alleles per locus, and the number of possible genotypes (*after* Wallace, 1968). Even with only two alleles at each of these loci (and we know that there are often many more than two), the number of possible human genotypes is 3^{8000}, a number impossible to visualize or represent (although this is obviously not meant to suggest that all these genotypes would be viable combinations). The total number of the human race at the present time is approximately 3^9 and so clearly every human being alive now or who has ever lived possesses a unique genotype; a genotype, moreover, which probably differs from all others at a very large number of genetic loci.

But we have so far discussed mainly those genic variants present in human populations in reasonably large proportions; large enough for them to be considered as polymorphisms. There are, however, an enormous number of rare inherited conditions which show variation in frequency throughout the world. Albinism, for instance, appears to occur at a frequency of about 1/10,000 to 1/20,000 in some European countries, but in West Africa it is probably three to five times as high. The overall rate of occurrence of congenital malformations such as anencephaly, spina bifida, club-feet, hare-lip, and congenital hip dislocation seems to be about 1 to 1·5 per cent among Europeans, American Negroes, and Japanese, but there is evidence that these populations differ in the incidence of particular defects, although the reasons for such differences are likely to be only partly genetic (*see* Chapter 6: ss. 6.2 and 6.5).

Although, perhaps, these rare genetic variants are of less significance for racial differentiation than the true polymorphisms we have already considered, they are still symptomatic of some genetic difference and we cannot predict whether any of them are likely to be important with respect to racial differentiation in the future, and in many respects it is genetic variation of this latter sort which, as we have already mentioned, is of prime importance in studies of racial mixing.

Under these circumstances we can already begin to see that racial variation is only a small part of the total genetic variation spread through the human species as a whole. Any concept of the pure race also automatically dissolves when it is realized what an enormous amount of genetic diversity every human race automatically possesses—far greater in magnitude than the relatively few genetic differences which define the various races in the first place.

But even all this is only part of the genetic variation present in the human population.

2.1.b. Continuous Variation

Characters such as height, weight, rate of development, intelligence, and so on, show a complete spectrum of variation which, in most cases, follows a

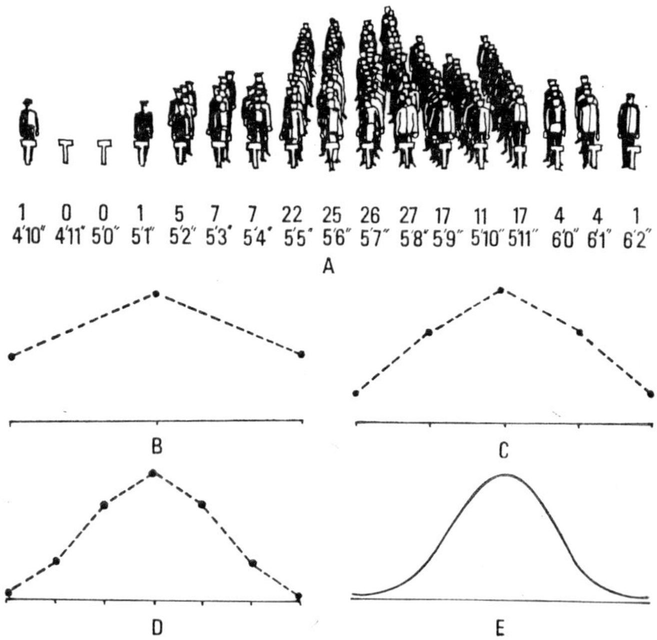

Fig. 2.1.—A, The distribution according to height of a company of American soldiers, and possible genetic situations which could bring this distribution about. B, 2 alleles, 3 phenotypes. C, 2 pairs of alleles, 5 phenotypes. D, 3 pairs of alleles, 7 phenotypes. E, Very large number of alleles, continuous phenotypes. (*From Principles of Human Genetics*, 2nd ed., 1960 (Curt Stern). Copyright, London: W. H. Freeman & Co.)

normal or bell-shaped curve of distribution. *Fig.* 2.1 shows the distribution according to height of a company of American soldiers and possible genetic situations which could bring this distribution about. As is made clear by the figure, normally distributed traits such as this are usually determined by a considerable number of collaborating genes each with rather minor effects, although, because of the blurring effects of environmental influence, in some cases a relatively small number of loci may account for a continuous trait. Thus investigating the genetics of human height a little further, *Fig.* 2.2 shows a

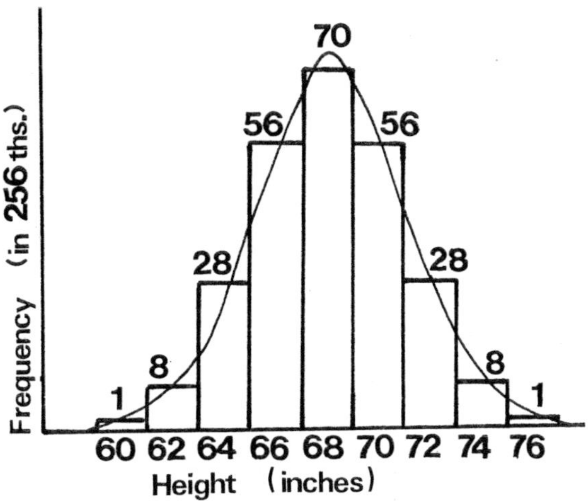

Fig. 2.2.—A distribution curve for human height supposing that only two loci, each with three alleles, affects this character; one allele decreasing stature by 2 in., a second determining average stature of 68 in., and a third increasing stature by 2 in.

distribution curve for human height assuming that only two loci, each with three alleles, affects this character; one allele decreasing stature by 2 in., a second determining average stature of 68 in., and a third increasing stature by 2 in. As can be seen, a satisfactory approximation to a normal curve can be obtained. The inheritance of human skin colour has of course received a great deal of attention, and although skin colour obviously does not usually follow a normal distribution because of racial polarization, in interracial hybrids, as we shall see, skin coloration does approach a normal distribution. Estimates of the number of genes involved vary between 2 and 6, each with a restricted number of alleles.

Other examples of these many quantitative or multifactorial traits are blood-pressure, dermatoglyphic ridge counts on finger and palm prints, and refractive index of the eye. In fact multifactorial traits are far more numerous than genic ones. Their contribution to human genetic variation is considerable. Analysis of dermatoglyphic ridge counts has been particularly useful in some cases of racial hybridization and warrants a little more detailed description.

On the surfaces of the hands and feet are numerous fine epidermal ridges which form regular but complex patterns. Francis Galton in 1894 was one of the earliest biologists to try to classify finger-prints and to investigate their genetic control and inheritance. Most finger-prints can be classified as one of a few major types comprising circles, loops, whorls, and double loops. There is

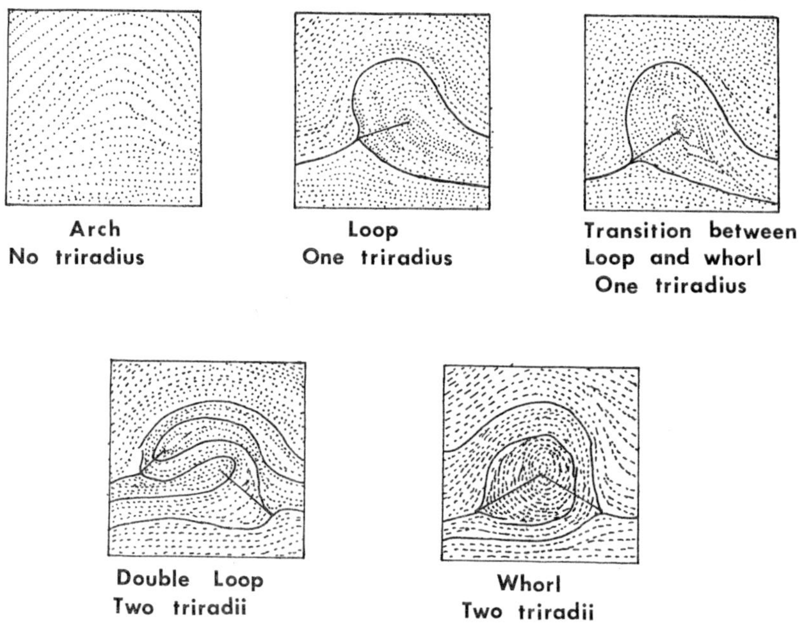

Fig. 2.3.—Some of the principal types of pattern seen in finger-prints. (*From* Harrison, Weiner, Tanner, and Barnicot, 1964.)

so much individual variation in detail that no two individuals have identical patterns, not even identical twins. The main finger-print types are shown in *Fig.* 2.3.

The principal individual variations in finger-print patterns are certainly under genetic control, but cannot yet be described in terms of any simple mode of inheritance. Differences between the dermatoglyphs of populations therefore reflect variations in the distributions of genes, but of course the details cannot yet be specified. Some of the more extreme variations in human populations are shown in *Fig.* 2.4. The dermatoglyphics of human hybrids are quite normal and fall within the range of types shown in *Fig.* 2.4. This in itself is an important indicator of the normality of development in these individuals, since finger and palm prints are quite sensitive indicators of biological stability (Holt, 1968).

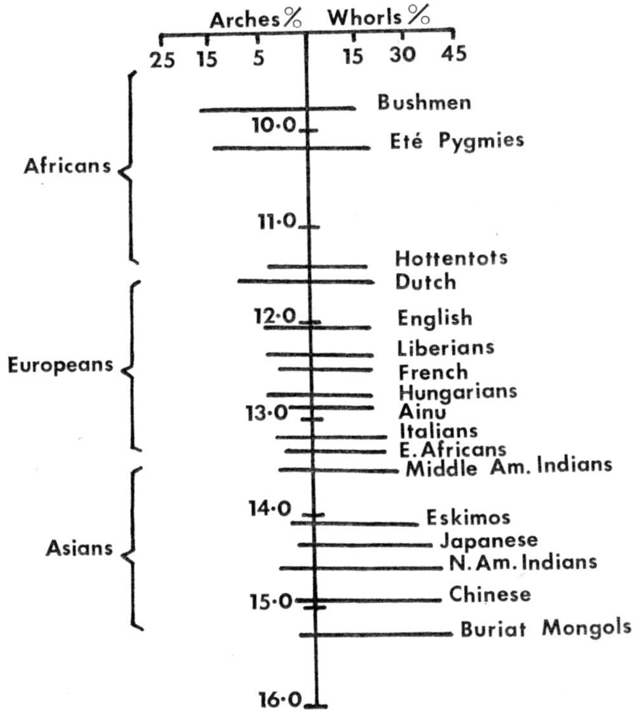

Fig. 2.4.—The range of proportions of arches and whorls and the pattern intensity index in various human populations. (*Adapted from* Cummins and Midlo, 1943.)

2.1.c. *Epigenetic or Quasi-continuous Variation*

As mentioned in the introduction of this section, the underlying genetic phenomena controlling the various types of observed variation are essentially similar, but the variation is expressed in different ways and in different systems in the phenotype. The variation we are discussing in this section includes that where essentially discontinuous characters are controlled by many genes of small effect so that there results, perhaps, a threshold whereby the character either is or is not expressed as a result of a small change in the genetic background. Alternatively, environmental effects during development may play a similar part in changing the nature of the expression of the particular character concerned. Thus in the terms that C. H. Waddington (1957) has used, these characters may be regarded as being 'incompletely canalized', exhibiting a marked departure from the mean genetically controlled expression of a character.

H. Gruneberg and his collaborators (1950, 1952, and later papers) have described over 50 discontinuous variants occurring in inbred strains of mice, and other workers (Berry, 1963; Berry and Searle, 1963; Hedges, 1969) have since shown that a similar number of variants (not all the same) can be found in more than a dozen wild rodent species and higher mammals, such as seals (Berry, 1969a) and of course man (Berry and Berry, 1967). One of the particular characters concerned is the absence of third molars in mice, and Gruneberg (1951) showed clearly that a threshold mechanism is involved in this type of variation, splitting an originally continuous distribution into sharply distinct phenotypes. What is inherited is the size of the tooth itself; if the tooth falls below a certain critical size, the tooth fails to develop. There are genes which

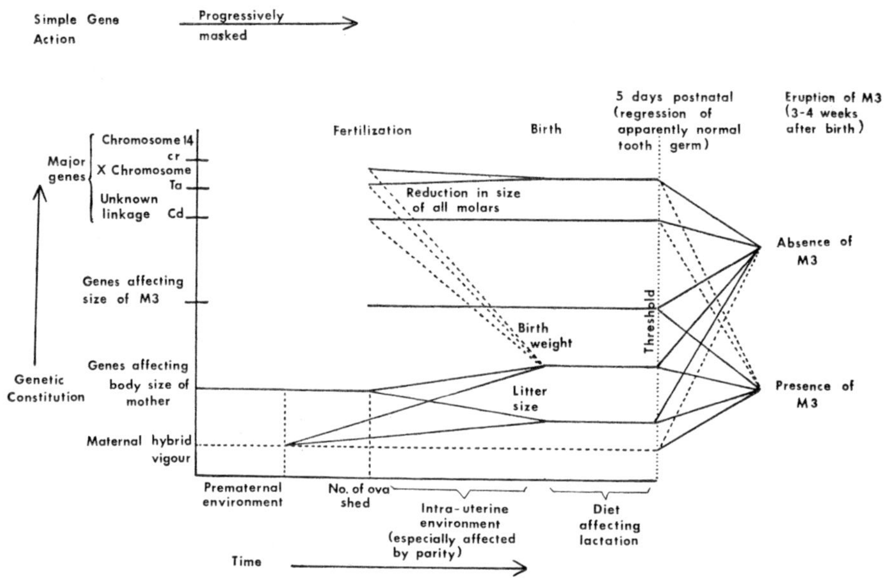

Fig. 2.5.—The interaction of genetical and environmental factors in the determination of third molar loss in the mouse. (*From* Berry, 1968.)

control tooth size as such and tooth size is also highly correlated with general body size, but the final factors which determine whether a tooth will develop or not in an inbred strain are the environmental factors connected with maternal physiology. *Fig.* 2.5 summarizes the interaction of genetical and environmental factors in the determination of third molar loss in the mouse. This particular character has been described in some detail because it happens that the quasi-continuous variant in man which has been most fully investigated is the same character, namely agenesis of the third molar (Grahnen, 1962; Garn, Lewis, and Vicinus, 1963). Although a full explanation is not yet available, it seems likely that third molar loss in man is a complex process at least analogous to that in the mouse.

The epigenetic or non-metrical variants that occur in man are at least morphologically analogous to those which occur in rodents, and the little that is known about their inheritance agrees with them being inherited in the same way as in the mouse. The most extensive group of variants that have been studied are those of the human skull where Berry and Berry (1967) have classified nearly 600 skulls for 30 different non-metrical variants. *Fig.* 2.6 shows different aspects

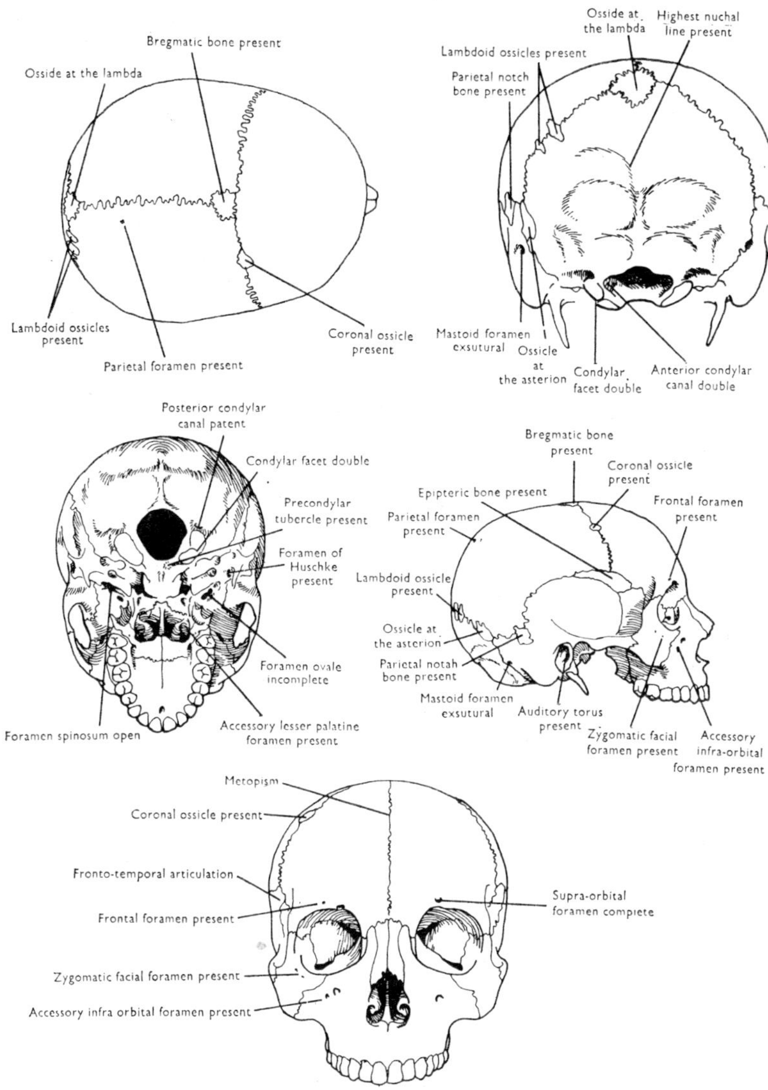

Fig. 2.6.—Some of the discontinuous minor variants which are commonly found in the human skull. (*Adapted from* Berry and Berry, 1967.)

of a human skull indicating a number of the variants which were scored. In addition to these 30 characters a number of similar variants in the vertebral column have been used, together with those of the skull, either as indicators of evolutionary change or of population divergence and racial history (Brothwell, 1965; Berry, and Ucko, 1967; Berry, Brothwell and Powers, 1968; Kellock and Parsons, 1970). Brothwell (1958) was able to construct a map showing the frequencies of metopism throughout the world on which he drew 0, 5, and 10

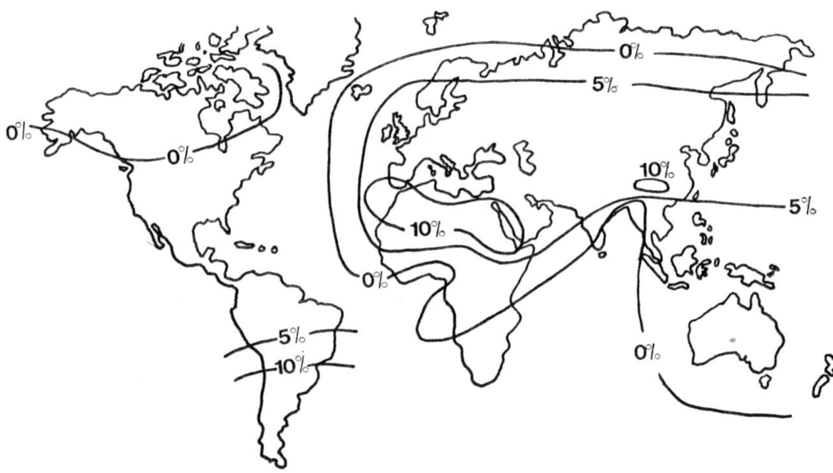

Fig. 2.7.—World variation in the incidence of metopism. (*From* Brothwell, 1958.)

per cent 'isovariant' lines exactly as is done with blood group frequency maps (*Fig.* 2.7). In addition A. C. Berry (quoted in R. J. Berry, 1969b) has used non-metrical skull characters as genetical markers to determine degrees of genetical relationship between the populations of Iceland, Shetland, different regions of Norway, and part of Scotland, and found evidence of relationships agreeing with those suggested on the basis of the incidence of multiple sclerosis. Multiple sclerosis itself, on the basis of this and other evidence, appears likely to have a significant genetic component in its causation. Many other developmental abnormalities such as cleft palate, congenital pyloric stenosis, anencephaly, spina bifida, and most structural heart defects are also essentially non-metrical variants (Falconer, 1965).

Penrose (1957) has described the incidence of anencephaly, a severe congenital defect in which the brain is grossly under-developed, in different parts of Europe. In Ireland and Western Great Britain the frequency varies from 1·0 to 0·32 per cent, whereas it is less than one-tenth of this in Continental Europe, and is even lower in certain regions of South-eastern France. In South Africa the incidence of anencephaly among native Africans appears to be only about one-quarter of that in Europeans. In Singapore Searle (1959) discovered a frequency of 0·77 per cent, but there was great heterogeneity between the racial groups with frequencies ranging from 0·0 per thousand among Hakka Chinese to 6·5 per thousand in Sikhs. This particular defect is closely associated with spina bifida, which appears to show similar variation, and it merely exemplifies the sorts of results which have been obtained with other characters of this type. We shall examine the racial incidence of characters of this sort in more detail in Chapter 6.

2.1.d. *Variation in the Incidence of Disease*

Differential mortality is one of the prime factors in natural selection and gene frequency changes stemming from it. In many examples of genetic variation in man it is possible to show that the presence of a certain gene or genes leads directly to a pathological condition and therefore geographical or racial variation in the incidence of certain diseases. Sickle-cell anaemia (*p.* 49), porphyria variegata (*p.* 64), and Tay-Sachs disease (*p.* 46) are examples of this.

But the causation of a large number of diseases such as TB, diabetes, cancer, cardiovascular disease, and others is much less clear. It is thought that there is a genetical component in many of them, partly because of racial differences and partly because of the tendency of many of them to run in families. The influence of environment and diet is probably also important in many cases, and it seems likely that genetical diversity may be playing some part in determining individual and group responses to the many environmental factors causing complex diseases such as these. *Table* 2.6 shows comparisons in the incidence of various types of

Table 2.6.—NUMBER OF CASES OF CANCER IN BANTU OF JOHANNESBURG (1935–55) AND EXPECTED NUMBERS IN THREE OTHER POPULATIONS WHEN STANDARDIZED FOR AGE (*From* Higginson and Oettlé, 1960.)

SITE OF CANCER	BANTU (observed)	U.S. WHITE	U.S. NEGRO	DANES
Buccal and lip	2	21·3	2·0	24·3
Stomach	41	89·0	128·4	119·4
Colon	7	75·7	54·1	43·6
Liver	114	13·8	25·7	1·4
Lung	40	103·0	104·3	51·1
Skin	14	212·7	18·2	49·5

cancer in a number of racial groups appropriately standardized for age distribution. In some cases the similarities can be explained, e.g., Bantus and American Negroes both have low incidence of cancers of mouth, lips, and skin attributable to the protective effects of their pigmented skin (*see* pp. 56–62). *Fig.* 2.8

suggests that human diseases can be represented as lying on a continuous spectrum from being entirely genetic in origin to entirely environmental. Certainly differences in the incidence of disease in different races is evidence of genetic variation and can also be seen as a force in producing further racial differentiation. Here, too, we see that there is an enormous amount of genetic variation present, some of which is involved in racial differentiation and much

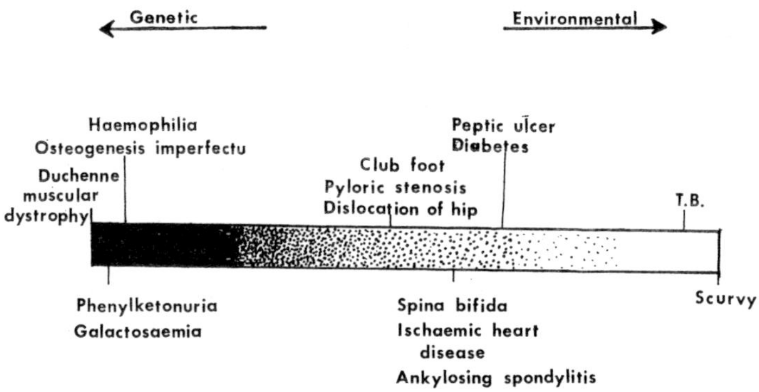

Fig. 2.8.—Human diseases represented on a spectrum ranging from those which are largely environmental to those which are entirely genetic. (*From* Emery, 1971.)

of which is common to the whole gene pool. *Table* 6.49 in section 6.5.4 shows some racial differences in disease incidence illustrating how widespread is the phenomenon, and Goldstein (1969) discusses what an important part it has played throughout the evolution of man.

2.1.e. *Miscellaneous Genetic Variation*

In addition to the above genetic variation there are a number of other racial variations whose presence, so far, defies explanation but which are persistent, just as characteristic of different races, and under some measure of genetic control. They have not been subject to the same analysis as the visible characters or the biochemical differences or the continuous genetic variation, but they are likely to receive an increasing degree of attention in the future as the links between genetics, demography, and epidemiology become stronger (Neel, Shaw, and Schull, 1965; Dyer, 1969b; Cavalli-Sforza and Bodmer, 1971).

The first of these variations is the human sex ratio at birth, which normally varies between 104 and 107 (male births per hundred female births). This can be confirmed from data in the United Nations *Demographic Yearbook* for 1959 and 1965, particularly from those countries in which the registration of births is said to be relatively complete. There are statistics of birth from 80 different populations for the decade 1949–58 and 50 of them fall within the range 104–107.

There are still a number of exceptions, however, and *Table* 2.7 shows the sex ratio at birth in those cases that fall significantly below 104 and significantly above 107.

For the purposes of comparison, the sex ratio in England and Wales over this period is 106·00 and in Scotland 105·76. The magnitude of the difference

Table 2.7.—EXTREME SEX RATIOS IN DIFFERENT COUNTRIES, 1949–64
(*From* Dyer, 1969b.)

COUNTRY	NUMBER OF BIRTHS	SEX RATIO
1. Sex ratio significantly below 104·00		
Monserrat	5,869	94·34
St. Lucia	44,884	98·77
St. Kitts	28,592	100·06
Antigua	28,228	100·13
Grenada	50,358	100·20
S. Africa (coloureds)	922,963	102·04
Bermuda	17,893	102·07
2. Sex ratio significantly above 107·00		
Hungary	2,754,122	107·15
Poland	10,753,456	107·17
Hong Kong	1,405,479	108·04
Gambia	13,050	110·18
Aden	60,852	120·31

between highest and lowest sex ratios is surprising. Most biologists since Darwin have argued that strong selective forces exist to keep the sex ratio closely approximating to 1 : 1. Yet here we have several departures in either direction from this. What might be the reason? Examination of the levels of prenatal mortality in the countries with aberrant sex ratios shows no significant correlations with either high or low sex ratio, and beyond noticing that the Negroid races, especially those in the West Indies, have a propensity for a lower sex ratio at birth, we can make little comment. We know, furthermore, that most of these ratios are reasonably stable. For instance, the sex ratio for all births between 1878 and 1950 in Grenada was 101·4 and in Jamaica was 102·3, while among U.S. Negroes between 1915 and 1948 the sex ratio at birth was 102·9.

Although this variation in sex ratio may appear to be of merely curiosity value, it does have more significance as an indicator of biological stability that we shall return to in Chapter 4. The heterogametic sex, the XY sex, is often poorly represented in interspecific crosses and may be sterile. This is because the contribution of the two parents is exactly equal in the XX sex, but the XY sex lacks a piece of chromosome from one parental species. Inbreeding may also depress the XY sex, as has been demonstrated experimentally in mice and as has been suggested by analogy in the case of man, e.g., Darlington (1966). There is really little evidence to support this suggestion and Morton, Chung, and Mi (1967) conclude there is really no hybridity effect on sex ratio in Hawaii.

Among the races represented in Hawaii are the Koreans and these have a very high sex ratio which persists in interracial crosses involving Korean fathers. This suggests that the high sex ratio, at least in this case, could be due either to a primary segregation disturbance or gametic selection. It is not likely to be due to differential viability of male and female foetuses, since Morton and others state that there is no suggestion of an effect in Korean mothers.

The second largely inexplicable variation concerns frequencies of multiple births. The statistics here are of two sorts: First, there are the total numbers of live children resulting from single, double, or multiple births. Secondly, there are the total frequencies of twins among all births, these twins being divided into either monozygotic or dizygotic twins (i.e., the products of one egg or two). Both of these statistics show considerable variation. The proportion of live-born children resulting from twin births varies from 2·57 per cent in Jamaica (1652 in a total of 64,338), 2·46 per cent in Northern Ireland (802 in a total of 32,505), and 2·35 per cent in Scotland (2456 in 104,355), down to 1·33 per cent (1739 in 130,958) in Bulgaria, 1·19 per cent (2595 in 216, 531) in Ecuador, and 1·06 per cent (17,587 in 1,659,521) in Japan. The frequencies of children from multiple births (three or more) are equally variable. For instance, Ecuador has only 4 per 100,000 births, whereas Jamaica has 4 per 1000. The British Isles exhibit more medium values, but even they show curious variation: 5 per 10,000 in Scotland, 3 per 10,000 in England and Wales, and 1 per 10,000 in Northern Ireland. What the reason for this variation might be, we cannot tell.

If twins are of different sexes they must, of course, be dizygotic and among twins of the same sex a similar number to the number of different sexed twins can be assumed to be dizygotic. The remainder of the single-sexed twins are therefore monozygotic. When we divide the total incidence of twin births in this way, making due allowance for the age structure of the population (because the frequency of twinning is known to be age dependent), the results in *Table* 2.8 can be derived from national natality statistics. Monozygotic twinning appears to be of a fairly uniform frequency of 3–4 per 1000, but dizygotic twinning is very variable. The lowest rates are among Mongoloid races, which are markedly lower than those among white Caucasoids. The Negroes have higher rates still. The racial hybrids, both first generation and later, show the marked effects of racial admixture. The Mongoloid/Caucasoids with a very low rate of dizygotic twinning, the Negroid/Caucasoids with a rather high rate intermediate between the two parental races.

A third source of mystery is differential fertility. Between 1955 and 1964 in England and Wales the number of children born every year increased from less than 700,000 to more than 900,000. The number of these children born to mothers aged more than 50 years varied between 12 in 1955 and 1 in 1964.

Over the whole decade about 8 million children were born in England and Wales, only 50 of them to women aged 50 years or more.

In France the figures are slightly higher. There are about 800,000 children born every year, and of these about 50 annually are to women more than 50 years old. No other country in western Europe has higher figures than those of

Table 2.8.—TWINNING RATES IN DIFFERENT POPULATIONS (*Number of twin births per* 1000 *live births.*)

COUNTRY	YEARS	DIZYGOTIC	MONOZYGOTIC	TOTAL
Caucasoid				
U.S.A.	1964	5·7	3·8	9·5
Spain	1951–53	5·9	3·2	9·1
France	1946–58	7·1	3·7	10·8
England and Wales	1957–61	8·5	3·5	12·0
Sweden	1946–55	8·6	3·2	11·8
Italy	1949–55	8·6	3·7	12·3
Mongoloid				
Hawaii	1948–58	2·5	4·5	7·0
Japan	1926–31	2·7	4·2	6·9
Japan	1956	2·3	4·1	6·4
Negro				
Antigua	1857–1951	11·5	3·9	15·4
Jamaica		13·4	3·8	17·2
W. Nigeria (Yorubas)		49·8	4·4	54·2
M-W. Nigeria (Edos, Urhobas)		28·7	4·6	33·3
E. Nigeria (Ibos, Efiks)		36·6	5·1	41·7
First Generation Racial Hybrids				
Mongoloid/Caucasoid (Hawaii)	1948–58	2·1	5·0	7·1
Other Hybrids				
Negroid/Caucasoid				
U.S. Negro	1964	9·8	3·9	13·7
Salvador mulatto		13·6	4·0	17·6

France and most are rather lower. The overall figure for children born to women aged 50 years and more in western Europe lies between about 10 and 60 for every million births. Clearly these are rather infrequent events.

In at least two countries in eastern Europe, however, the situation is very different. Tiny Albania is the most striking. There are only about 60,000 children born every year in this country but between 300 and 500 of these are to women more than 50 years old. Therefore, women over 50 years are involved in about 5 in every 1000 births, or 5000 in every million; this is an astonishing contrast with the 10 in England and Wales. In Yugoslavia, 1–2 births in every 1000 are to women over 50 years old, which is also very different from western Europe. The reasons for this most curious prolonged fertility in these two countries is just not known. The number of girls under 15 years of age who became mothers in Albania or Yugoslavia is no higher than any other country in Europe and rather lower than many, so clearly it is not just a matter of greater overall fertility at all ages. *Table* 2.9 shows some of the worldwide variation of

Table 2.9.—FERTILITY OF WOMEN AGED 50 YEARS AND OVER.
(*From* Dyer, 1969b.)
(*United Nations Statistics for* 1965.)

COUNTRY	BIRTHS TO WOMEN AGED 50 YEARS AND OVER	NUMBER PER MILLION BIRTHS
Europe		
England and Wales	2/862,725	2·3
France	56/862,333	64·9
Yugoslavia	433/408,158	1060·9
Albania	359/ 65,692	5464·7
North America		
Canada	11/373,626	29·4
Panama	17/ 49,394	344·2
El Salvador	778/137,950	5639·7
Asia		
Japan	18/1,823,697	9·9
Indonesia	2825/3,182,617	887·6
Africa		
S. Africa		
White	7/75,569	92·6
Coloured	75/75,171	997·7

this feature. In some cases, particularly from South America, the birth-rate is higher among women at all ages from 15 to 50 years; in others, as we have seen, this is not so.

This example of differential fertility of one particular age group is certainly only a part of the total genetical variation for this trait present in the human species. This again is a trait for which it is difficult to isolate all the environmental factors concerned. Infant mortality, deliberate restriction of family size, and the lack, until recently, of sufficiently accurate statistics make comparisons very difficult. Nevertheless the importance of this particular subject is obvious enough in the present context and we return to it later, particularly in Chapters 4, 6, and 7.

2.2. The Origin of Racial Diversity

As we have seen, man, in common with almost all other organisms examined to date, possesses an enormous amount of genetic variability. It is obviously very important in studies of racial integration to discover how this variation arose and in particular how the races have become differentiated with respect to this variation.

2.2.a. *Mutation*

The origins of the racial diversity are many and various but ultimately all new variation must arise as new mutations in the genetic material. New mutations are the raw materials of evolutionary change. They are the bases on which selection operates and through which other processes such as genetic drift, gene migration, and so on, are made effective. A mutation is defined as a permanent heritable change in that small part of the deoxyribose nucleic acid (DNA) which codes for one polypeptide or protein unit. When a mutation occurs in a germ cell of an individual which is subsequently passed on to the offspring generation, the genotypic frequencies of that generation are obviously slightly changed.

In organisms commonly studied in the laboratory it is possible, by experiment, to obtain accurate estimates of mutation rates at a fairly large number of representative genetic loci: dominant and recessive mutations can be studied, those with visible effects and those that are lethal, those that are autosomal, and those that are sex-linked. Mutations are found to be relatively infrequent events and this accentuates the difficulties posed for the study of mutation rates in man. Nevertheless several estimates of mutation rates in man have been obtained by a variety of methods and, by making use of judicious comparisons with other organisms, much useful information can be gained.

An examination of *Table* 2.10 shows that the mutation rates which have been determined for man are broadly similar to those in most other organisms and therefore the rate of origin of new variation must also be fairly similar. It is interesting to note that many examples of variation in mutation rates at particular loci are known amongst experimental organisms and it is therefore at least possible that similar variation exists in man, thereby contributing to racial differentiation.

To what extent do new mutations contribute to the pool of genetic variability in man? From *Table* 2.10 it appears that an average mutation rate might be 1 or 2 mutations per 100,000 individuals per locus per generation. Now on the basis of estimates in *Drosophila*, man, as we have mentioned, is thought to possess at least 50,000 genetic loci. Therefore, between 10 and 20 per cent of all gametes carry a new detrimental mutation. Since each individual is the product of 2 gametes, this means that perhaps 30 per cent of all human beings contain a newly

Table 2.10.—SPONTANEOUS MUTATION RATES AT SPECIFIC LOCI FOR VARIOUS ORGANISMS

ORGANISM	TRAIT	MUTATIONS PER 100,000 GAMETES
E. coli K 12	To streptomycin resistance	0·00004
	To arginine independence	0·0004
	To phage T1 resistance	0·003
Neurospora crassa	To adenine independence	0·0008–0·029
Drosophila melanogaster males	y^+ to yellow	12·0
	bw^+ to brown	3·0
	e^+ to ebony	2·0
	ey^+ to eyeless	6·0
Zea mays	sh to shrunken	0·12
	su to sugary	0·24
	I to i	10·60
Mus musculus	+ to brown	0·9
	+ to dilute	1·2
	+ to agouti	3·0
Homo sapiens	Autosomal dominants: Epiloia	0·8
	Autosomal dominants: Acridia	0·5
	Autosomal dominants: Amyotrophic lateral sclerosis	3·0
	Autosomal dominants: Neurofibromatosis	13·25
	X-linked recessive traits: Duchenne muscular dystrophy	7–9
	X-linked recessive traits: Haemophilia A	2–4
	X-linked recessive traits: Haemophilia B	5–10

mutated gene that is more or less detrimental. Suppose that the true figure is only 10 per cent, then this would mean that nearly 5 million people in Great Britain, for instance, carry a detrimental gene not carried by their parents; and in the whole human population perhaps 350 million people carry such new genes. As high as this figure is, there are strong indications (Dyer, 1971c, and below) that the overall mutation rate, that is including genes with very small effects or no effects at all, is considerably greater than this. Bearing in mind the 10,000 or so generations that *Homo sapiens* has existed, it is apparent that a great deal of new genetic variation has been created in this way.

In addition, of course, mutant genes that persist for several generations must accumulate in a population. The frequency of genes that persist for 10 generations will be 10 times their mutation rate and that of genes which persist for 100 generations will be 100 times their mutation rate. The net result, it has been estimated, is that there are about 50 times as many detrimental mutants present in the human population as would be produced by new mutation. Of course a number of variable factors affect the rate of accumulation or elimination of mutant genes—these include the severity of their effects, their degree of dominance, the population size, and the breeding structure of the population. Each of these has its part to play in the genetics of racial differentiation.

The above remarks have applied almost exclusively to the origin of genic variation by mutation. But we have already seen that continuous and quasi-continuous variation is equally important in human populations and therefore we must consider the origin of this variation. Considering first the genetics of continuous variation, Dr. T. Mukai in 1964 studied the mutation rate of viability polygenes on the second chromosome of *Drosophila melanogaster* and estimated the lower limit to be 14·1 per cent per chromosome per generation. This is nearly 30 times as high as the normal recessive lethal rate on this chromosome. Estimates of the spontaneous mutation rate at gene loci affecting broadly continuous characters such as the number of abdominal or sternopleural chaetae in *Drosophila* also suggest that they are somewhat in excess of those at loci controlling so-called major characters.

Estimates of mutation rates of genes affecting quasicontinuous characters have come mainly from sub-line differentiation in laboratory strains of mice. Although the uncertainty of the relationship between the character and the number of genes which control it render estimates uncertain, it is fairly clear that here, too, mutation rate is fairly high. Professor H. Gruneberg (1970) has recently questioned whether the observed changes are in fact due to mutation—rather that they might be due to viral infection, but even if this were true (for the other side of the argument *see* Beardmore, 1970) the consequences, from the present point of view, are basically unchanged.

Dr. M. S. Grewal in 1962 tabulated the incidence of 27 skeletal variants in various sub-lines of the C57BL strain of mice. He found that, considering 10 different sub-lines, there was a good correlation between the degree of divergence and the number of generations the lines had been separated. The average divergence per character was found to be 0·003. Now in the British sub-lines of C57BL which Grewal tabulated, Deol, Gruneberg, Searle, and Truslove (1957) had found that 21 'mutations' affecting the 27 variants had occurred. This represents a rate of 0·01 mutations per variant per generation or 1 mutation per 100 generations. It was shown by these authors that the average effect of these mutations amounted to 0·6 standard deviation = 0·30 expressed in terms of variance. Hence this should give a divergence of 0·01 × 0·36 per character per generation. In other words these mutations can account for the whole of the divergence actually found.

A. G. Searle (1964) estimated that 20 mutations among 23 variants occurred in a total of 40 generations in the sub-lines he studied—in other words, more than 2 mutations per 100 generations, which is over twice that estimated by Deol and others (1957). These apparent mutation rates are 1000 or 2000 times greater than those recorded for any other type of character in the mouse (Schlager and Dickie, 1971).

Of course, because of the nature of the characters investigated and the lack of similar studies in man, the significance of results such as these for the origin of human genetic variation cannot be accurately assessed. Nevertheless the rates of skeletal and genetic changes in an isolated colony of African green monkeys are important indicators as to their evolutionary importance in man. The rate of change during the 300 years that this monkey colony has been isolated appears to be 100 times greater than the average rate of evolution in the horse during the Tertiary period (Dyer, 1968a), on which J. B. S. Haldane's and G. G. Simpson's classic figures on evolutionary rates were based (Haldane, 1949; Simpson, 1953). Clearly both continuous and quasicontinuous variation originates at a high rate and is an important contributor to racial differentiation. The degree of difference between the teeth of the St. Kitts and African green monkeys is similar to that which now separates the African green monkey from other cercopithecoids which, on the basis of coat coloration and geographical distribution, are usually given distinct subspecific or specific status.

2.2.b. Natural Selection

Natural selection is responsible for the spread of new mutations, and therefore the establishment of new variation, and also for the maintenance of stable polymorphisms. In this section we will examine its role in bringing about genetic change. Because of the very long time span normally involved in genetic change, much of our knowledge in this field comes from the theoretical and mathematical work of such geneticists as Sewall Wright in the U.S.A. and J. B. S. Haldane and Sir Ronald Fisher in England. Nevertheless, there are some examples known of selective changes in gene frequency in man indicative of the rates at which such changes can take place.

The first example concerns genetic changes among certain Jewish populations. At least three culturally and ethnically distinct Jewish groups are recognized in our times. These date back principally to the first century A.D. when the conquest of Jerusalem by the Romans instigated the mass exodus of Jews from Palestine. One group, the Ashkenazi, are those Jews who dispersed in central, eastern, and western Europe (and subsequently went to North and South America, South Africa, and Australia). Another group, the Sephardi, lived in the countries around the Mediterranean, including the European part of Turkey. A third group, the Oriental Jews, live in Asia Minor, Iraq, Iran, and the Yemen. The three groups have been genetically isolated from each other, and from the non-Jews (to varying extents), for this period of 2000 years, or between 60 and 70 generations. In this situation, then, the fact that the three groups of Jews have rather different frequencies of many well-known genes, and that in particular the Ashkenazi Jews have a very high frequency of certain genetic disorders compared with the Sephardic and Oriental Jews and with non-Jews, provides a fine opportunity to investigate genetic evolution in fairly well-defined circumstances.

Table 2.11 shows the results of a number of surveys indicating the sort of genetic divergence which has occurred between these populations. Presumably

Table 2.11.—GENE FREQUENCIES (PER CENT) AMONG DIFFERENT JEWISH GROUPS

	ABO Blood Group			MN Blood Group		Rh^- Individuals	PTC Tasters	Colour Blindness	G6PD Deficiency	Hp^1 Gene
	A	B	O	M	N					
Ashkenazim	26	12	62	64	36	9	21	8	0·1	30
Sephardim	32	15	53	50	50	7	22	6	2	38
Oriental										
Kurdistan	32	14	54	53	47	4	13	6	58	31
Iraq	30	20	50	50	40	6		4	25	27
Iran	20	19	55	60	40	5	16	6	15	29
Yemen and Aden	18	10	72	76	24	10	18	5	5	25

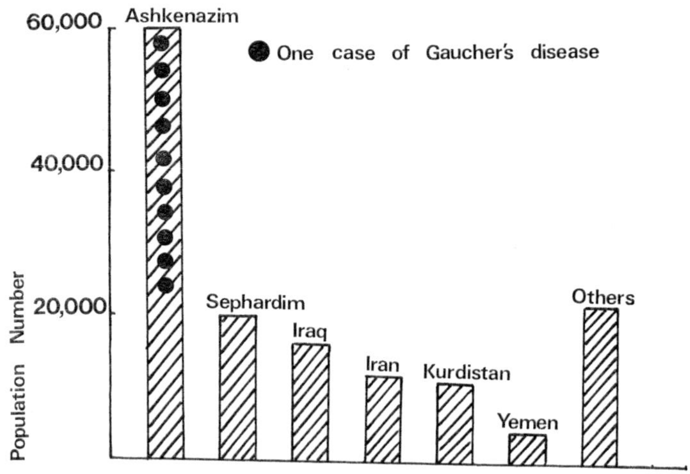

Fig. 2.9.—The incidence of Gaucher's disease among different ethnic groups in Jerusalem in 1957.

the original Jewish population in Palestine was fairly homogeneous for most of these genes. Of course, some of this divergence may be due to a degree of intermarriage with local non-Jewish populations and it is especially relevant, therefore, that certain characters in the Ashkenazi Jews are significantly different both from other Jews and from non-Jews, thereby excluding this possibility. The most noteworthy of these characters are Gaucher's disease, Tay-Sachs disease, and Niemann Pick disease, three lipid-storage diseases possibly sharing similar polymorphic properties and subject to the same selective forces. *Fig.* 2.9 shows the distribution of Gaucher's disease among ethnic groups in Jerusalem in 1957. Clearly some differentiation has taken place, but a more instructive example concerns Tay-Sachs disease.

Tay-Sachs disease, also known as 'infantile amaurotic familial idiocy', is a hereditary disease of lipid storage resulting in a progressive degeneration of cerebral function which commences soon after birth and ends in death usually within the first or second years of life. The disease is due to the homozygous condition of a single autosomal recessive gene. It occurs once in approximately 6000 Jewish births and once in approximately 500,000 non-Jewish births, thereby indicating that 1 in 40 Jewish persons and 1 in 380 non-Jews is heterozygous for the TSD gene. The birth incidence of TSD among the Sephardic and Oriental communities of the Middle East and North Africa appears to be even lower than that found in non-Jewish Europeans and Americans. It appears, therefore, as if there exists a selective advantage to heterozygous carriers of the TSD gene among the Ashkenazi, sufficient to raise the frequency of this gene from 0·0013 at the end of the first century A.D., when the mass emigration of the Jews began, through 60 generations to a new figure of 0·0126 in the twentieth century.

If the frequency of the TSD gene is q and the fitness s, then

	TT_2	Tt	tt_2
Frequency	p	$2pq$	q
Fitness	$1 - s$	1	0

and the appropriate calculations suggest that the selective advantage, s, of about 4·5 per cent in heterozygotes would be sufficient to raise the gene frequency to its present level among Ashkenazi Jews. Studies on the relative fertilities of grandparents of children showing Tay-Sachs disease, and who must therefore have been heterozygotes, suggest that they have a net reproductive advantage of about 6 per cent. In this case, therefore, demographic studies provide a neat confirmation of a genetic hypothesis.

The problem still remains, however, as to whether the incidence of Tay-Sachs disease is at equilibrium among Ashkenazi Jews or whether it is still ncreasing in frequency. In fact a selective advantage of about 5 per cent will, as *Fig.* 2.10 shows, lead ultimately to an equilibrium value of over 2000 per

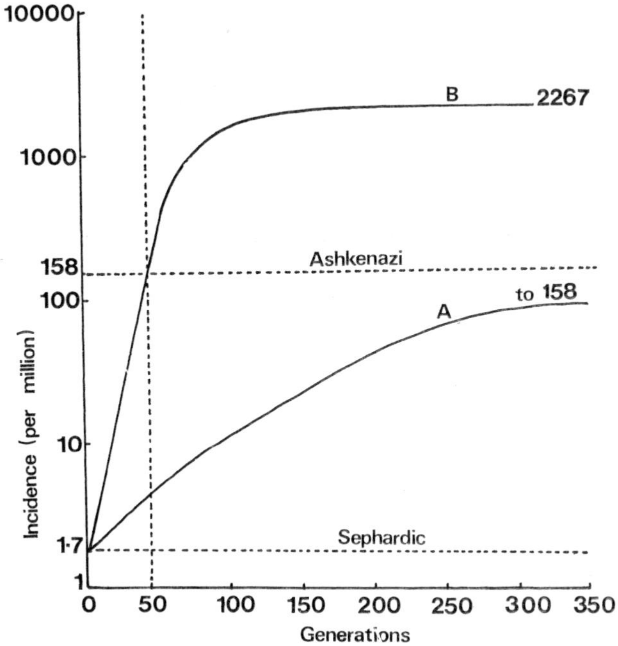

Fig. 2.10.—The relationship between the present incidence of Tay-Sachs disease in Ashkenazi Jews and possibly heterozygote advantages giving rise to this incidence. *A*, Fitness Tt 1·01295; *B*, Fitness Tt 1·05265.

In order to reach the observed frequency among Ashkenazi, Tay-Sachs heterozygotes must have more than 5 per cent reproductive advantage. The disease must therefore be increasing at an observable rate. (*From* Dyer, 1970a.)

million births and this indicates that the incidence of the disease must at present be increasing rapidly among Ashkenazi Jews. It appears that the rate of change in the incidence of the disease at the present is about 7·5 per cent per generation or between $\frac{1}{4}$ and $\frac{1}{3}$ per cent per year. A change of this magnitude should be measurable in an appropriate study. It seems possible, in other words, actually to detect evolution occurring in large human populations by demographic means.

There are a number of other examples in which heterozygous advantage, and in particular fertility differences, have been postulated to explain the relative high frequency of homozygous detrimental genes and the fact that the gene in question might be increasing in frequency. One such case is that of cystic fibrosis of the pancreas, which in various surveys of Caucasian populations has an incidence of about 4·0/10,000. Two recent surveys (Danks, Allan, and Anderson, 1965; Knudson, Wayne, and Hallett, 1967) have shown that heterozygous carriers of cystic fibrosis have a higher number of offspring than non-carriers. Knudson and others found that grandparents of cystic fibrosis patients had an average of 4·34 offspring compared to 3·43 for controls ($P < 0{\cdot}01$). The magnitude of the relative disadvantage s of the normal homozygote compared with the heterozygote is given by

$$\frac{1 - s}{1} = \frac{\text{average size of sibship for normals}}{\text{average size of sibship for heterozygote}}.$$

The value of s obtained in Australia by Danks was 0·10 and in the U.S.A. by Knudson was 0·21. These figures are very large and the Caucasian population cannot be in equilibrium with respect to this gene if either of them are correct.

Knudson and his co-workers calculate that a heterozygous carrier advantage which began operating 23 generations ago with a coefficient of selection of 0·1 could explain the observed incidence. If this is so the gene frequency for cystic fibrosis will continue to rise.

A third important example of genetic evolution occurring in human populations is less clear cut, partly because the genetic constitution of the founders, unlike that of the Jews, is not accurately known, and also, again unlike the Jews, racial amalgamation and gene flow has occurred on a fairly large scale. It concerns the frequency of those genes, one of whose functions is to confer some protection for its carrier against the malarial parasite, and how the frequencies of these genes are changing in populations removed from areas of endemic malaria or which are in areas where malaria has now been eradicated.

One of the principal genes giving such protection is the sickle-cell gene. This gene produces an altered form of haemoglobin, the iron-containing red pigment in the red blood-cells responsible for carrying oxygen. It has long been known that individuals heterozygous for the sickle-cell gene show no morbid effects because half of their haemoglobin is normal and carries oxygen efficiently. Individuals homozygous for this gene, however, develop severe anaemia and almost always die young from distortion of their red cells by crystallized haemoglobin. Despite this almost inevitable elimination of the gene in the homozygous condition it appears that the frequency of heterozygotes is stable at about 40 per cent of the population in certain African tribes. It is reasonably certain that the reason for this high frequency of genes is that the heterozygotes have a much greater resistance to certain forms of malaria than those without the gene at all. If 75 per cent of the homozygotes for the sickle-cell gene die without reproducing, the heterozygotes must have a fitness of about 1·26 that of the normal homozygote in order to maintain an equilibrium level of 40 per cent heterozygotes. Clearly the sickle-cell gene will only remain at this level in the population if this heterozygote advantage remains. If these populations are removed from malarial areas, for instance, or if malaria is eradicated, then this stable situation will be upset and we will expect gene frequency changes to ensue. This is just what happened among those Negroes removed from Africa to the U.S.A. during the slave trade. It has been calculated that the average incidence of the sickle-cell gene in the original slave population was 22 per cent. These Negroes, in the U.S.A. at least, have now been diluted by between 20 per cent and 25 per cent White and Indian admixture and therefore a sickle-cell frequency of perhaps 16–17 per cent would be expected. The actual overall incidence of the sickle-cell trait in the United States Negroes is about 9 per cent. Such a change between expected and observed gene frequencies would, it is calculated, have occurred in about 12 generations (300–350 years) and this fits very well with the known details of the slave trade, from malarial Africa to the non-malarial western hemisphere, which commenced during the sixteenth century. The opposite direction of change, namely an increase in the sickling allele, has occurred in parts of West Africa and has been related to man's destruction of the tropical rain forest, which in turn has provided new habitats for the mosquito vector (*Anopheles gambiae*) and the plasmodium of falciparum malaria (Livingstone, 1965).

Although it was stated that this example was less clear cut than that of gene frequency changes in various Jewish populations, there is one confirmatory piece of evidence which, although on a somewhat smaller scale, is just as striking. The two Dutch territories of Surinam and Curaçao imported Negro slaves for a rather short time 200 to 300 years ago from what was then the Gold Coast. These two territories have broadly similar environments except that malaria is absent from Curaçao but highly endemic in Surinam. The present frequency of the sickle-cell gene in Southern Ghana, where malaria is prevalent, is 20 per cent. In Surinam this frequency has remained at 20 per cent, but in Curaçao, where in the absence of malaria the selective superiority of the heterozygote is obviously removed, the sickle-cell gene has dropped to 5 per cent (Firschein, 1961). Selective gene frequency change has occurred in one locality but not in the other.

As we have seen, the type of mutations which can occur, and therefore the nature of the raw material on which natural selection can operate, can be very varied. They can be fully dominant or recessive or they may have some heterozygous effects which can be harmful or beneficial to varying extents. *Figs.* 2.11 and 2.12 are designed to show the lengths of time which a range of genes would

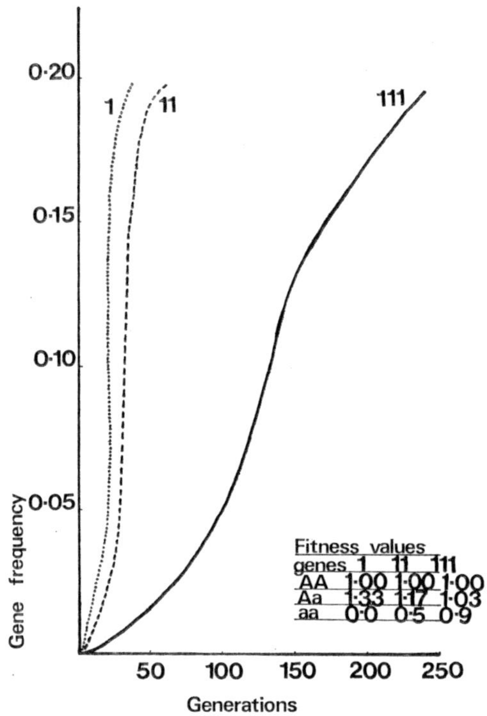

Fig. 2.11.—Relative fitness values of homozygotes and heterozygotes for one segregating gene which will give the same final equilibrium value but will take different lengths of time to reach final equilibrium. (*From* Livingstone, 1964.)

take to change in frequency under certain specified conditions. *Fig.* 2.13 showing calculated possible changes in the frequency of sickle-cell haemoglobin towards its equilibrium value is shown for comparison. The species *Homo sapiens* is more than 300,000 years old, or more than 10,000 generations; all of the theoretical examples given in these figures show that this period of time is more than sufficient to allow an enormous amount of genetic differentiation, enough, in fact, to allow for all the racial differences now observed to have originated subsequent to the separation of the species itself from its ancestor *Homo erectus*. As we shall see this is of some importance in considering the genetic organization of the whole human gene pool and that of the different races in particular. It is on the whole theoretical evidence, merely indicating the possibilities rather than what actually has happened. The existence of at least one undoubted case of the species-wide spread of a mutant in roughly this period of time would be significant confirmatory evidence of the correctness of

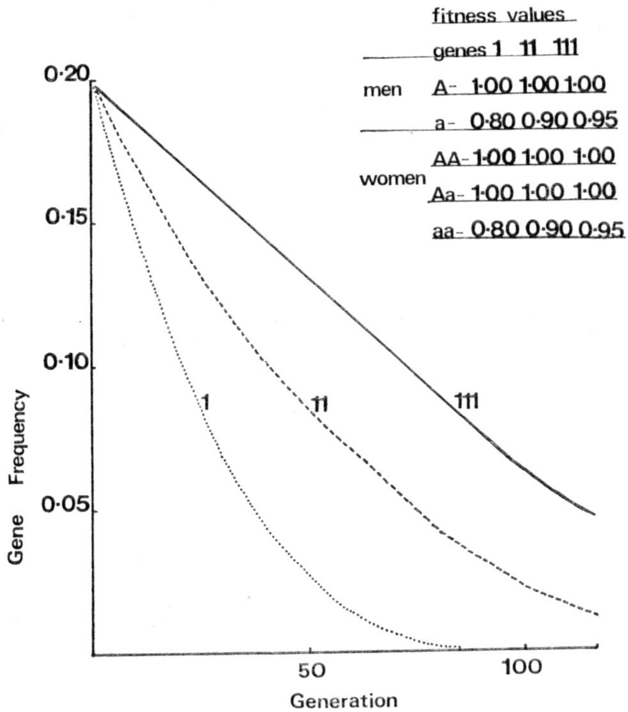

Fig. 2.12.—The reduction in frequency of a sex-linked gene under various fitness conditions, showing various periods of time to elimination. (*From* Livingstone, 1964.)

these theoretical speculations, although because of the difficulty of obtaining fossil evidence on this point such a case would be difficult to find.

The serum of human blood contains a complex mixture of many different proteins, one of which is haptoglobin. Its function is to bind with free haemoglobin to form a soluble complex, possibly to minimize the loss of iron in the body which would otherwise occur by excretion as free haemoglobin. There are three commonly occurring haptoglobin types in man—Hp 1-1, Hp 2-1, and Hp 2-2. Electrophoretic analysis shows a number of structural differences at the level of the polypeptide components of the protein molecule (*see Fig.* 2.14). The α polypeptides produced by the genes Hp^{1F} and Hp^{1S} each contain 83 amino-acids and differ only in a single one (Black and Dixon, 1968). The α polypeptide produced by Hp^2 has 142 amino-acids and appears to be a linear fusion of the Hp^{1F} and Hp^{1S} α polypeptides, although with a sequence of 24 amino-acids missing at the site of fusion.

This almost certainly arose as the result of some kind of unequal crossing over in an individual heterozygous for Hp^{1F} and Hp^{1S} as indicated in *Fig.* 2.15. The characteristic polymerization of the haptoglobin molecule brought about by the Hp^2 polypeptide is easily detected by electrophoresis in man, but has

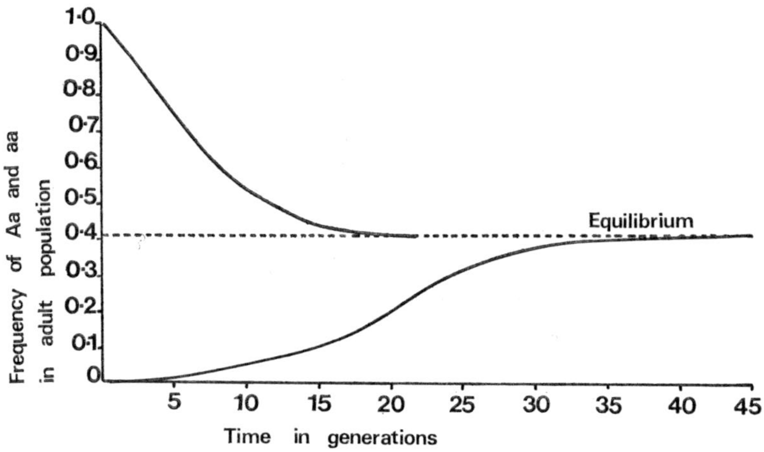

Fig. 2.13.—The change in sickle-cell homozygote-heterozygote frequency in the adult population which would occur from a high or a low level when the fitness of the normal homozygote, sickle-cell homozygote, and heterozygote are 0·95, 1·19 and 0·30 respectively. (*From* Allison, 1955.)

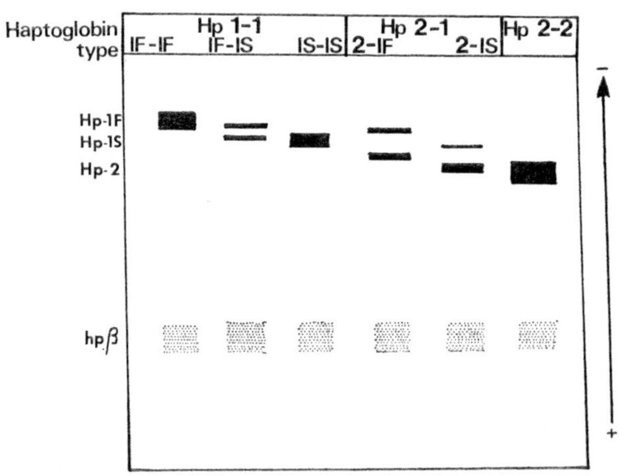

Fig. 2.14.—Electrophoretic separation of Hpα chains in different haptoglobin types. (*From* Harris, 1970.)

never been seen in higher apes. It is probable, therefore, that the event giving rise to the Hp^2 allele occurred only after the evolutionary separation of the hominid line. And yet today it is the commonest of the three alleles in most human populations (*see Fig.* 2.16) and must, therefore, be conferring some selective advantage on the individuals carrying it. We do not know, of course, just when this mutational event occurred during hominid evolution but the widespread pattern of Hp^2 in *Homo sapiens* suggests that its spread within this species must have been very rapid.

Unfortunately our knowledge of the formal genetics of man is limited, and a great deal of the observed morphological and physiological differences

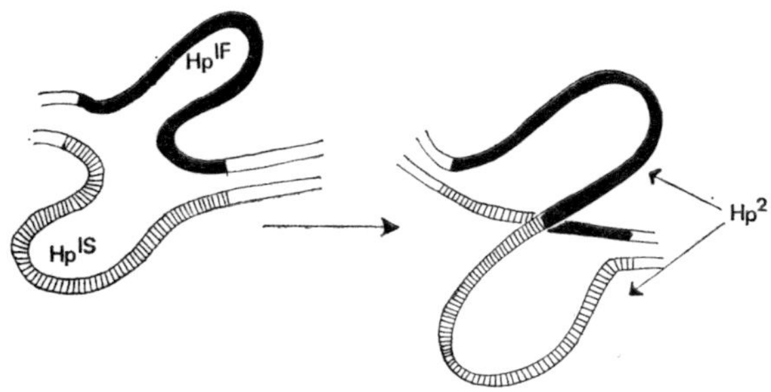

Fig. 2.15.—Diagram illustrating postulated formation of Hp^2 allele as a result of unequal crossing and between Hp^{1F} and Hp^{1S} alleles. (*From* Harris, 1970.)

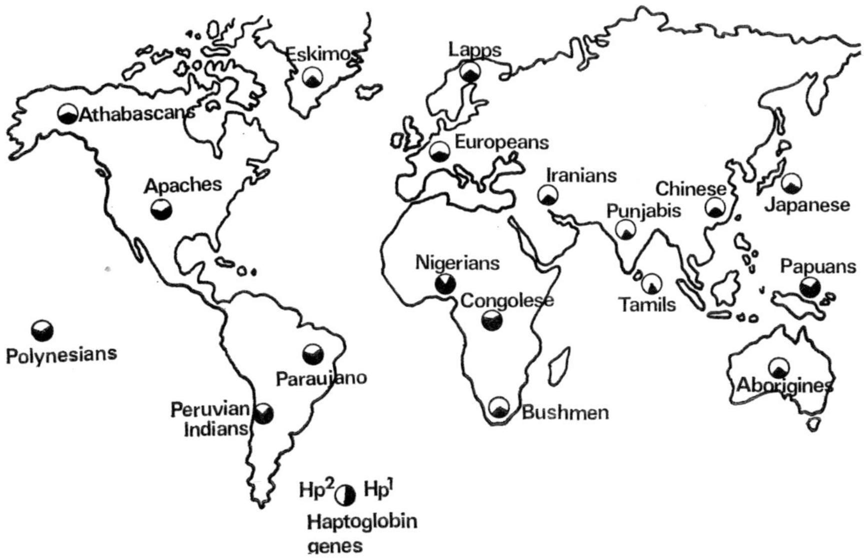

Fig. 2.16.—The world distribution of the Hp^1 and Hp^2 alleles. (*From* Kirk, 1968.)

between the races cannot yet be fully described in genetic terms. Nevertheless, some of the obvious differences between different races and people can be explained in general as being due to the effects of natural selection on genetically controlled characters. We can briefly consider some of these under the headings of adaptation to altitude, cold, and heat, although much of the evidence is necessarily speculative at this stage.

Most of the adaptations to altitude known are simple physiological alterations, merely indicative of man's overall phenotypic plasticity, most of which can be gained or lost in a fairly short period after transition from one environment to the other (Lasker, 1969). Among races living permanently at high altitudes, however, some differences can be recognized which are probably genetically controlled and these differences apparently confer selective advantage on the peoples concerned. Thus, for instance, among people living in the Peruvian Andes at a height of more than 15,000 ft., growth and skeletal maturation is retarded. The consequent relative stunting is probably advantageous. Chest measurements do not follow this trend toward small dimension though. Indian mountain dwellers of South America, while generally developing more slowly than coastal people, develop a larger thorax and greater lung capacity. This clearly is of some advantage in the rarified oxygen at high altitudes. Among animals the llama, the vicuna,and Bolivian goose, all native to high altitudes, have high values for red-cell count and increased affinity of their haemoglobin for oxygen. These characters are retained by first-generation llamas born at sea level. If this sort of genetic character can be fixed by natural selection, it is possible that similar changes could be effected in man.

So far as adaptation to cold is concerned, M. T. Newman (1953, 1961) has applied to man the two rules which express the relationship between bodily physique and temperature. Bergmann's rule states that, in bodies of the same shape the larger one has relatively smaller surface area: cold adapted animals therefore tend to be large. Allen's rule states that short extremities further increase the ratio of mass to surface area and that cold adapted animals have relatively short limbs. In man general body size is, on the average, positively correlated with climate, especially the climate in the coldest month of the year, and this association is best shown by the contiguous populations inhabiting the large land areas of the northern hemisphere in which there is a gradient from large average size in the north to smaller in the south. There are many exceptions to both Bergmann's and Allen's rules, but clearly there is some association of body form with climate such that metabolic efficiency is, on the whole. maximized (Roberts, 1953; and *see* Chapter 6, s. 6.*2.a*).

Although there are clear-cut differences in the ability of different racial groups to withstand low temperature, to escape frostbite, and so on, in almost all cases these are acquired abilities which can usually be developed within the space of a few weeks or months. The same is broadly true of adaptations to heat tolerance. There is one trait, however, in which the differences in body size between peoples living in different climates is consistently manifest and appears almost certainly to be due to genetic factors. This difference is in mean birth weight. That these differences are genetic is indicated by the fact that there is no immediate effect of climate on birth weight when first generation Chinese, Indian, or European babies are born in climatic regions other than those in which they normally live. Further data on human birth weight and the forces of selection operating on it are discussed in section 6.*2.c.* There it is shown that there appears always to be a strong element of stabilizing selection acting on these different birth weights so that the optimum birth weight in different populations is very different, and has probably been so for a number of generations. The relatively small degree of directional selection serving to change birth weight which is normally operative, is further evidence in support of this view.

Other morphological features may also be related to thermoregulation. Nose form, for example, seems to be adapted to the degree of need to moisten and warm the air inhaled. Noses are certainly narrow in colder zones and areas of excessive dryness, and the average ratio of width to the length of the nose, the nasal index, is highly correlated with climate, especially with vapour pressure.

Table 2.12.—Nasal Index and Climate
(Harrison and others, 1964.)

Nasal Index (146 groups living) correlated with	Correlation Coefficient	Standard Error
Dry-bulb temperature	0·63	0·050
Relative humidity	0·42	0·068
Dry-bulb temperature and relative humidity	0·72	0·040
Wet-bulb temperature	0·77	0·034
Vapour pressure of the air	0·82	0·027

Table 2.12 shows some examples of this association of nose shape and climatological factors. Both Mongoloid and Negroid races have depressed nasal roots and it has been suggested (Gates, 1956) that this is the original primitive condition, with the long, thin, high nose of Caucasoid populations a later development in response to living in cool temperate climates of the northern hemisphere.

A more general hypothesis has been advanced by Coon, and others (1950) that the Mongoloid face shows a number of features particularly adapted for living in conditions of extreme cold. These include the reduction of brow ridges, and with them the frontal sinuses, and the flattening and widening of the eye orbits and malar regions, to permit fat padding, and the reduction of the prominence of the nose. All this, though, is very much an unproved hypothesis.

2.2.c. *Natural Selection and Skin Colour*

There are two factors above all which are involved in controversy in the field of race relations. One is the possibility of differences in racial intelligence and the other is the obvious differences in skin colour. These are, therefore, both worthy of special study throughout this book. In this section we shall try and examine the role that natural selection has played in the origin of racial differences in skin coloration. A fuller analysis of the genetics of skin colour is reserved until later, but it is appropriate now to consider how the observed differences might have been brought about by natural selection.

The colour of human skin depends on two main factors; the blood in the smaller vessels of the dermis and the amount of the dark pigment, melanin, in the epidermis. The melanin is formed by specialized cells, the melanocytes, situated in the basal layers of the epidermis. In Negroes the pigment granules formed by these melanocytes are both more numerous and larger than in Europeans. Now the ultraviolet radiation contained in sunlight, which is damaging to tissues, is strongly absorbed by melanin and, since dark-skinned people are in general found in tropical regions, it is usually considered that this distribution is due to natural selection in favour of pigmented skin as a protection against strong sunlight. If, then, we argue that dark skin is present as a protective agent, we must suggest a mechanism whereby it is advantageous to a large proportion of the human race to become depigmented. Why, in other words, are there any White races at all? One of the effects of ultraviolet light is to stimulate the production of vitamin D in the skin, and it is suggested that people with dark skin might be at a disadvantage in this respect, particularly in northern latitudes, and therefore it is advantageous to lose the excess melanin.

There are in fact three possible harmful results from exposure of the human skin to ultraviolet light: sunburn, suntan, and skin cancer. The changes in the skin associated with sunburn are rather complex but protection can be achieved by a thickening of the corneal layer and increase in the melanin content of the skin, and migration of this melanin from basal cells towards the surface. The damage to the skin by sunburn appears to involve destruction of the sweat glands, and during the period of maximum reddening and blistering an individual's heat-regulatory ability may be severely disturbed. The biological hazard is quite a serious one and it is probably of considerable selective importance that the threshold for sunburn is much higher in Negroes than in White races (Blum, 1959).

There is a great deal of experimental evidence that ultraviolet light can damage DNA, the hereditary material, and there is also some experimental work in mice and in men suffering certain genetic defects that this UV-induced damage is directly implicated in the production of various skin cancers and tumours (Dyer, 1969f). There is also a certain amount of circumstantial evidence for the role of ultraviolet light in causing cancer of the skin in man.

In very sunny parts of the world such as South Africa, Australia, and parts of southern U.S.A., the incidence of skin cancer among White individuals is considerably higher than it is among the pigmented inhabitants of these regions. Secondly, among these White populations 95 per cent of the skin cancer occurs on the two principally exposed areas of the body, the face and the back of the hand. Thirdly, within White populations the incidence of skin cancer is greatest in those regions of the world receiving most sunlight and therefore most UV, particularly that of the shorter wavelengths most likely to be responsible for carcinogenesis. *Fig.* 2.17 shows one aspect of this relationship in the U.S.A.,

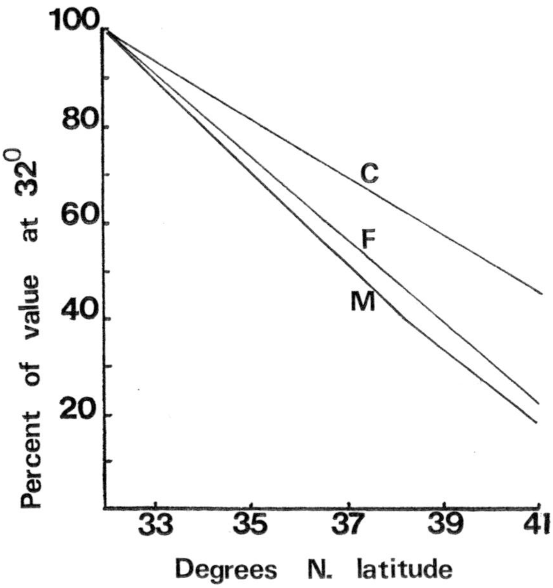

Fig. 2.17.—Variation with latitude of the carcinogenic radiation; and incidence of cutaneous cancer. (*From* Blum, 1959.)
C, Estimated maximum variation of annual carcinogenic radiation with latitude.
F, Variation of skin cancer in White females in the U.S.A. with latitude.
M, Variation of skin cancer in White males in the United States with latitude.

and it is also found that, for instance, skin cancer is about twenty times more frequent in Queensland than in England.

Thus it seems reasonably certain that the geographical and racial distribution of dark skin colour is controlled by the protective effect of dark pigmentation, both from sunburn and cancer. While the darkest skinned individuals are found in the regions of higher UV intensity and also where its shortest wavelengths penetrate the atmosphere, it does seem possible that dark pigmentation may have arisen independently more than once. The dark Caucasoid people of south India and Arabia, the Oceanic Negroes, and native Africans appear to be rather distantly related according to most other genetic characters, but all possess dark pigmentation. If this hypothesis of multiple origin of dark pigmentation is correct, it raises a number of problems regarding the rate of spread of colour genes, in particular whether the skin colour distribution observed today can be explained in terms of the anthropological evolution of man. Would it, in fact, be possible for either the Black or White races to evolve their present skin colours during the period of existence of *Homo sapiens* as a species? Recent theoretical work by Dr. F. B. Livingstone, of the University of Michigan, provides some important evidence on this point (Livingstone, 1969a, b, d).

When we examine those areas of the skin which are not normally exposed to sunlight, it appears that the amount of pigmentation is one of those characters least affected by environmental factors. Although there are many genetic loci known for the production of melanin in mammals (Searle, 1968), a number of recent studies in man (Stern, 1953, 1970; Harrison and Owen, 1964; Harrison, Owen, da Rocha, and Salzano, 1967) suggest that most of the observable variation in skin colour can be explained on the basis that only four principal loci are involved. A more detailed consideration of the genetics of skin pigmentation in man is given in Chapter 6, but we may for the moment take this as our working hypothesis.

If these genes are unlinked and possess two alleles with equal and additive effects, there will be nine possible phenotypes, ranging from homozygosity for all alleles giving the whitest possible skin, to homozygosity for all alleles producing the maximum amount of melanin and therefore the darkest skin. Disregarding, for the moment, genetic complications concerning dominance relations and the details of selective differentials, it is clearly possible to plot the paths of genetic change by supposing that the genotypes associated with a particular skin colour contribute disproportionately to succeeding generations, and writing a computer programme to simulate the many generations of reproduction, genetic recombination, and natural selection necessary to show up any phenotypic change. This is the approach which Dr. Livingstone has recently adopted (Livingstone, 1969d). Among the situations which he investigated were the changes in phenotype in populations originally at a midpoint value which were selected for either darker or lighter skin colour; the selection of either light or dark phenotype starting from the opposite extreme; and the evolution of dark skin from light supposing dominance to operate. The detailed procedure was as follows: for each combination of gametes the amount of melanin (MEL) was established and the fitness of this combination was set equal to 1-FIT(WOPT-MEL)^2, where WOPT is the melanin production of the optimum phenotype and FIT is a factor to adjust the amount of selection. When FIT was set equal to 0·004, 0·002, or 0·001, the maximal amount of selection was 25·6 per cent, 12·8 per cent, and 6·4 per cent respectively against one extreme phenotype when the opposite was the optimum phenotype, and the amount of melanin in the population varied from 0·0 to 8·0.

Livingstone's results are represented here as *Figs.* 2.18–2.22. The number of generations before a really stable situation is reached varies from about 150 up to 1500 or so, depending on the selective values and the degree of dominance

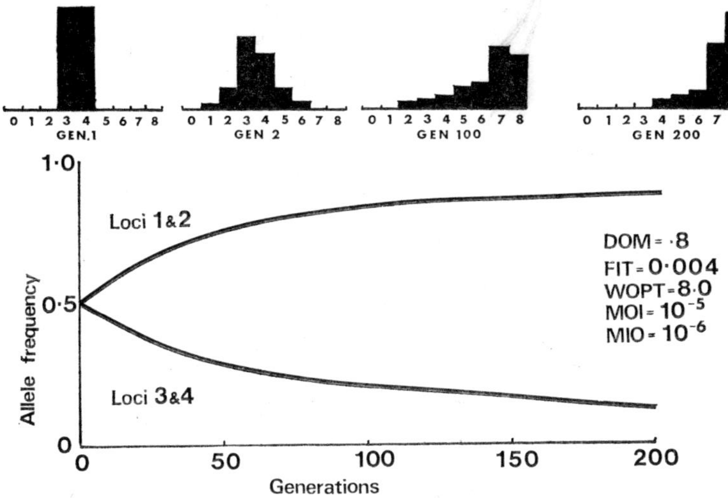

Fig. 2.18.—Change in allele frequencies and the distribution of phenotypes in a population which began as the cross between two populations homozygous for alleles contributing maximally to an absence of melanin and for alleles producing a maximum amount of melanin respectively. (*From* Livingstone, 1969d.)

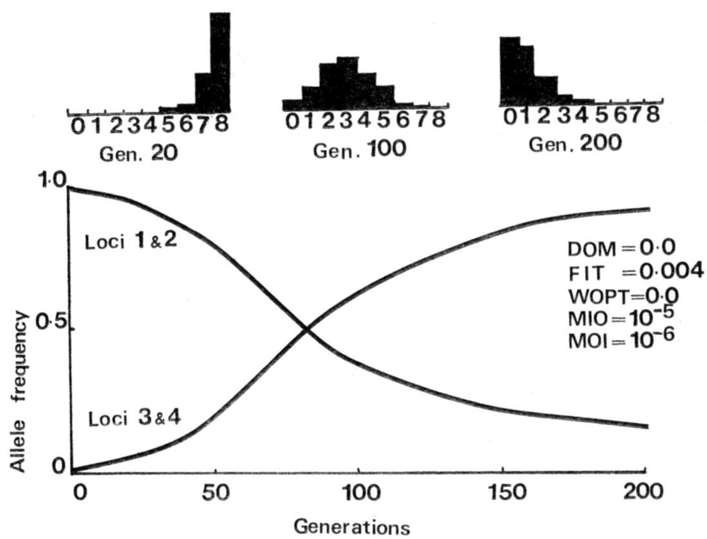

Fig. 2.19.—The change in allele frequencies and the distribution of phenotypes during the evolution of light skin colour with no dominance. (*From* Livingstone, 1969d.)

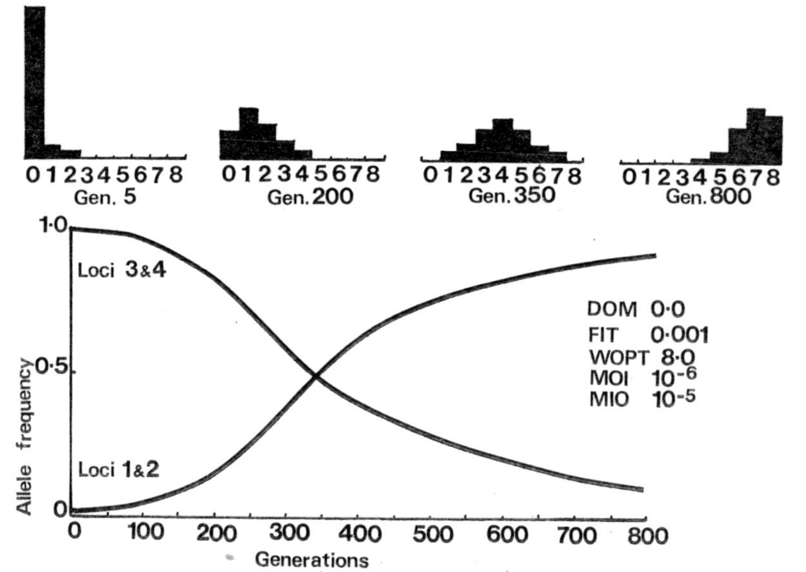

Fig. 2.20.—The change in allele frequencies and the distribution of phenotypes during the evolution of dark skin colour with no dominance. (*From* Livingstone, 1969d.)

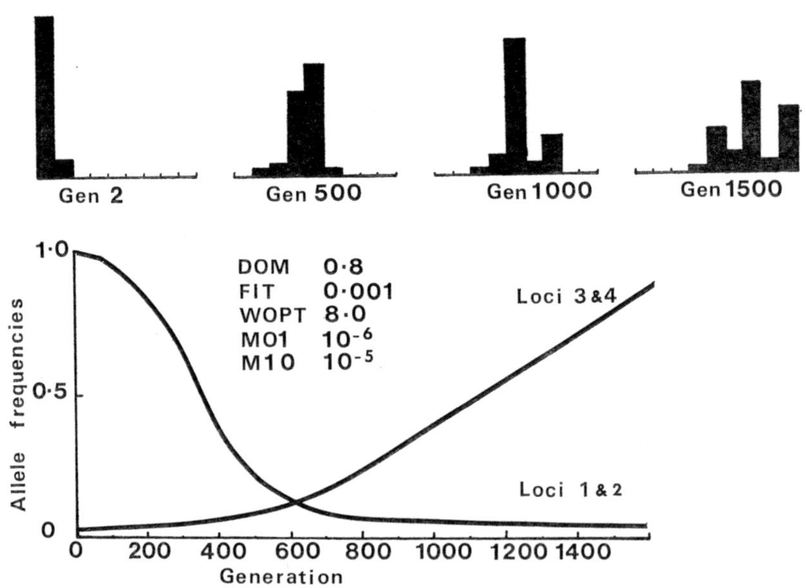

Fig. 2.21.—The change in allele frequencies and the distribution of phenotypes during the evolution of dark skin colour with dominance. (*From* Livingstone, 1969d.)

chosen. In other words, it appears possible that a dark skin can evolve from a light one in as few as 4500 years, although normally it would be expected to take longer than this.

The validity of predictions such as these can only be checked against the situations as we see them today. Firstly, there are a number of exceptions to the rule that those people with the darkest skins live nearest the equator. The American Indians living in the tropics, for instance, do not have the very dark skins characteristic of indigenous tropical populations of Africa and India.

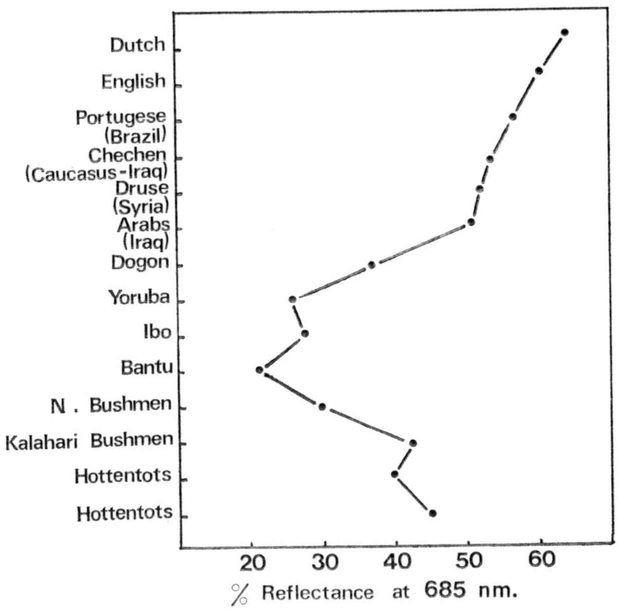

Fig. 2.22.—The north–south variation in skin colour in Europe, Western Asia, and Africa as measured by percentage reflectance at 685 nm. (*From* Livingstone, 1969d.)

Now it seems likely that most of these Indians crossed into South America between 25,000 and 15,000 years ago, after the establishment of the land bridge from North America. J. B. S. Haldane (1956) therefore concluded that it must take at least this period of time, between 500 and 1000 generations, for very dark skin colour to evolve. In Livingstone's model (*Fig.* 2.21), a population which began with very light skin colour would in this time have only evolved the dominant alleles which contribute to dark skin colour and not the recessive ones. In 500 generations a population with intermediate skin colour would be expected, such as that represented by the Aguarana Indians of Peru analysed by Weiner, Sebag-Montefiore, and Peterson (1963). In the old world there have certainly been more than 1,500 generations or 45,000 years for *Homo sapiens* to disperse, and therefore for very dark skin to evolve. This would be expected to result in a complete gradation of skin colours from pole to equator. Such a north–south variation in skin colours does in fact exist as we have already mentioned (*Fig. 2.22*). There are a number of other situations which show that skin colour is one of the more rapidly evolving human traits. For instance, among the dark-skinned semitic peoples of the middle east, the Oriental Jews are normally significantly darker than the Sephardic and Ashkenazi Jews, although these peoples have only been separated for about 2000 years. In India the major Caucasoid influx was probably about 8000 years ago and now most Indians are very much darker than most other Caucasoids. The unknown factor in this situation, however, is the amount of mixing which took place with Indian aboriginal tribes who are very dark skinned.

One final check on the reality of these models is possible: the intensity of selection. The models suggest that with a maximum of only about 6 per cent selection the skin colour difference among human populations could certainly arise within the time which seems to be available from fossil and archaeological evidence. We have already mentioned selective values of 4·5 per cent in connexion with Tay-Sachs disease, of either 10 per cent or 20 per cent in connexion with cystic fibrosis, of 26 per cent in connexion with sickle-cell anaemia, and in s. *2.3* we discuss further levels of natural selection which operate in man, some greater and some less than 6 per cent. All in all there seems no doubt that the differentiation of the human species with regard to skin colour could easily have taken place within a period of time rather shorter than that of the existence of the species as a whole.

2.2.d. Genetic Drift

'Genetic drift' is a single name which is used to describe at least three different phenomena. The first involves those circumstances in which the size of the breeding population becomes very small, so small in fact that the random accidents of sampling which accompany the processes of gamete formation and fertilization become as important, if not more important, than natural selection in determining the genetic constitution of the succeeding generations. Because random processes are inevitably involved at these two stages of reproduction, genetic drift in this sense must occur. Its long-term importance, however, is still disputed, since some authorities suggest that the population size at which drift can become operative must be so low as to make such populations of negligible importance on an evolutionary scale. Natural selection, it is argued, will in almost every case be the deciding factor in determining gene frequencies.

One variant of this form of genetic drift is the so-called 'founder principle'. By this is simply meant the fact that when a new population is founded on an island, say, or in a new ecological habitat by a relatively small number of migrants, these few migrants are unlikely to carry with them the full range of genetic variability contained within the parental population and therefore this new population will be genetically distinct. Again, this phenomenon must occur, but its importance is also the subject of debate.

A rather different phenomenon is the possible existence of genuinely neutral alleles at some genetic loci. Such alleles, by definition, have no selective values either positive or negative and might, therefore, change in frequency in a purely random way. Although the existence of such selectively neutral alleles is still a matter for debate, supporting evidence comes from the enormous amount of genetic variation normally present in human populations, suggesting that some of it at least must be selectively neutral, and also from the apparently random way in which individual protein molecules such as haemoglobin vary in many of the amino-acids of which they are composed (Arnheim and Taylor, 1969; King and Jukes, 1969; Dyer, 1971c). For the opposite point of view *see*, for example, Clarke (1970), Maynard Smith (1970), and Angseesing (1972).

The final phenomenon associated with the term 'genetic drift' is the possibility that different combinations of alleles and genes in different individuals are in fact effectively equivalent in terms of fitness under natural selection. Genes not only act but interact, and although each individual gene in any one organism may normally be subjected to favourable or unfavourable selective forces, many different combinations of genes, when brought together in one individual, may give rise to phenotypes which are of absolutely equal fitness.

If man had ever lived in one single interbreeding population then genetic drift would certainly never have been very important. But man was for a long time merely a collection of isolated food-gathering tribes, of which the Aborigines of Australia were, until fairly recently, living examples. The total number of men alive was also very small until historical times and therefore there was, as well as physical separation between different populations, considerable fluctuation in their size as a result of famines, epidemics, and wars. All of these circumstances, when considered in conjunction with the considerable migrations among human populations which are known to have taken place, are just those in which genetic drift may have occurred. It seems likely *a priori*, therefore, that genetic drift may have played a part in racial differentiation and in fact there are a number of examples known today which suggest that this is indeed so.

The Blood and Blackfeet Indians of North America differ not only from other North American Indian groups but from all other populations in containing the A allele of the ABO blood group in frequencies of more than 50 per cent. This is probably the result of genetic drift in initially small populations. The Naskapi Indians have a serum albumin variant at a level of 25 per cent, even though this variant is unknown anywhere else in the world. Other probable examples of drift are known in certain Eskimo tribes in northernmost America. Most of the Aleutians and Eskimos who live along northern America from western Alaska to eastern Greenland have frequencies of the A allele of around 30 per cent and frequencies of the B allele of about 6 per cent. For the MN group most Eskimos have frequencies between 78 per cent and 91 per cent of the M allele, but Eskimos living in Baffin Island and in Labrador differ in both of these systems. They have only 7 per cent A allele and 1 per cent B allele, and in the MN system they have only 56 per cent M. Eskimos of Alaska have a frequency of up to 25 per cent of a silent allele for serum cholinesterase, whereas in the rest of the world it is very rare. These examples are merely inferential. Bentley Glass (1954, 1956), however, has investigated a number of other situations where genetic drift has almost certainly occurred, and he describes one situation in particular in which the gene frequencies were found to have changed over the short span of only three generations, almost certainly as a result of genetic drift. He investigated the genetics of the German Baptist Brethren, or Dunkers, who emigrated from Germany to the U.S.A. in the nineteenth century and have remained isolated ever since, with a population size of around 60,000. Their A blood group frequency at 60 per cent was found to be different both from West Germany (45 per cent) and from U.S.A. (40 per cent) and they also differed in the MN blood group. In both West Germany and the U.S.A. the frequencies are 30 per cent M and 20 per cent N, the rest being the MN heterozygotes. Among the Dunkers there are 44·5 per cent M, 13·5 per cent N, and 42 per cent MN heterozygotes. Not only has there been a marked shift in the frequencies of the M and N phenotypes but the frequency of heterozygotes has declined as would be expected in a small population isolated from all others.

One clear-cut example of the operation of the founder principle is that of the spread of the gene causing porphyria variegata in South Africa. This is a very unfavourable gene since it causes a defect in the synthesis of porphyrin, part of the haemoglobin molecule, and affected individuals are liable to develop blisters and abrasions on those parts of the skin which might be exposed to sunlight. Sulphonamides and barbiturates cannot be used on carriers since they become paralysed, often fatally. In this particular case, by carefully tracing the family histories of all known porphyrics, it has been possible to trace all of the carriers back to one of either two immigrants, who arrived in South Africa in 1685 and 1687, were married in 1687, and had eight children, four of whom inherited porphyria. This single mutant allele has spread to the extent that it is now carried by 8000 people. In the rest of the world the gene is rare; among the Afrikaner population it has a frequency of about 1 per cent. This major difference is due simply to the presence of one carrier among the very few original immigrants in the seventeenth century. Pedigree studies have also been used to trace the origin of the gene causing hereditary tyrosinaemia in an isolated French Canadian population living in Quebec. This normally fatal condition occurs with a frequency of 2 per cent in this isolate and has been traced back to a French immigrant of the seventeenth century (Laberge, 1969).

Another condition which confers high sensitivity to sunlight is albinism. Long thought to be inherited in a simple abnormal recessive manner in man, it is now known to be a heterogeneous condition due to a number of different genes (Witkop, 1971). It is curious that some forms of albinism, contrary to expectation, may be of higher frequency in Negroes than Whites (*Table* 6.12). Furthermore in different tribes in the Belgian Congo albinism varies between 1 in 400 and 1 in 40,000 (Waardenberg, 1961); variation which is presumably indicative of the effects of genetic drift. Some American Indian tribes of central America also show complex interactions between selection and drift since the frequency of albinism shows a peak incidence of about 1 in 140 in southwestern United States, falls to about 1 in 6500 southwards in Guatemalan Indians, and rises again to about 1 in 143 further south still in Honduras. The Trobriand Islanders, just north of New Guinea, are an example of a Melanesian population with a relatively high incidence of albinism. This is presumably also due to genetic drift or the founder effect, since albinos, although looked after as children, are not normally allowed to mate in this society.

Another example of the possible operation of the founder effect concerns the genetical differentiation of the colonizers of the northern Atlantic Islands Orkney, Shetland, Faroe, and Iceland. These islands differ with respect to blood groups, gene frequencies, certain non-metrical skeletal differences, and in the incidence of certain diseases such as anencephaly and multiple sclerosis. These differences are discussed by Dr. R. J. Berry (1967, 1969b), who concludes that the islands were colonized by genetically different and probably atypical Viking populations during the late eighth and ninth centuries.

These examples concern mainly single gene differences in fairly small populations. On a somewhat larger scale Drs. M. Nei and Y. Imaizumi have investigated the variability of a number of Japanese populations with respect to the ABO blood groups and found a good agreement with a general theoretical model which suggests that the observed variability is largely the result of the operation of genetic drift in the populations into which the Japanese were subdivided (Nei, 1965; Nei and Imaizumi, 1966a, b). Dr. L. Cavalli-Sforza and his co-workers have demonstrated this phenomenon in action in Northern Italy. They have shown (*see Fig.* 2.23) that genetic variation in frequencies of blood groups between one village and another was greatest in isolated upland hamlets and declined as population density increased farther down the valley in the hill towns, on the plain, and in the city of Pavia (Cavalli-Sforza, 1969b).

The extension of the concept of random genetic drift to larger populations over longer periods of time, however, is somewhat more controversial. Nevertheless, it is instructive to compare the world-wide variation of gene frequencies with those calculated on the assumption that genetic drift alone and not selection is responsible for the variation. This comparison based on the work of Cavalli-Sforza and co-workers (Cavalli-Sforza, Barrai, and Edwards, 1963, 1967; Cavalli-Sforza and Edwards, 1964, 1967 is made in *Fig.* 2.24. Various curves are shown for the increase of variation in gene frequencies with time. This is in fact a measure of the extent to which genes increase in frequency in some races and decrease in others.

Alongside these curves are shown the standard variances of some of the principal genes used in differentiating human populations. It is obvious that

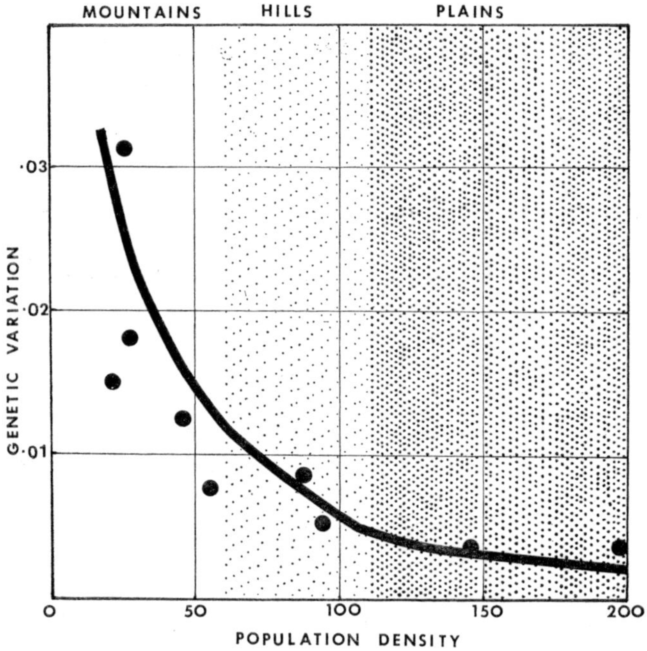

Fig. 2.23.—The relationship between geography, population density, and genetic variation in the province of Parma, northern Italy, showing the importance of isolation and genetic drift in this situation. (*From* Cavalli-Sforza, 1969b.)

even the greatest variance observed would be compatible with a large isolate size of N = 10,000 and a separation time of 300,000 to 400,000 years, i.e., that of the time of existence of *Homo sapiens*. The fact that the normal isolate size in the past was likely to be rather less than 10,000 lends further credence to the viewpoint that genetic drift has been an important process in man's evolution.

Of course, natural selection is known to play some part in fixing the equilibrium values of many of these genes and, for instance, the ABO blood groups and PTC sensitivity are true selectively balanced polymorphisms since they are known to occur in other primates and are therefore of long standing. On the other hand some genes with high variances have most probably been subject to different selective conditions in the different environments in which populations carrying them have lived, so that extra variance must have been added to that due to drift. Nevertheless, the fact that the origin of man's genetic diversity can, by random processes alone, be so easily accommodated within the time of existence of *Homo sapiens* is of importance in racial studies.

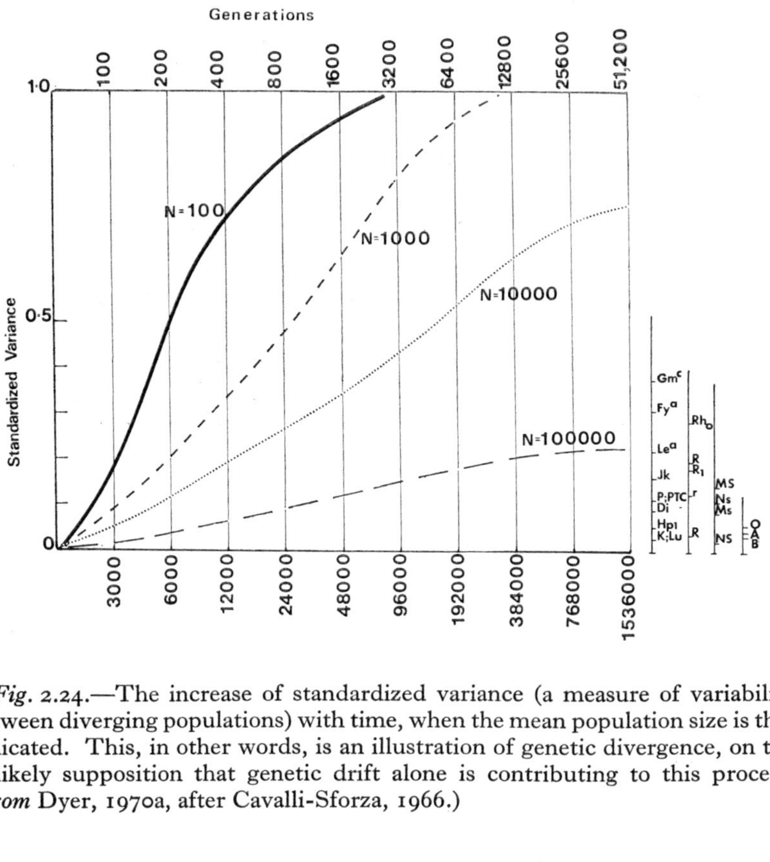

Fig. 2.24.—The increase of standardized variance (a measure of variability between diverging populations) with time, when the mean population size is that indicated. This, in other words, is an illustration of genetic divergence, on the unlikely supposition that genetic drift alone is contributing to this process. (*From* Dyer, 1970a, after Cavalli-Sforza, 1966.)

2.2.e. *Relaxation of Selection*

There is one further phenomenon which may be adding to the degree of racial diversity in man: the relaxation of selection. Man's environment has certainly changed during his evolution from the stone age to modern civilization. Undoubtedly this has meant that the force and direction of selection, and also the characters selected for, have changed. In so far as different races have progressed farther along the pathway from a so-called natural environment, this is a further contributory factor in racial diversification.

Our knowledge of human gene frequencies and selective forces is not yet sufficient to allow direct estimates of the effects of reversal, suspension, or relaxation of selection on various phenotypic traits. However, comparisons have been made of the incidence of various anatomical or physiological defects in primitive societies, in which these defects might be thought to have selective disadvantage, with their occurrence in more advanced societies where it is supposed that selection has been relaxed. Red-green colour blindness is one such trait. Among Eskimos, Australian Aborigines, and other hunters and gatherers the frequency among males is about 2 per cent. Among some American Indians and African Negroes the frequency is about 3·3 per cent. Populations which have evolved even further from the hunting stage and which have been examined for this trait such as Europeans, Filipinos, and some Arabs, show a frequency of 5·1 per cent. (More comprehensive data are shown in *Table* 2.13.) It seems, therefore, that selection no longer operates against eye

Table 2.13.—FREQUENCY OF COLOUR BLINDNESS IN MALES OF DIFFERENT POPULATIONS

POPULATION	RACE	NUMBER	PERCENTAGE
Dutch	European	97	15·5
Todas	Indian	320	12·8
U.S.A. Whites	European	803	11·4
Brahmins	Indian	100	10·0
Russians	European	1343	9·3
Belgians	European	1243	8·0
Hindus	Indian	142	8·1
Norwegians	European	9040	8·0
Scottish	European	464	7·8
Germans	European	1000	7·5
Tonga Islanders	Polynesian	67	7·5
Chinese	Asian	1164	6·9
Turks	European	473	5·3
Marshall Islanders	Micronesian	268	5·3
U.S.A. Negroes	African	2019	3·9
Aborigines	Australian	378	3·3
Navaho Indians	Amerindian	163	2·4
Bagandas	African	537	1·9
Congolese	African	929	1·7
Fiji Islanders	Melanesian	200	0·0
Brazilian Indians	Amerindian	230	0·0

defects of this sort in modern societies and that this divergence in gene frequencies is of fairly recent origin.

It is in fact possible to calculate the differences in selection levels which might account for these differences in gene frequency. If selection occurs entirely in males (since the frequency of colour-blind females must be negligibly small and there are no known heterozygous effects), then the gene frequency at equilibrium $\hat{q}$, which is a function of the mutation rate per locus per generation u

and the selection coefficient against the mutant gene s, where the fitness of the mutant hemizygote is $1 - s$, can be defined as: (Li 1955),

$$\hat{q} = 3u/s$$

If $u = 10^{-5}$ then $s = 0{\cdot}003$ for $\hat{q} = 0{\cdot}01$, $s = 0{\cdot}0006$ for $\hat{q} = 0{\cdot}05$ and $s = 0{\cdot}0001$ for $\hat{q} = 0{\cdot}1$.

Therefore it takes six times the selective forces required to keep the gene frequency at 0·1 to reduce it to 0·05 and thirty times to reduce it to 0·01. It was therefore suggested by Post (1962) and Neel and Post (1963) that natural selection under the rigorous conditions experienced by primitive hunters and gatherers holds their mutant gene frequencies low, whereas the less rigorous environment in agricultural and advanced populations allows a relaxation of selection.

We can take this argument a little further by considering deuteranomaly alone and supposing that selection against it was suspended 4000 years ago. If the average generation time is 25 years, then the mutant gene for deuteranomaly would have been accumulating for 160 generations. We might assume that the mean frequency in 2000 B.C. was 1·25 per cent, which is midway between the 0·7 per cent of the Bantus in Uganda and the 1·8 per cent of the Papuans in New Guinea, which are the only two large samples of primitive peoples to have been accurately tested by anomaloscope. Nine European samples similarly tested have an average of 4·72 per cent. Therefore if this increase of 3·46 per cent has been due entirely to the accumulation of new mutants, then the rate of mutation has been about 0·0346 ÷ 160 or 0·00022/gen. ($= 22 \times 10^{-5}$). This is rather high compared with the values of mutation rates of autosomal dominants shown in *Table* 2.10, but the assumption of 4,000 years since the suspension of selection and the presumed initial value of 1·25 per cent could both be in error, which would reduce this value somewhat to more normal levels. In fact, if we substitute this mutation rate into the equation $s = 3u/\hat{q}$ to obtain the coefficient of selection as indicated above, we find that $s = 0{\cdot}0528$. This seems a remarkably high value for what is apparently a rather minor defect. Nevertheless, even though it may be too large on the basis of the rather uncertain assumptions involved, it indicates that the relaxation of selection may lead to rather rapid divergence of different races.

We can return to the *micro-races* of Great Britain for further illustration of both this point and the degree of racial mixture in that country. Post (1963, 1971) reports that 123,414 recruits for the Royal Navy were tested for colour blindness at 77 stations around Britain. The highest frequencies of colour blindness were found in Cornwall and along the English Channel, which were all in excess of 10 per cent. There was a progressive decrease northwards and eastwards, the lowest values being found in the east of Scotland, which averaged 4·9 per cent. These 77 recruiting stations could be arranged into geographical areas which were homogeneous, each within its area, but heterogeneous amongst themselves. A decreasing cline northwards and eastwards is consistent, says Post (1971), with the pattern of successive cultural invasions of southern Britain from the earliest Neolithic time onwards right through to the Saxons, Danes, and Normans, who had shorter cultural histories than the older inhabitants. These older inhabitants were pushed ever westwards by each wave of invaders, and we have already examined some of the mosaic of *micro-races* in Wales which resulted partly from this succession of invasions. As a final point we may note that the average colour blindness frequency of these naval recruits in Britain was 7·5 per cent compared with 9·0 per cent found in France among 6635 students.

Deformities of the nasal septa (the long bony membranes of the nose) have also been studied from this point of view (Post, 1966, 1969a, 1971). It seems likely that among hunters the need for oral instead of nasal breathing would be a distinct disadvantage, whereas in primarily agricultural communities it would not be so much of a handicap. Certainly such selective factors go some way towards explaining the different frequencies of abnormal septa among many different populations of the world. The data gathered by Dr. R. H. Post, in fact, provide a more searching test of the hypothesis of relaxed selection, because, of the 26 populations he has analysed, 16 are from simple palaeolithic cultures of hunters and food gatherers and there are also samples of European Neolithic skulls and Egyptian Neolithic skulls which can be compared directly with contemporary European and White American populations. Two extracts of Post's data shown in *Tables* 2.14 and 2.15 show how well the hypothesis is

Table 2.14.—NORMAL AND ABNORMAL NASAL SEPTA IN SKULLS OF COMPARABLE NEOLITHIC AND CONTEMPORARY POPULATIONS (*From* Post, 1966. Copyright, University of Chicago Press.)

	No. of Skulls	Percentage Deformed
Egyptians Neolithic	103	12
Egyptians Contemporary	30	37
European Neolithic	192	30
European Contemporary	42	64

supported. In fact, the supposed rate of change between neolithic and contemporary populations seems embarrassingly high to be explained by relaxed selection alone. Five thousand years is probably less than 200 generations and, even for a complex polygenically controlled trait, such as a deformed nasal septum almost certainly is, the mutation rates and rates of phenotypic change must be very rapid. This, though, is in line with evidence mentioned earlier regarding very rapid rates of change for many continuously varying characters (*see* s. *2.2.a* and Dyer, 1968 a, b, 1971c).

Post (1971) mentions a number of other traits which may have differentially altered in frequency in recent years among the races. Among these, myopia and differences in size of the external opening of the tear duct are reasonably

Table 2.15.—FREQUENCIES OF SEPTAL DEFORMITIES IN CONTEMPORARY 'PRIMITIVE' AND 'CIVILIZED' POPULATIONS (*From* Post, 1969a. Copyright, University of Chicago Press.)

POPULATION	NUMBER	PERCENTAGE DEFORMED
Bushmen, South Africa	12	0
Negritos Philippines	3	0
Eskimo*	271	10
American Coloured*	144	24
Negroes West Africa	37	27
American White*	263	56
Czechoslovakia, Bohemia 1880's	86	58
Armenia 1880's	15	67
Crete 1880's	17	76

* Percent of vomer deformities only.

well established (Post, 1969b). Others include some cancers, possibly associated with a reduction in breast feeding, and some congenital abnormalities.

2.3. THE MAINTENANCE OF GENETIC VARIATION

We have already stated that many races in species other than man are incipient new species and therefore must exist as recognizable distinct entities for fairly lengthy periods of time. Even those races not destined to give rise to new species must still exist for sufficient periods of time to be recognized and therefore the genetic variation on which the races are classified must be maintained for similar periods of time. There are, therefore, two aspects to the study of the maintenance of genetic variation: firstly, how it is preserved at individual genetic loci and, secondly, how the arrays of genetic variation characteristic of each race are maintained together.

2.3.a. *Selection and Genetic Polymorphism*

The discovery in the last decade of very high levels of polymorphism in a very wide range of species has posed some interesting problems for theoretical and population geneticists. The nature of this problem will emerge after a discussion of some of the classic examples of polymorphism known in man.

Haemoglobin, the protein which is found in the red blood-cells of man and which carries oxygen around the body, is known to exist in a number of aberrant forms and it has already been mentioned that one of these forms Hb^S, sickle-cell haemoglobin, is virtually lethal when homozygous because of its effects in distorting the red blood-cells. Despite this lethality among homozygotes, many different African tribes carry high frequencies of the Hb^S allele among heterozygotes and groups of people with high, medium, and low frequencies of this allele have also been found in southern India, Greece, and Italy.

A

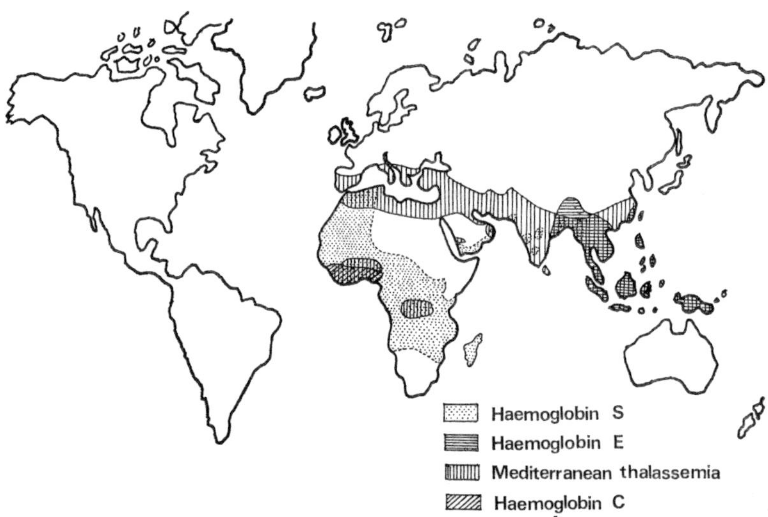

B

Fig. 2.25.—A, The world distribution of falciparum malaria before 1930. B, The distribution of some abnormal haemoglobins showing their relationship with falciparum malaria.

Now as *Fig.* 2.25A and B shows, there is a positive correlation between the occurrence of the sickle-cell gene and the incidence of malaria in many parts of the world. This suggested the hypothesis that we have already mentioned, that Hb^A/Hb^S heterozygotes are more resistant to falciparum malarial infection than Hb^A/Hb^A homozygotes. A small experiment was carried out in which volunteers of the genotype Hb^A/Hb^A proved to be much more susceptible to infection by the malarial protozoan parasite *Plasmodium falciparum* than the heterozygotes Hb^A/Hb^S (Allison, 1954a, b; 1955). Further confirmation is provided by a series of morbidity statistics collected in various parts of Africa, in which it appears that of 104 deaths among African children only 1 was a sickle-cell heterozygote, although if mortality from malaria is the same in the sickle-cell heterozygotes as in other members of these populations, then 23 would have occurred (*Table* 2.16).

Table 2.16.—Incidence of Sickle-cell Trait among African Children whose Deaths could be attributed to Malaria (*From* Harris, 1970.)

Locality	Deaths due to Malaria	Number with Sickle-cell Trait	Incidence of Sickle-cell Trait	Expected No. with Sickle-cell Trait if no Selective Differential
Uganda (Kampala)	16	0	0·16	2·6
Congo (Leopoldville)	23	0	0·235	5·4
Congo (Luluborg)	23	1	0·25	5·7
Ghana (Accra)	13	0	0·18	2·13
Nigeria (Abadan)	29	0	0·24	7·0
Totals	104	1		23·0

Blood group frequencies figure importantly in racial genetics and for some of these, also, selective differences are known to play a part in the maintenance of the observed variation, although the direction of selection pressure may well be different in different parts of the world or environmental conditions or even for pre- and postzygotic stages within the same population. In one survey in Japan, for instance, it was found that significantly fewer children were born to O mothers from A fathers than to A mothers from O fathers (Matsunaga, 1955). This was an important selective mechanism the magnitude of which can be seen in *Table* 2.17. However, later results in a different Japanese community enjoying more favourable economic and social circumstances did not show this phenomenon at all. Hiraizumi (1964) suggested that there is a reduction in viability of children due to ABO incompatibility but it is only manifest in poor environmental conditions. He also suggested that the proportion of O carrying sperm from AO and BO fathers is higher than the expected 50 per cent. The balance of selective

Table 2.17.—NUMBER OF O AND A CHILDREN FROM COMPATIBLE AND INCOMPATIBLE O × A MATINGS (*From* Matsunaga, 1955.)

MOTHER	FATHER	NUMBER OF O CHILDREN Obs.	NUMBER OF O CHILDREN Exp.	NUMBER OF A CHILDREN Obs.	NUMBER OF A CHILDREN Exp.
A	O	282	284·5	430	427·5
O	A	320	275·3	369	413·7

forces maintaining the ABO polymorphism must therefore be a very complex one, a situation further illustrated by a series of studies on susceptibilities to particular diseases among persons of different blood groups. For many diseases there are no significant associations with different blood groups, but as *Fig.* 2.26

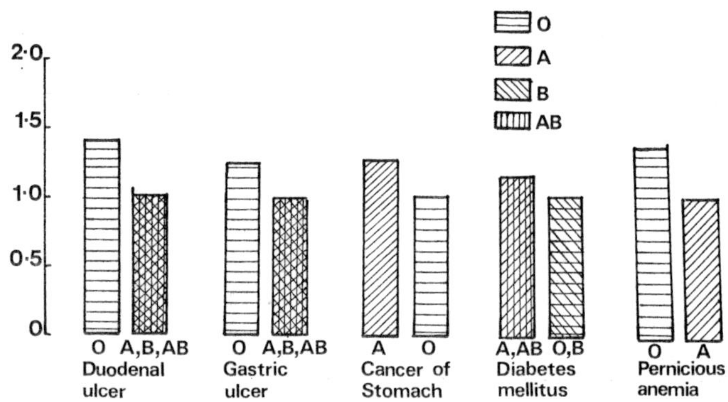

Fig. 2.26.—Relative incidence of five different diseases in patients of different blood groups. (*From Principles of Human Genetics* (Curt Stern), 2nd ed., 1960. Copyright, London: W. H. Freeman & Co.)

shows, some associations have been found. The effective strength of many of these selective processes is no doubt limited but, coupled with the known environmental components in the causative factors of many diseases, their possible part in the maintenance of genetic variability can be appreciated.

The very fact that the allele frequencies of the ABO system occupy, on a world basis, a very restricted part of the possible range of frequencies (*Fig.* 2.27), suggests that there are strong selective pressures serving to keep the allele frequencies stable.

The Rhesus system is a blood group in which the important effects of selection are obviously apparent. Although the system is a very complex one genetically, about which there is still controversy, we need consider here only three genotypes: Rhesus-positive homozygotes (DD), Rhesus-negative homozygotes (dd), and the heterozygote (Dd). A marriage between a Rhesus-positive man (DD or Dd) and a Rhesus-negative woman is, in a genetic sense, an incompatible

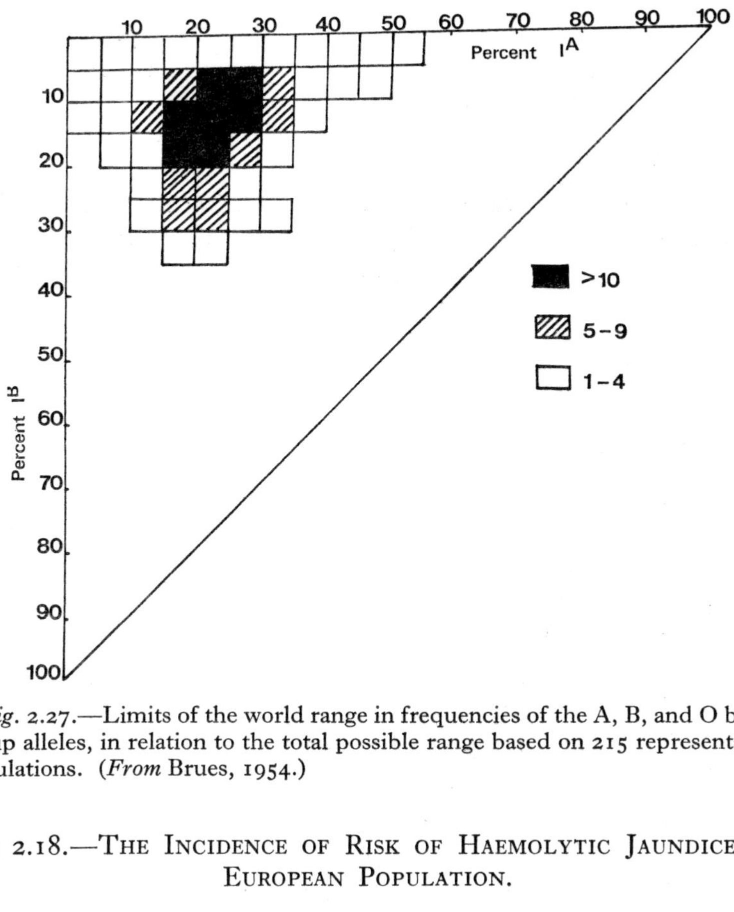

Fig. 2.27.—Limits of the world range in frequencies of the A, B, and O blood group alleles, in relation to the total possible range based on 215 representative populations. (*From* Brues, 1954.)

Table 2.18.—The Incidence of Risk of Haemolytic Jaundice in a European Population.

	Frequency of the Paternal Genotype		
	DD 0·36	Dd 0·48	dd 0·16
DD 0·36	12·96	17·28	5·76
Dd 0·48	17·28	23·04	7·68
dd 0·16	(5·76)	(7·68)	2·56

one. The first heterozygous child suffers no ill effects but its mother produces antibodies against the D antigen during the pregnancy. Subsequent heterozygous children are exposed to these antigens, which diffuse through the placenta, and they are liable to develop a type of haemolytic jaundice which is lethal if not treated. *Table* 2.18 shows the incidence of risk in a European White population remembering that not all of the approximately 14 per cent incompatible matings will produce a diseased child. The possibilities are in fact as follows:

First child	dd	no effect
First child	Dd	no effect
Subsequent children	dd	no effect
Subsequent children (after a Dd child)	Dd	jaundice

It is clear that a population whose D and d alleles are of equal frequency will lose both equal numbers and equal proportions of the two alleles, but if the allele frequencies are initially unequal, the proportional losses of the two

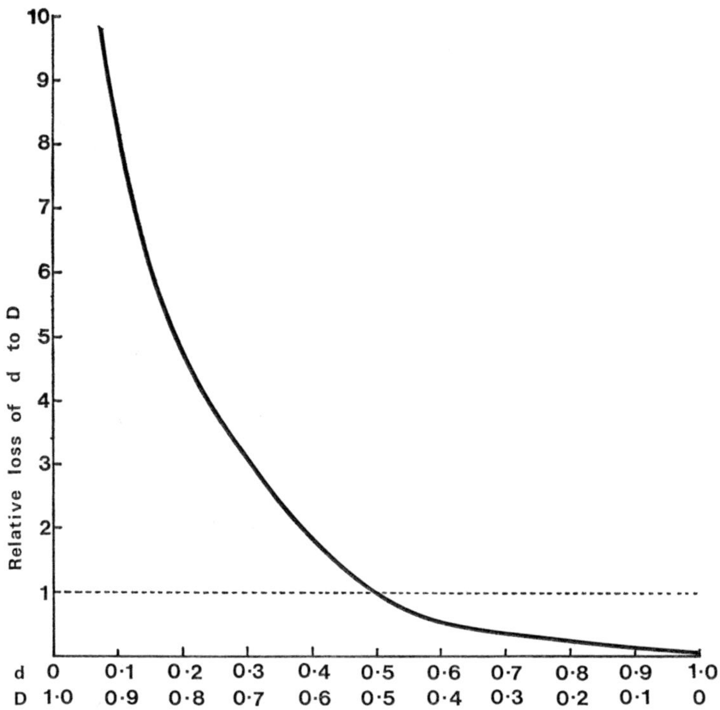

Fig. 2.28.—The loss of d alleles relative to D alleles due to fatal Rhesus incompatibilities during pregnancy at different allele frequencies. (*From Principles of Human Genetics* (Curt Stern), 2nd ed., 1960. Copyright, London: W. H. Freeman & Co.)

alleles will also be unequal. *Fig.* 2.28 shows the loss of d alleles relative to D alleles at different relative frequencies.

Now many of the populations of the world differ greatly in their frequencies of the Rhesus alleles and clearly, therefore, there must be differential selection serving to increase or at least maintain these differences. In fact, because of an overcompensation effect in affected families and other possible modes of selection, it is not at all clear in which direction selection is proceeding for Rhesus alleles in Caucasoid or other populations. It does seem, though, that selection may be increasing the difference between the races with regard to this trait.

In *Table* 2.19 are shown some data from Glass (1950) on the fertility of Rh^+

Table 2.19.—MEAN NUMBERS OF PREGNANCIES AND LIVING CHILDREN OF Rh-POSITIVE AND Rh-NEGATIVE WOMEN (*From* Glass, 1950.)

GROUP OBSERVED	MEAN NUMBER OF PREGNANCIES	MEAN NUMBER OF LIVING CHILDREN (per woman)
Whites	$t = 4.65 \quad P < 0.001$	$t = 3.05 \quad P < 0.01$
Rh+	1.777 ± 0.017	1.378 ± 0.016
Rh−	1.908 ± 0.023	1.454 ± 0.020
Non-sensitized	1.83	1.43
Sensitized	2.46	1.62
Negroes	$t = 1.35, 0.10 < P < 0.2$	$t = 2.41, 0.01 < P < 0.02$
Rh+	2.865 ± 0.048	2.412 ± 0.042
Rh−	2.725 ± 0.085	2.195 ± 0.074
Non-sensitized	2.61	2.11
Sensitized	3.6	2.92

and Rh^- women. In both Negroes and Whites, women who have produced anti-Rh antibody had significantly higher mean numbers of pregnancies and of living children than their non-sensitized Rh^- counterparts. In Negroes the mean number of live births per Rh^+ woman was significantly higher than Rh^- women. Among Whites, on the other hand, the mean number of living children per Rh^- woman exceeded that per Rh^+ woman.

The selective factors involved in the maintenance of many of the other human polymorphisms are mostly either unknown or incompletely understood. For instance there is a relationship between the ability to taste phenylthiocarbamide (PTC) and certain types of thyroid disease called 'nodular goitre'. PTC is chemically related to substances which produce goitre and a significantly higher frequency of PTC non-tasters has been found among those affected with goitre than among non-affected. The exact mechanisms involved are not yet clear, but the ability to taste PTC is another polymorphism which shows world-wide variation in frequency.

2.3.b. *Non-random Mating*

Hitherto we have considered the establishment of genetic variation. If mating was completely random throughout the whole human population, this would merely spread the existing variation out fairly evenly, whereas in fact, of course, the various races of man possess characteristic and unique arrays of different alleles. The notional concept of a completely random mating population rests on the assumption that a male chosen at random will select his spouse from a fixed universe of women, every one of whom he has an equal probability of marrying. Distance alone rules this out for man, but even at the local level there are many other constraints on random mating, some of which are associated with already established racial variation, and others such as age, kinship, religion, polygamy, and polyandry which, on the whole, are not.

The possible genetic effects of various demographic factors in human populations have been considered from a number of points of view (Lasker, 1952; Sutter and Tran-Ngoc-Toan, 1957; Cavalli-Sforza, 1958; Freire-Maia and Freire-Maia, 1964; Mange, 1964; Küchemann, Boyce, and Harrison, 1967; Cavalli-Sforza and Bodmer, 1971). From the point of view of the maintenance of genetically distinct isolates and races the most important considerations are: (1) the size, density, and spatial distribution of the population; (2) the amount of migration; and (3) the nature of mate selection within the population. We must therefore briefly consider each of these factors.

It appears that, at least in the past, the likelihood of marriage between two individuals is a decreasing exponential function dependent on the distance between their homes. *Fig.* 2.29 shows the results of some work by Boyce, Küchemann, and Harrison (1967) on the relationship between distance and the contribution of mates to the 1861 breeding population of Charlton-on-Otmoor in Oxfordshire. The type of curve they described for this relationship is one found very frequently in measurements of gene flow in both animals and plants and has been shown to fit the distribution of *Drosophila* dispersion (Wallace, 1966), windborne pollen dispersal (Bateman, 1950), and other examples of the spreading of living things. Of course the geographical population as recognized by the ecologist, anthropologist, or demographer is not necessarily the same as the Mendelian one recognized by the geneticist, but often there is a fairly close correlation, since geographical distance and barriers are, or were until recently, the main isolating mechanisms in man just as they are in other animals.

It seems certain that the increasing density of the human population, together with the change in many areas from simple agricultural communities to industrial and advanced societies, is having considerable repercussions on genetic and therefore racial structures. Initially when man lived in simple communities of hunters, gatherers, pastoralists, and cultivators, the total population could be considered as being composed of a large number of these clusters between which there was only limited migration or gene exchange. As society has changed the population has become distributed more diffusely in space and the individuals composing it have become rather more mobile. From a theoretical

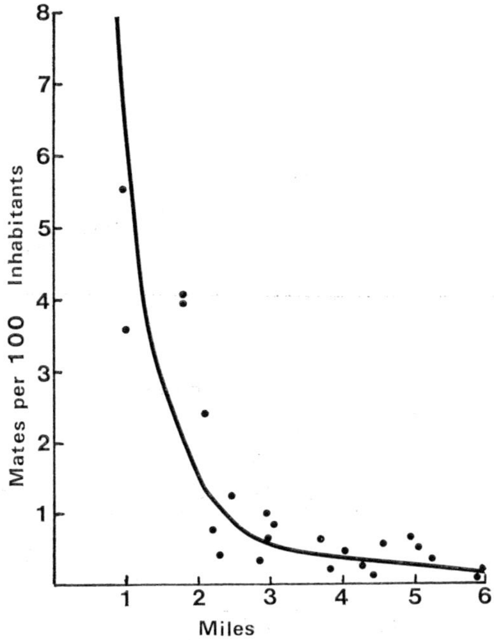

Fig. 2.29.—Observed and expected distribution of marriage distances. The observed distribution is based on the enumerators' returns for the 1861 census of the Oxfordshire parish of Charlton-on-Otmoor. Each dot represents a village which contributed mates to the 1861 breeding population of Charlton. The expected distribution, the solid line, is the fitted curve $y = 4{\cdot}75x^{-1{\cdot}88}$. (*From* Boyce and others, 1967.)

point of view, changes of this sort affect our estimates of the rates of evolution and race formation quite considerably.

The distance over which spouses are commonly drawn is also a factor in the likelihood of a consanguineous marriage, and a number of studies (Hajnal, 1963; Cavalli-Sforza, Kimura, and Barrai, 1966; Yasuda and Morton, 1967) have investigated this. But whatever the reasons, the fact that the proportions of consanguineous marriages vary considerably from one society to another and between one race and another, plays an important part in the maintenance of racial variation. In most of Europe, for instance, less than 1 per cent of all marriages are between cousins, among the Parsees of India 13 per cent of marriages are between cousins, and in the state of Andra Pradesh on the east coast of India up to 25 per cent of all marriages are between uncle and niece.

Here, too, the general change in the structure of man's society has led to a general decrease in the incidence of consanguinity. In Japanese feudal families, for instance, as many as 20 per cent of the marriages were consanguineous, a frequency which has been drastically reduced in modern Japan. *Fig.* 2.30

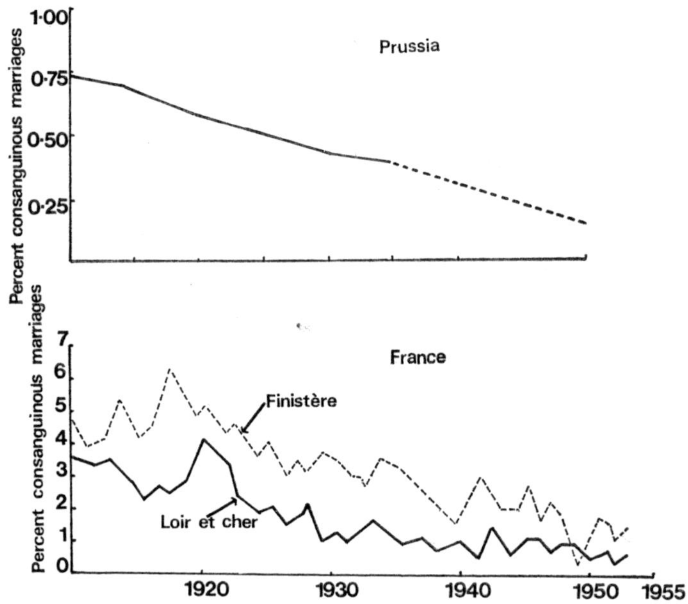

Fig. 2.30.—The decline in consanguineous marriages in Europe in the last century.

shows dramatic decreases in the frequency of consanguineous marriages in two European countries over recent decades, and there is no doubt that these are merely examples of a world-wide gradual breakdown of isolates due to increased human mobility, both geographical and social. This phenomenon must have continuing and considerable repercussions on the rate of racial integration, owing to the release of a wide range of genetic variability into many of the racial gene pools. We deal with this problem at greater length in Chapter 5.

Finally, we turn to the possibility of mating based on phenotypic similarity—assortative mating or the converse. By this is meant simply the greatly increased likelihood that, for instance, a tall man will marry a tall woman, or that people of similar intelligence are likely to marry each other, or that people with different skin colours are rather unlikely to marry.

The level of assortative mating can be measured in these continuously varying characters by calculating the statistical correlation between mates, and, with regard to all-or-none traits, by comparing the proportion of marriages observed in which the partners share the given trait with that expected on the basis of chance. The following three examples (*Table* 2.20, *a*, *b*, and *c*) show such correlations for rather different kinds of characters with varying degrees of

Table 2.20.—NON-RANDOM MATING IN MAN: EXAMPLES FROM A CAUCASOID POPULATION

a. SIMPLE MENDELIAN CHARACTERS

			Husbands Dark Eyes	*Husbands* Blue Eyes
Wives	Dark Eyes	Obs.	92	117
		Exp.	73	136
	Blue Eyes	Obs.	77	197
		Exp.	96	178

b. PHYSIQUE AND PERSONALITY TRAITS

Trait	*Correlation between Spouses*
Memory	0·57
Intelligence	0·47
Neurotic tendency	0·30
Stature	0·28
Eye colour	0·26
Weight	0·21

c. FACIAL AND CONFORMATION CHARACTERS

Trait	*Correlation between Spouses*
Ear lobe length	0·40
Ear length	0·40
Waist circumference	0·38
Hip circumference	0·22
Neck circumference	0·20

genetic basis and also racial association. Even though these examples come from within one race, they give an idea of the great extent to which assortative mating is responsible for the preservation of racial diversity.

2.4. HOW MUCH NATURAL SELECTION IN MAN?

Natural selection occurs whenever two or more genetically different types of individuals pass their genes to later generations in different proportions from those they hold in the parental population. The rate at which the carriers of one genotype reproduce their genes relative to other genotypes is a measure of their selective value. Now there are at least two important factors contributing to the selective value, firstly differences in fertility and secondly differences in infant or pre-adult mortality. We have already mentioned that it is very difficult to assess variability of these characters in genetic terms. It is, nevertheless, possible to show that there is still great opportunity for natural selection to occur in man despite considerable reductions in pre-adult mortality and the number of children born; trends which have become very marked in many human populations in recent generations.

Dr. J. F. Crow (1958a, b; 1961) derived an index to measure natural selection in man:

$$I = V/\bar{W}^2$$

where V is the variance (the sum of squared deviations from the mean, divided by the number of observations) of the number of children born per person and $\bar{W}$ the mean number of their children. As I increases, the opportunity for natural selection to operate also increases. As we have already mentioned, however, I is the sum of two components, Im due to mortality and If due to variations in fertility. Im is the ratio of the proportion of children who die before reaching maturity (Pd) to those who survive (Ps), i.e., $Im = Pd/Ps$. The fertility component, $If = Vf/\bar{x}_s^2$, where Vf is the variance of the number of children born per person who survive to maturity and $\bar{x}_s^2$ the mean number of children per survivor.

Clearly, the greater the mortality and variation in fecundity, the greater is the opportunity for selection, although it is not apparent which is the most important. Dobzhansky (1962) reports some work of Crow (1961) and Spuhler (1961) which shows that *If* actually tends to increase as the number of children born per family decreases (*Table* 2 21). The reason is that the variances of the

Table 2.21.—THE INDEX OF SELECTION IN SOME HUMAN POPULATIONS (*From* Dobzhansky, 1962.)

POPULATION		CHILDREN BORN ($\bar{W}$)	*If*
Rural Quebec	Canada	9·9	0·20
Hutterites	U.S.A.	9·0	0·17
Gold Coast	Africa	6·5	0·23
N.S.W.	Australia 1898–1902	6·2	0·42
U.S.A.	born in 1839	5·5	0·23
U.S.A.	born in 1866	3·0	0·64
U.S.A.	alive in 1910	3·9	0·78
U.S.A.	alive in 1950	2·3	1·14
Ramah Navajo Indians		2·1	1·57

number per family do not grow as rapidly as do the squares of the mean number of children, thus increasing *If* in populations with fewer children per family. Spuhler (1961) calculated the mortality and fecundity components (*Im* and *If*) and the total index (*I*) for a series of populations. It appears that in families where fewer children are born but a greater proportion survive, the reduction of *Im* is more than compensated for by an increase in *If*, resulting in a higher total *I*. It is also worth remembering that even if childhood mortality could be eutirely eliminated, natural selection could still operate on those genotypes influencing the probability of an individual marrying and wanting and being able to have children.

To turn to some specific examples of the action of selection on an individual character and how this can and does differ between different racial groups, we can consider selection and human birth weight, first investigated in Europeans by Karn and Penrose (1951) with later studies in different races by other workers (Fraccaro, 1956; Jayant, 1964, 1966; Hollingsworth, 1965; Van Valen and Mellin, 1967). The data analysed by Karn and Penrose showed that of 6693 female births during a certain period in a London hospital, 6419 were still alive 1 month later. The proportion of survivors at various birth weights are shown in *Fig.* 2.31, which also shows the total frequencies of births at different

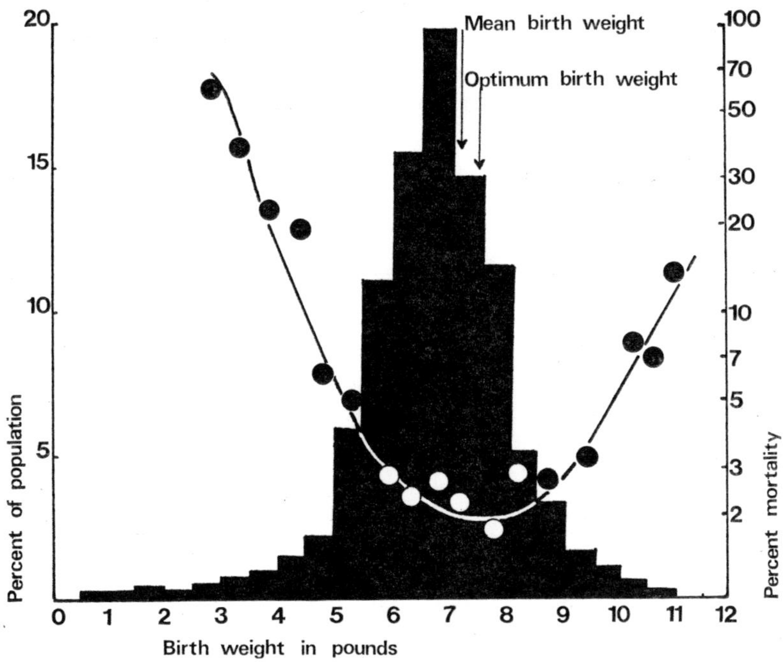

Fig. 2.31.—The distribution of birth weight among 13,730 children and the rates of early mortality of the various birth weight classes. An example of stabilizing selection in man.

weights. The optimal birth weight is exactly 8 lb. and departures from this value at birth increase the likelihood that the infant will die within this 1-month period.

The overall mortality was 274/6693 (4·1 per cent), but the mortality of the optimum 8 lb. class was only 1·2 per cent. Therefore 4·1 percent — 1·2 per cent is a measure of the effect of selection for an optimum birth weight. The average weight of all these children at birth, i.e., before selection, was 7·06 lb. and the average birth weight of all these children who survived at least 1 month was 7·13 lb., which represents a change in average birth weight of 1 per cent. The main factor, though, is a reduction in the standard deviation of genetic variability of 10 per cent, and this type of selection is therefore termed 'stabilizing'. In a study of birth weight and survival of African babies (Hollingsworth, 1965) the overall mortality was somewhat higher at 11·6 per cent for females and the optimum birth weight was somewhat lower at 6·75 lb., at which weight there was only 8·4 per cent mortality. In other words, the force of selection for an optimum birth weight was 3·2 per cent. The mean birth weight before selection was 6·16 lb. and of the survivors was 6·22 lb., showing that for Ghanaian babies the direction of selection was the same as that for European babies and of a similar magnitude—about 1 per cent. The results from a number of similar investigations are in *Table* 2.22, which shows the difference in mean

Table 2.22.—MEAN BIRTH WEIGHT IN DIFFERENT HUMAN POPULATIONS.

		MEAN BIRTH WT. (all births)	MEAN BIRTH WT. (survivors)	DEGREE OF DIRECTIONAL SELECTION (per cent)
England	♀ ♂	7·06 lb.	7·13 lb.	0·99
India	♀	6·32 lb.	6·40 lb.	1·26
(upper class)	♂	6·51 lb.	6·60 lb.	1·38
India	♀	5·48 lb.	5·64 lb.	2·91
(lower class)	♂	5·84 lb.	5·67 lb.	2·99
Ghana	♀	6·16 lb.	6·22 lb.	0·97
	♂	6·44 lb.	6·49 lb.	0·77
New York	♀	3·067 kg.	3·109 kg.	1·36
	♂	3·160 kg.	3·198 kg.	1·20

birth weight in different races both before and after selection and also the different levels of selection which are operative.

We have considered so far in this section the overall magnitude of selection on human populations and the amount of selection on a continuously varying character. We consider at various places elsewhere in this book selection against certain characters such as the sickle-cell trait, but it is also appropriate to present in summary some results of a demographic approach to measuring selective differential. This approach can be used to calculate the selective differentials associated with a number of traits which have hitherto not been amenable to genetic analysis. The methods used in this approach are fully explained by Cavalli-Sforza and Bodmer (1971). We give in *Tables* 2.23–2.25 some results for Huntington's chorea, schizophrenia, and IQ. Among the interesting features shown by these tables are some of the undoubted changes which are occurring in selective coefficients at the present time, and the

Table 2.23.—UPPER AND LOWER BOUNDS TO THE SELECTIVE DISADVANTAGE OF PEOPLE WITH HUNTINGTON'S CHOREA (*From The Genetics of Human Populations* (L. L. Cavalli-Sforza and W. F. Bodmer), 1971. Copyright, London: W. H. Freeman & Co.)

	NET REPRODUCTIVE RATE (R_0)	APPROXIMATE SELECTIVE DISADVANTAGE (per cent)
U.S.A., 1940. Total Population	1·0485	—
Choreics		
Assuming fertility stops at onset of disease	0·7854	26·3
Assuming fertility is normal until death	1·0308	1·8

Table 2.24.—NET REPRODUCTIVE RATES AND APPROXIMATE SELECTIVE DISADVANTAGES OF SCHIZOPHRENICS ASSUMING THAT 1940, 1950, AND 1960 U.S.A. BIRTH-RATE DISTRIBUTIONS APPLY BEFORE ONSET OF THE DISEASE (*From* Bodmer, 1968.)

	NET REPRODUCTIVE RATES (R_0)			SELECTIVE DISADVANTAGES		
	1940	1950	1960	1940	1950	1960
Males	0·56	0·77	0·92	0·86	0·61	0·51
Females	0·76	1·04	1·25	0·40	0·27	0·22

Table 2.25.—THE RELATIONSHIP BETWEEN I.Q. AND FITNESS. (*From* Bajema, 1963. Copyright, University of Chicago Press.)

IQ RANGE (m)	INTRINSIC RATE OF NATURAL INCREASE ($r_m \times 10^{-3}$)	AVERAGE GENERATION LENGTH T_m	RELATIVE FITNESS MEASURED BY $e^{r_m T}$, WHERE T IS THE AVERAGE GENERATION LENGTH
≥120	8·9	29·4	1·00
105–119	3·9	28·9	0·87
95–104	0·3	28·4	0·78
80–94	7·5	28·0	0·96
69–79	−10·0	28·8	0·58
Total sample	3·9	28·5	

U-shaped relationship between IQ and fitness which is in disagreement with other predictions regarding this trait.

The relationship between intelligence and fertility seems to be changing fairly rapidly at the present time. Bajema (1963) points out that in the United States there has in recent years been a decrease in the proportion of unmarried individuals, childless families, and one-child families, as well as a decrease in the proportion of families having five or more children. Bajema (1963) found that the intensity of natural selection in relation to intelligence in the population he was studying (about 1000 native White inhabitants of Michigan born in 1916 or 1917) was 0·13, where the optimum phenotype was the IQ $\geqslant$ 120 group. The fitness, or reproductive performance of the population he studied, was therefore 13 per cent less than what it would be if all the IQ's were in the optimum range.

The whole point of this section has been to show that natural selection is playing a very prominent part in maintaining and enhancing the genetic variety which is present in such abundance in man. It is also clear that it is affecting the whole range of variable characters, not just the more obvious range of sharp phenotypic discontinuities.

2.5. How Much Genetic Difference between Races?

Although the total amount of genetic variation in man and other species can now be estimated, the problem of the total amount of genetic identity between species and races and the amount of genetic reorganization accompanying speciation have been, and still are, the subject of some controversy. Even now, the evidence we have is rather limited and comes principally from the fruit-fly *Drosophila*, although with some supporting evidence from man himself. The collecting of evidence on this point did not really begin until 1965 when Drs. J. L. Hubby and L. H. Throckmorton investigated 10 different species of the *Drosophila virilis* group using electrophoretic techniques, which allowed them to compare the mobilities, and therefore the structures, of a large number of the flies' constituent proteins. Since the proteins are direct gene products, this analysis is also one of the genetic material. They found that within one species subgroup the species shared, on average, 71·6 per cent of their proteins with each other, with a maximum figure of 78·9 per cent. Of 38 proteins investigated electrophoretically in *Drosophila virilis* only 2·6 per cent were unique to that species, and of 38 in *Drosophila americana* only 5·3 per cent were unique. In other words, all the other proteins had been inherited apparently unchanged from their precursor species. Although other *Drosophila* species have higher frequencies of unique proteins, it seems clear that specific differences *can* be established without major genetic reconstruction, although of course it is by no means certain that they in fact always are. In a later report (Hubby and Throckmorton, 1968) the results of studies on a number of sibling species of *Drosophila* are presented. Sibling species are species which are almost identical in phenotype and may, therefore, be very closely related. If indeed this is so they would provide particularly telling comparisons from our present viewpoint. In addition to each pair of sibling species, a closely related but non-sibling species was tested, and altogether Hubby and Throckmorton were able to assemble and test nine such species trios. They found that about 50 per cent of the proteins of the sibling pairs, on average, showed identical electrophoretic ability, and that one sibling pair shared 85 per cent of proteins with identical mobility. The morphologically distinct relatives shared, on average, a rather lower proportion of their proteins. It appears that, at least in this case, sibling species are more closely related than non-sibling relatives, and the conclusion from these results again is that speciation *may* occur as a result of only small amounts of genetic change.

This work on *Drosophila* provides virtually the only quantitative data on the amount of genetic difference between species that we have. It is supported in general terms by the work of Manwell and Baker on a number of animal groups, but particularly a group of closely related vespertilionine bats (Manwell and Kerst, 1966) and anseriform geese (Baker and Hanson, 1965). These two groups, 4 bat species and 11 geese species, are of relatively recent evolutionary divergence and the genes controlling a large number of blood-serum proteins and enzymes show remarkable similarity in the various species. In these cases it appears that speciation has been and is being effected with changes of no more than, and probably less than, 20 per cent structural genes. This conclusion is closely confirmed by Selander and others (1969) working on two subspecies of the house mouse (*Mus musculus domesticus* and *M. m. musculus*) from six geographic regions of the Jutland Peninsula, Denmark. They studied a total of 41 polymorphic loci in these two subspecies and found that, for the 24 monomorphic loci and the 4 polymorphic loci at which the same alleles are present in approximately equal frequencies in both *M. m. musculus* and *M. m. domesticus*, the electrophoretic mobilities of bands are identical in the two subspecies. Thus these semispecies of mice shared 68 per cent proteins with identical mobility as opposed to 50 per cent in full species of *Drosophila*. The conclusion reached by Selander and others is, in fact, slightly different from that of Hubby and Throckmorton. On the basis of other evidence presented in their paper they suggest that 'reorganization and differential coadaptation of the genomes of allopatric populations may occur before intrinsic barriers to interbreeding have been developed, and, thus, before the populations have reached the full species level' as conventionally defined (Mayr, 1963); in other words, species may be genetically established before they are reproductively isolated.

The alternative approach is to examine homology, defined in chemical terms, between samples of isolated DNA. The overall composition of DNA varies widely in some groups but is rather constant among mammals. McLaren and Walker (1965, 1968, 1970) have made a number of comparisons between the

DNA of a number of rodent species, using as their measure the amount of binding which takes place between isolated DNA samples in agar under a number of experimental conditions. As McLaren and Walker (1970) state: 'It seems unlikely that sequence complementarity . . . will prove useful in the near future as a quantitative measure of taxonomic relatedness'. Nevertheless for general interest it seems worth presenting some of their recent results (*Fig.* 2.32) which show that, relative to a within-species binding percentage

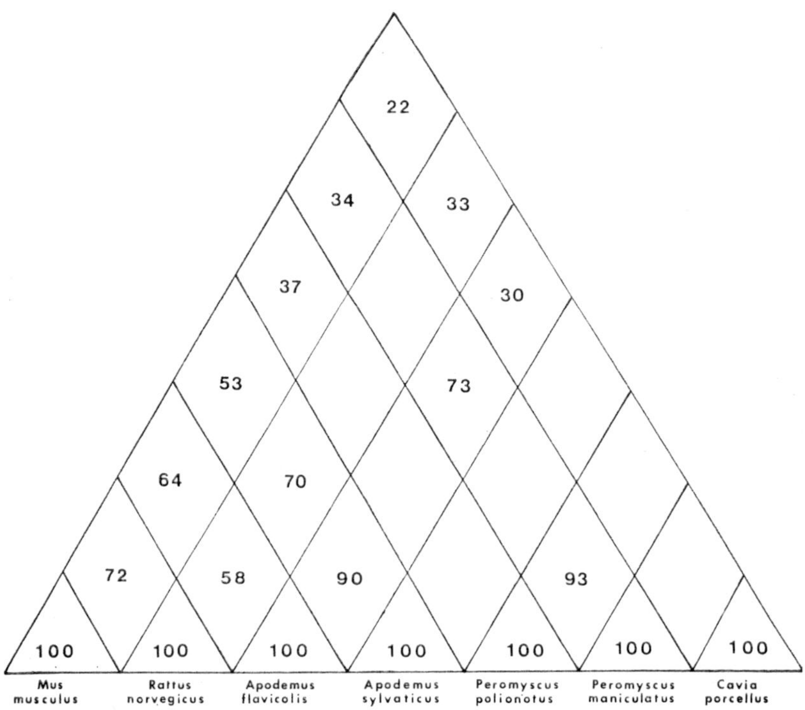

Fig. 2.32.—The average between-species binding percentage of extracted DNA expressed as a percentage of the average within-species percentage. The binding percentage is determined by *in vitro* experimentation. (*From* McLaren and Walker, 1970.)

fixed arbitrarily at 100 per cent, two species of the same genus gave binding percentages of 90 per cent for *Apodemus* and 93 per cent for *Peromyscus*. These figures are to be compared with estimates of between 22 per cent and 33 per cent for binding percentages between difference suborders. The main conclusion from this work, in line with that from orthodox genetic studies, is that species may be much more closely related genetically than other taxonomic categories.

Our estimates of the amount of genetic change which have taken place in human races rest on various mathematical analyses, principally by Drs. L. L.

Cavalli-Sforza and A. W. F. Edwards (1963, 1967). They used the overall gene frequencies at such loci as the major blood groups and widely scored biochemical differences in a number of different populations, to compute the

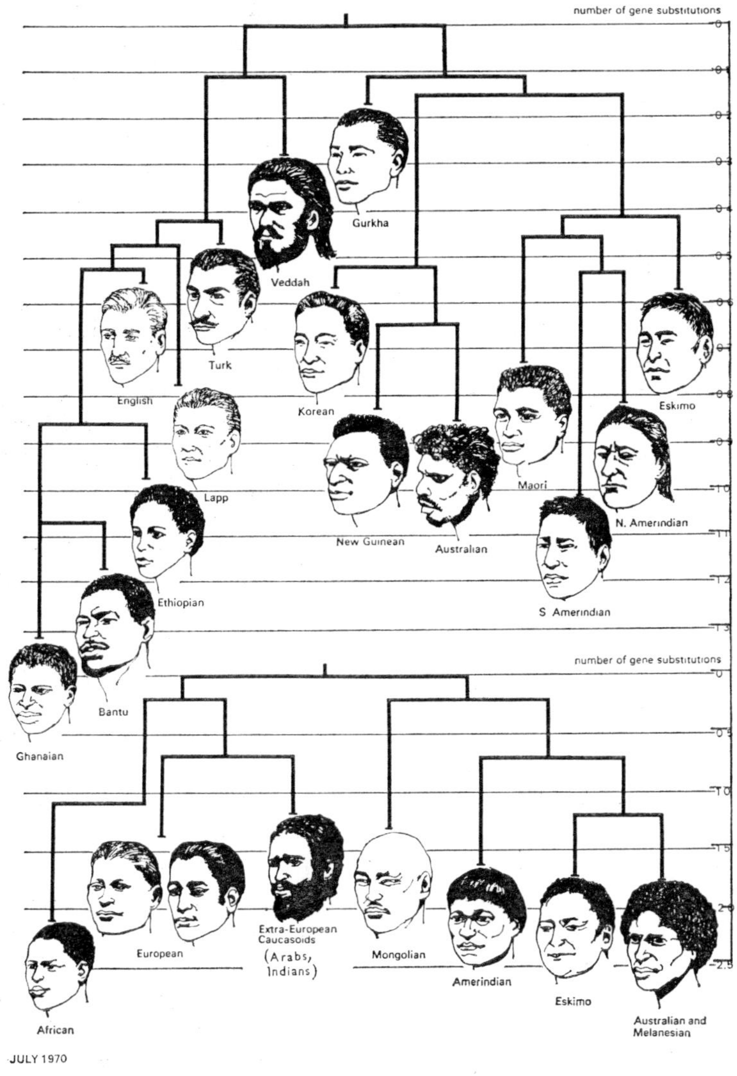

Fig. 2.33.—Two reconstructions of human evolution. The upper reconstruction is based on comparisons of five genes in fifteen different populations. The lower diagram is based on comparisons of thirteen genes in seven ethnic groups. The scale shows the overall number of substitutions for the genes considered which took place during the divergence of the races. (*From* Dyer, 1970a, *after* Cavalli-Sforza, 1966.)

total number of partial gene substitutions, often referred to as the 'genetic distance', which have taken place between the populations compared in all possible pairwise combinations. *Fig.* 2.33 shows two reconstructions of human evolution which they derived by this method (Cavalli-Sforza, 1966). These reconstructions are purely genetic; their approximation to generally accepted racial evolutionary groupings is an indication of their validity. In each part of the figure the length of the arms represents the number of gene substitutions. Either of these two reconstructions shows that, although some of the races of man have completed up to 20 per cent substitutions or, in other words, show up to 40 per cent gene differences in a paired comparison among those genes which were used in carrying out the calculations, most races show rather less than 10 per cent substitutions and the average difference between races is of the order of 10 per cent. It is much as expected to find that the average amount of gene substitution is as low as this since, as we have just seen, with more than 20 per cent gene substitutions, we are approaching the level of species differences, and we already know that man is definitely one biological species.

The racial groupings which Cavalli-Sforza and Edwards use are clearly major *geographical races* but including possibly some of the more clearly differentiated *local races*. More recent work, however, has extended their technique and has compared it with others on some *micro-races* of man. Fitch and Neel (1969) and Neel and Ward (1970) present data on the number of genetic substitutions which have occurred between some of the villages and tribes of Indians inhabiting South and Central America. Fitch and Neel utilize six genetic loci (MN, Rh, Kidd, Duffy, Diego blood groups and serum haptoglobins) to investigate tribes which were chosen on the basis of having non-Indian admixture estimated at less than 5 per cent and a sample size greater than 200. They found that the maximum divergence between any two tribes for these six systems was 1·37 gene substitutions and that, on average, the number of gene substitutions was considerably less than this, being of the order of 0·6 gene substitutions. Now these 12 tribes are rather widely separated geographically. It is likely that they trace back to a common ancestral population either 15,000 years ago when the central American Isthmus became established, or possibly 25,000–30,000 years ago when the Bering land bridge between the Asian and American continents was available. (These dates may be subject to some modification in the light of recent archaeological evidence—*see* Chapter 3, 5s. 3.4 and 3.5). We have, then, a figure of about 10 per cent genetic difference attached to racial differences of the order of possibly *local races* or maybe larger *micro-races*. Neel and Ward (1970) have taken their analysis a step farther by investigating the genetics of these same six loci in a number of villages of two of these Indian tribes, the Makiritare and the Yanomama. They used a rather different method of comparing genetic distance but they were able to show that the mean genetic distance between Indian villages is well over half the mean distance between Indian tribes. Since the mean distance between these Indian tribes appears, in turn, to be approximately half the mean distance between the fifteen populations which Cavalli-Sforza and Edwards investigated, our conclusions as to the amount of genetic difference between the races of man are substantiated by internal consistency.

In all discussions of racial differences and racial integration it is perhaps as well to remember that the genetic differences involved for all but the most distantly related races are most probably around 10 per cent of the total genetic complement of man, and that even this may be an over-estimate, because the genetic techniques available to us in the study of man do not allow us to test a random sample of genes, as perhaps we are doing in *Drosophila*. The genes used in these calculations in man are necessarily those which we know to be polymorphic and highly variable. The inference is that if we could obtain a truly random sample of genes, including these which are invariant in *Homo sapiens* our estimate of average genetic differences between the races would be rather less than 10 per cent.

CHAPTER 3

THE ORIGIN AND HISTORY OF THE RACES OF MAN

3.*1*. INTRODUCTION

THE problem of the origin of the races of man is essentially an insoluble one because in the first place there is, as we have seen, little general agreement on just how many races of man exist, let alone sufficient evidence for tracing their origins. Secondly, the nature of racial classifications is based essentially on population concepts which can hardly be satisfactorily developed on the rather scarce material provided by palaeontological or archaeological investigations.

The best we can do in this situation is to present the historical and prehistorical evidence bearing on racial origins, to relate it to the known present distribution of man and the genetic variability which he currently exhibits, and then do our best to deduce racial origins and relationships from this rather varied evidence. To present the evidence in outline in this way will not lead to a racial classification, but will firstly make clear the sort of evidence on which existing classifications are based, and secondly, and more important, expose the true biological relationships of the different racial groupings. Since we have already decided that the term 'race' has no formal status and that different races are not necessarily biologically equivalent, these twin aims are of major importance in human studies.

All the races of man living today belong to one species, *Homo sapiens*. This species, and other fossil species of the genus *Homo*, together with the immediately preceding genus *Australopithecus*, are placed by competent taxonomists in the subfamily Hominidae. This subfamily, together with two or three other families of apes, is placed in the superfamily Hominoidea, one of three superfamilies in the suborder Anthropoidea. The details of this classification are continually being changed somewhat as new fossils are discovered and new biochemical evidence of affinities or otherwise between surviving species is found. *Fig.* 3.1 shows a scheme of classification indicating man's evolutionary

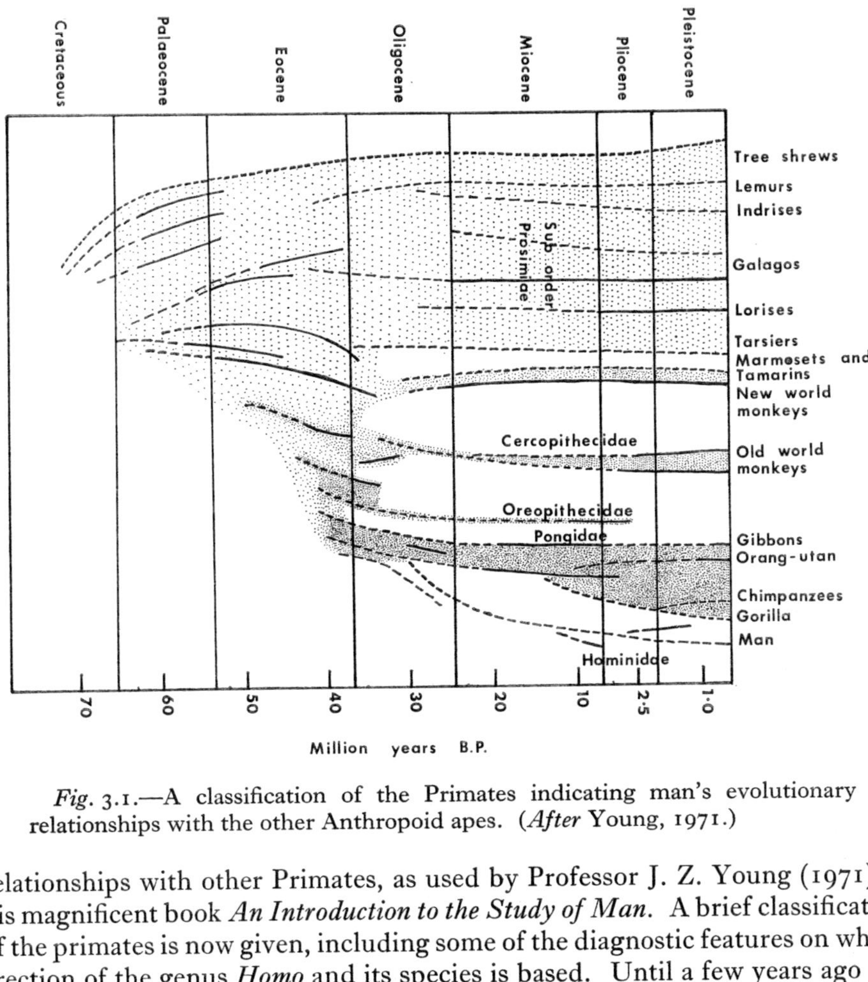

Fig. 3.1.—A classification of the Primates indicating man's evolutionary relationships with the other Anthropoid apes. (*After* Young, 1971.)

relationships with other Primates, as used by Professor J. Z. Young (1971) in his magnificent book *An Introduction to the Study of Man*. A brief classification of the primates is now given, including some of the diagnostic features on which erection of the genus *Homo* and its species is based. Until a few years ago the taxonomy of the Hominidae was in utter confusion owing to the gay abandon with which new genera and specific categories were erected by palaeontologists. Today the position is simpler, but the consequence of this earlier irresponsibility can still be seen in many earlier textbooks. This book largely eschews early nomenclature and recognizes only three genera within the Hominidae and no more than two or three species within the later two important genera *Australopithecus* and *Homo*. The classification used by J. Z. Young is as follows:

ORDER: PRIMATES
- Suborder 1. Prosimii
  - Infra order 1. Lemuriformes
  - Infra order 2. Lorisiformes
  - Infra order 3. Tarsiiformes
- Suborder 2. Anthropoidea
  - Superfamily 1. Ceboidea
    - Family 1. Cebidae
    - Family 2. Callithricidae
  - Superfamily 2. Cercopithecoidea
    - Family 1. Parapithecidae
    - Family 2. Cercopithecidae
  - Superfamily 3. Hominoidea
    - Family 1. Oreopithecidae (Oligocene)
    - Family 2. Pongidae (Oligocene–Recent)
    - Family 3. Hominidae (Pliocene–Recent)
      - *Ramapithecus*
      - *Australopithecus*
      - *Homo.*

FAMILY: HOMINIDAE

The family Hominidae, one of three families within Hominoidea is characterized by: large brains with greater development of frontal lobes than the great apes; rotation of the foramen magnum; small supra-orbital ridges; marked reduction of the facial skeleton; a relatively simple pattern of the molar teeth and reduction of canines; an upright posture; increased length of lower limb relative to trunk and upper limb; completely opposable thumbs and unopposable big toes.

GENUS: *Homo*

There is no real agreement about the number of genera within the family or about the characteristics which specifically define the genera. Cranial capacity within the genus *Homo* ranges from 900 to 2000 c.c. Members usually lack a diastemata (space between canines and premolar teeth); temporal ridges never reach the midline of the skull; brow ridges are highly variable.

SPECIES: *Sapiens*

The species *sapiens* is characterized by: increased reduction of the facial skeleton, brow ridges, and palate size; long bones that are straighter and more slender than those of *H. erectus*.

In most respects the details of the Hominoid classification which are adopted are much less important than agreement on the actual evolutionary steps by which the line or lines leading to modern man became distinct. Unfortunately information is scanty for many of these evolutionary steps, and therefore disagreement is rife. In fact, it is hardly surprising that this is so when we consider the number of variables that must be taken into account when constructing an evolutionary framework and the hazardous, and therefore infrequent, process of fossilization which can provide data on these variables. C. S. Coon, in his book *The Origin of Races* (1963), discusses the variables of evolutionary relationships under four headings—Time, Space, Grade, and Line—and this is a suitable model for us to follow. We have, in other words, to discuss the chronology of the separation of the Hominid from the Pongid line and the subsequent evolution of man through the Australopithecine stages, through the stages of *Homo habilis* and *Homo erectus*, up to and within the species *Homo sapiens*. This, of necessity, involves us in examining the level of biological organization reached by each fossil stage, and in the later Hominid succession this is mainly a matter of rather subtle changes in dentition, brain size, cranial relationships, and technological achievement. We have, in addition, to consider the geographic radiation of all the species and subspecies concerned, particularly with reference to the distribution of races as we see them today. Finally, when discussing line, we are considering every racial, subspecific and specific lineage that we can distinguish, and linking them up as completely as we are able.

Such a procedure is obviously essential in any consideration of the significance of race crossing. If major races are really as old as some authorities believe, in particular Weidenreich (1945) and Coon (1963), who believe they pre-date the evolution of *Homo sapiens* himself, then crosses between them will be potentially more biologically interesting than if races are only a few tens of thousands of years old. To introduce some of the changes, particularly in skull morphology, which have occurred in man's evolution *Fig.* 3.2 illustrates some of the most important skulls to which we will subsequently refer, arranged in temporal sequence. This figure, in which the skulls are not drawn to scale, can be used as a reference for putting time and morphological change into perspective. The time span within which hominid evolution is believed to have taken place is also illustrated in this figure. This puts into perspective the ancestry of all the very early finds and we will now briefly examine them.

3.2. The Time Element in Human Evolution

Virtually the earliest fossil form which can be classified as a Hominid is dated by the potassium/argon radioactive method as being about 14 million years old. It was found in 1961 by Dr. Louis Leakey at Fort Ternan in Kenya and named by him *Kenyapithecus wickeri* (Leakey, 1962, 1967). The material consists of small pieces of both sides of the upper jaw, together with the first two upper molars and the last premolar. There is also one lower molar, a canine tooth, and possibly also an incisor which was found nearby. This material shows a number of similarities with both earlier and later forms, but the most striking comparison is with some pieces of jawbone discovered in the Miocene of north-western India by G. E. Lewis in 1934 and now known as *Ramapithecus punjabicus* (Lewis, 1934; Simons, 1963). Among the rocks in which these Indian specimens were found are none which are suitable for absolute dating, but they are probably between 14 and 18 million years old and it seems likely from other fossil evidence that they correlate quite closely with the late Miocene Strata of Kenya in which *Kenyapithecus* was found. It seems likely that these two forms are very closely related, even possibly belonging to one widely ranging species (Simons, 1963; Kinzey, 1970). Leakey (1967) has suggested that some fossil material from the Lower Miocene represents an earlier species of *Kenyapithecus*, which, if accepted as such, pushes the origins of the Hominidae back earlier still.

Between this period of time in the Miocene, the time of *Ramapithecus*, until the next groups of fossils are found there is a very long gap. Until recently there were almost no fossils from Pliocene deposits anywhere in the world which could be assigned to the Hominid line. The finding of a Hominid tooth in Early Pliocene deposits of northern Kenya (Bishop and Chapman, 1970) has provided us with our first important fossil material from the Pliocene. The sediments in which this tooth was found are dated at between 9 and 12 million years old and according to Dr. Leakey the tooth, probably a second upper left molar, has some affinities with *Kenyapithecus* and some also to both the later genus *Australopithecus* and even *Homo*. Apart from these recent finds, which cannot yet be really evaluated, one of the few other Hominid fossils from the Pliocene is the lower end of a humerus found at Kanapoi in north-western Kenya from rocks below a lava flow dated at 2·5 million years. This bone is thought to be about 4 million years old and to come from *Australopithecus africanus*.

The next Hominid fossils to be found are definitely members of the genus *Australopithecus* and they occur in rocks in Southern Africa of early Villafranchian age, which is about at the borders of the Pliocene and Pleistocene eras. From Taung in Botswana, Sterkfontein in the Transvaal near Johannesburg, and Makapansgat 200 miles north-east of Sterkfontein, has come fossil material assigned to the species *Australopithecus africanus*. This species has a small brain (only about 500 c.c.) but much of the dentition is characteristically hominid, although somewhat larger than *Homo sapiens* and, after initial controversies regarding R. A. Dart's original description of this form as a hominid, it is now universally recognized as such. Much of the exact dating of this material is rather difficult, but the general consensus is that it is between 1·5 and 2 million years old.

Recently a new series of hominid fossils has been found in the Omo Basin, Ethiopia (Howell, 1969), which contains teeth resembling South African *Australopithecus africanus* and also other forms possibly belonging to the closely related species *Australopithecus robustus* or another robust Australipithecine, *A. boisei*. In the preliminary report of the finds at Omo, Howell mentions ten fossil-bearing horizons each of which has been dated and which have yielded

Hominid fossil material as indicated in *Table* 3.1. It should be emphasized that these identifications are tentative and that most of the fossil material is very fragmentary, consisting of isolated teeth.

Yet another new Hominid fossil locality of Plio-Pleistocene age has recently been reported from East Africa near Lake Rudolph in Kenya (R. E. F. Leakey,

Table 3.1.—TEMPORAL RELATIONS OF THE PLIOCENE/PLEISTOCENE HOMINIDAE RECOVERED FROM THE LOWER OMO BASIN (*From* Howell, 1969.)

LEVEL	DATE AND LOCALITY*	FINDS
J		
I	K/Ar age = 1·81–1·87 m yr. B.P.	
H	Hominid locality 7	
	Hominid locality 74	*A. boisei*?
G	Hominid locality 28	} *A. africanus*
F	Hominid locality 26	
E	Hominid locality 9 and 64	*A. robustus*
D	K/Ar age = 2·37–2·56 m yr. B.P.	
	Hominid locality 2, 45, 51, 54	*A. africanus*
		{ mostly *A. africanus*
C	K/Ar age < 3·1 m yr. B.P.	possibly also robust
B	K/Ar age 3·75 m yr. B.P.	Australopithecines
A	K/Ar age > 4·0 m yr. B.P.	

*Locality numbers are those given by Howell on fieldwork.

1970). The potassium/argon date from volcanic tuffs at this locality is 2·61 ± 0·62 million years and two crania have been found at a horizon just below this tuff. One cranium resembles the robust Australopithecine *A. boisei* which is found in slightly later rocks from Olduvai Gorge and possibly also the Omo basin. The other cranium is very different from this and is either a gracile species of *Australopithecus* or a very early representative of *Homo*, and Richard Leakey in this early report suggests the latter.

The important conclusion which can be drawn from these preliminary results is that as early as three million years ago, at least two and possibly three Hominid species, *A. africanus*, *A. robustus*, and *A. boisei*, coexisted in one locality in southern Africa. It further appears that progress along the evolutionary road to *Homo* had proceeded much farther three, or maybe even ten million years ago, than has previously been thought.

On the basis of more extensive evidence from the finds at Kromdrai, Sterkfontein, and Swartkrans in South Africa, it is believed that *A. robustus* was somewhat larger and, as its name suggests, more robust than *A. africanus*, and it is also thought that the two species differed in ecological habitat. These two species are thought to have coexisted over a fairly wide area of southern Africa because at Swartkrans, very close to Sterkfontein, in addition to a large amount of fossil material clearly belonging to *A. robustus*, a lower jaw, some parts of an upper jaw and face, and a small piece of forearm have been found which are quite distinct. The upper jaw shows some features of both *A. robustus* and *A. africanus* but the dental material has resemblances to *Homo erectus*. It certainly appears that a form distinct from *A. robustus* and probably attributable to *A. africanus*, was living here between 1·5 and 1 million years ago, that is at a time corresponding to either the top of the early Pleistocene or the earliest middle Pleistocene.

Further specimens of *Australopithecus* and other hominids have been found at Olduvai Gorge in Tanzania, and this is one of the most important of all fossil sites for anthropologists, partly because of the number of absolute potassium/argon dates available from tests which have been carried out on suitable volcanic rocks. This site also allows representative faunal assemblages to be arranged in their true temporal order and this is of considerable importance in comparisons with equivalent series from other sites. It is at Olduvai, too, that more detailed studies on the relationships of the three named species of *Australopithecus* can be made.

The Olduvai Gorge is a river valley cut through Pleistocene lake sediments which are interspersed by volcanic material, and it is these volcanic rocks which have been dated by the K/Ar technique. Other means of dating the deposits have also been used and the date of the base of the sequence at 1·9 million years old is now very accurately known indeed. The very earliest fossil material to come from Olduvai is a juvenile mandible close to the base of Bed I, the lowest bed, and thought to be about 1·75 million years old. A number of fragments of skull parietal bones have also been found and some hand and foot bones. These, together with one or two other fragments from Bed I, were classified as *Homo habilis* by Drs. Leakey, Tobias, and Napier in 1964. As often happens in such cases, reappraisal of these fossils has now been made by a number of authorities (Wolpoff, 1969; Pilbeam, 1970) and, for reasons which we shall discuss later, it is suggested by some that this is a species of *Australopithecus*, either closely related to *A. africanus* or a subspecies of it. On the other hand Tobias (1967) believes that sufficient progress towards hominization had occurred by the lower Pleistocene for us to distinguish these early forms as *Homo habilis*.

Later representatives of this Hominid lineage have been recovered from Olduvai in later beds around 1·25 million years old in early Pleistocene, and 1 million years old in Bed II of lower Middle Pleistocene. The skull recovered from this later horizon in particular, although clearly related to *A. africanus*, has a number of features of its jaws and teeth which suggest that it has reached the level of *Homo erectus*. Certainly most authorities, whatever their position regarding the status of *H. habilis*, seem to agree that the transition from the genus *Australopithecus* of whatever species, to *Homo* of whatever species, had occurred in East Africa at or before the time of the faunal break marking the transition from lower to Middle Pleistocene—perhaps 1 million years ago or a little more.

There is one other group of fossils from this period which must be mentioned. In 1959 Dr. Mary Leakey found in the lower horizon of Bed I at Olduvai the skull of a robust hominid dated at around 1·75 million years. This specimen was originally named *Zinjanthropus boisei*, but a detailed study of this skull later showed that it has resemblances to both *A. africanus* and *A. robustus*. Professor Tobias believes that this form, which is now called *A. boisei*, may be ancestral to *A. robustus* (Tobias, 1967). Professor F. C. Howell at Omo has found similar forms which are also *A. boisei*, and which exist rather later in time than the first appearance of *A. robustus*. Thus, as we have concluded earlier, we are faced with the possibility of three distinct lineages of hominids in Africa at this time. Alternatively, it is possible that there were two subspecies of *A. robustus* in existence or even that *A. boisei* is the result of hybridization between robust and gracile hominids. On the basis of the evidence available to us at the present time, it is very difficult for us to distinguish between these possibilities. In *Fig*. 3.3 is shown a phylogenetic scheme suggested by Professor P. V. Tobias, who is probably the supreme authority on these fossils. He suggests that in fact *A. boisei*, although a forerunner of *A. robustus*, is also slightly off the main line of evolution from a general form of *Australopithecus* through *Homo habilis* to *H. erectus*.

We have already mentioned that some fossils from Olduvai and other Kenyan sites show some affinities with *Homo* as well as *Australopithecus*, particularly towards the top of Bed I and in Bed II. Fossil hominids of middle Pleistocene age are quite widely distributed in the old world and are found in Asia as well as in Africa. In particular a number of finds have been made in Java from beds of Middle Pleistocene between 0·5 and 1 million years old.

In the lowermost Djetis levels in Java dated at about 1 million years old two skulls and a mandibular fragment have been found which are assigned to *Homo erectus*. Tobias and von Koenigswald (1964) compare one of the composite specimens from this level with one of the finds of *Australopithecus* from Bed II at Olduvai, although there are certain important differences in shape which we shall discuss later. However, the time of evolution of *Homo erectus* in Java has been beyond doubt because from the Trinil beds, which are accurately dated at 0·7 million years at the bottom and 0·5 million years at the top, have come a number of skulls and limb bones which are undoubtedly of *Homo erectus*.

In north-west China at Lantien a similar skull and jaw, i.e., of *H. erectus*, have been found in beds said to correspond to the Djetis level in Java. But the widespread occurrence of *H. erectus* in Asia at this time is confirmed by the finding of a number of skulls from Choukoutien near Pekin, which can be assigned to *H. erectus* without doubt and which are dated at between 0·4 and 0·5 million years. At about the same time level as the Choukoutien skulls from

China, and similar to them, have come three mandibles from Ternefine in Algeria. The links between these African and Chinese fossils have been strengthened by the finding, associated with the Ternefine mandibles, of a small piece of parietal bone from the skull which is similar in much of its fine detail to the parietals of the Chinese material. *Homo erectus* is also represented

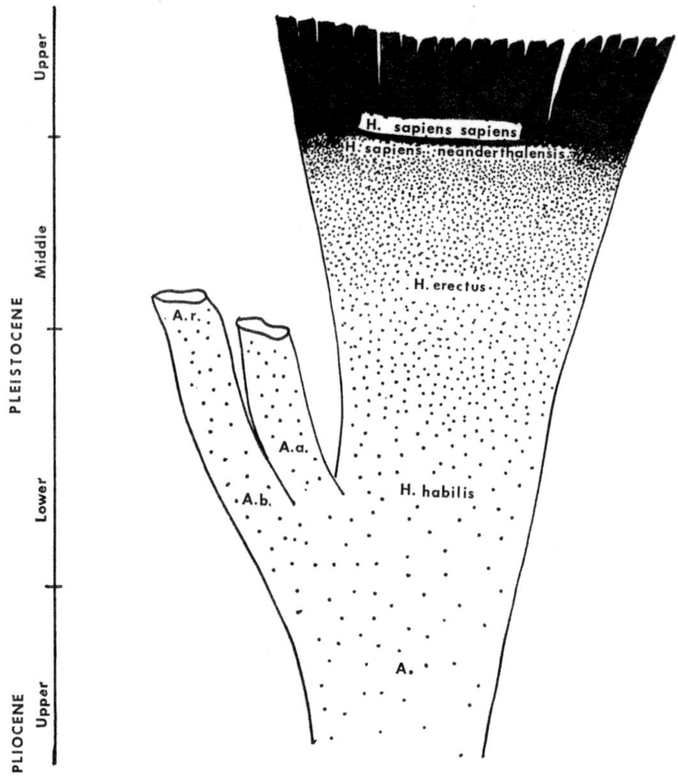

Fig. 3.3.—A provisional schema of hominid phylogeny from Upper Pliocene to Upper Pleistocene. Progressive degrees of hominization are represented by progressively darker shading.
A., Postulated ancestral australopithecine;
A.b., *Australopithecus boisei;*
A.a., *Australopithecus africanus:*
A.r., *Australopithecus robustus.*
(*From* Tobias, 1967.)

at Olduvai by a similar skull recovered from upper Bed II (Hominid 9) which is dated at about 0·5 million years. *Homo erectus* is thus a very widely distributed species.

The next fossil material in temporal sequence is that which is recovered from Europe. Fossil men are not known in Europe before the Middle Pleistocene, but from Heidelberg in Germany comes a lower jaw which probably dates from the Mindel glaciation somewhat over 400,000 years ago (*see Fig.* 3.8). At Verteszöllös in Hungary is another site around 400,000 years old, dating from inter-Mindel, which has yielded a hominid occipital bone and several teeth. The material from both these sites shows many resemblances not only to Asian *H. erectus* but also to later forms, and the Hungarian specimens are close to any boundary between *H. erectus* and *H. sapiens* that can be drawn, so close in fact that different authorities have classified them with both of these species.

The next hominids from Europe are definitely *H. sapiens* and date from about 200,000 years ago. Skulls found at Steinheim in Germany and Swanscombe in England are similar and are both rather primitive but with some similarities to later Neanderthal man. After this, though, there appears to be a distinct break in many parts of the world, both in the temporal sequence of the fossils and the anatomical progression which they show. This break is particularly marked in Europe because the next fossil men found in Europe begin to appear during the last interglacial, the Eemian, 70–100,000 years ago. These are representatives of so-called Neanderthal man and we will return to these later, describing their rather peculiar characteristics.

This evolutionary break, if such it is, is not in fact contemporaneous world-wide. From the Solo River in Java have come eleven skulls and two pieces of limb bones which also possess some of the characters of the Neanderthalers and these are dated as being about 150,000 years old; in southern Africa there are three rather younger sites which have yielded Neanderthaloid skulls and other material—Saldanha and Makapansgat in South Africa and Broken Hill in Zambia (Rhodesia Man). These skulls, although showing Neanderthal features, are also clearly related to *H. erectus* of Olduvai Gorge (Hominid 9), and it should be emphasized that merely because they show some Neanderthal features this does not necessarily imply that they were in fact related to the Neanderthals of Europe. These later South African fossils are dated at about 40,000–50,000 B.P. (before present). Since *H. erectus* is first found in Olduvai more than 500,000 years ago this would suggest that in Africa at least this line changed but slowly during this period of time. On the other hand the skull from Florisbad near Bloemfentein, dated at about 35,000 B.P., although primitive in many respects, is definitely not Neanderthaloid nor is it similar to the very early *H. sapiens* skulls of Europe. It suggests a certain heterogeneity among groups of *H. sapiens* inhabiting Africa at that time.

Some material dated just prior to these Neanderthaloid finds has recently been described in detail by Tobias (1971) comprising a mandible and some teeth from the Cave of Hearths, Makapansgat, Northern Transvaal, possibly 55,000 years old. Tobias finds that these remains have a number of features similar to both earlier hominids of N.W. Africa, including those from Ternefine, which are undoubtedly *H. erectus*, and to *H. sapiens neanderthalensis*. He suggests that a wide-ranging population, probably rather variable, inhabited most of Africa at this time, to which he assigns the name *H. sapiens rhodesiensis*.

Finally, we must mention a number of fossils found at Mount Carmel in Israel. One group of these fossils, dated at about 39,000 B.P., shows some characteristics resembling West European Neanderthals together with some features similar to the skulls from Southern Africa (Rhodesia Man). Another group of fossils, the skeletons from the Skuhl Cave, are known from the same area and these are obviously much more modern, although still related to the Neanderthals. They are thought to be a few thousand years later in time than the first group. Dobzhansky (1962) suggests this is an area of gene exchange between two subspecies or races of *H. sapiens*. This suggestion carries with it the implication that *H. sapiens* developed at some place outside Europe after the initial stages represented by the Steinheim and Swanscombe skulls and we shall find other evidence that this is probably the case.

Returning to consider the European Neanderthals, the earliest of them, from the Eemian interglacial between the Riss and Wurm glaciations, are a group of relatively unspecialized types apparently more modern looking than the classic Neanderthals from western Europe, who appear to have existed for a fairly restricted time period during the early part of the last glaciation between 50,000 and 70,000 years ago. This latter group, the classic Neanderthals, has a number of curious characteristics and may be migrants from outside Europe. About 35,000 years ago they were replaced in western Europe by types who were seemingly very modern in appearance, although the suddenness of this replacement has perhaps been overemphasized in the past. In other parts of the world the occurrence of men who were completely modern types of *Homo sapiens* appears to have been a little earlier in time, although we must remember that the Neanderthalers have not been found everywhere and were a very variable population anyway (Howell, 1957). An interesting idea was put forward by Virchow (1872), who suggested that the peculiar simian cast of these individuals was due to the fact that they suffered from rickets, a vitamin-D deficiency, caused by the poor diet and sunless environment available to them during the Wurm glaciation. Ivanhoe (1970) has recently resurrected this idea and suggested that the observed variability in the Neanderthal characters can be correlated with climatic factors changing with both latitude and longitude. Thus, Solo man from 7° is least affected and an attenuation of the Neanderthal facies can be seen in samples recovered from the interstadial. Although the final verdict on this suggestion has not yet been given, it seems rather an unlikely one and will not be considered further. Many authorities indeed believe that the true Neanderthal people were confined to Europe and that similar types from elsewhere were not in fact at all related.

It is instructive for the student of racial histories to consider the times of appearance of modern man on the various continents as closely as possible. In Europe, as we have seen, *Homo sapiens* first appears about 400,000 years ago and is also represented by some fossils dated at about 200,000 years B.P., from Steinheim in Germany and Swanscombe in England. At about 70,000 years B.P. the Neanderthals appear and these are for a time the only type of European Hominids. However, these appear to have been replaced at about 35,000 years B.P. by truly modern types.

In Africa the first skulls which can be classified as *Homo sapiens* are those of Saldanha and Broken Hill aged between 40,000 and 50,000 B.P., although they do show certain Neanderthal-like features. The mandible from Makapansgat at 55,000 B.P. is also probably of this group. A little later in time is the Florisbad skull, the relationship of which, as we have mentioned, is rather difficult to determine. There is very little other fossil material from any part of Africa to shed any light on the important racial evolution which undoubtedly took place there.

Turning to S.E. Asia and Oceania we find what is probably the earliest example of completely modern man, part of the skull of an adolescent boy from the Niah Cave in Borneo. Charcoal associated with this skull has been dated at about 40,000 years. The skull is very different from the Solo skulls of Java, and Brothwell (1960) suggests that its closest recent parallel may be the natives of Tasmania. Until fairly recently the earliest definite evidence we had of human occupation of Australia was dated at about 16,000 B.P. This consists of a skull, itself dated at 13,000 B.P., found in the gravels of the Keilor terrace of the Maribyrnong River near Melbourne. In the region of this skull have been found hearths at various levels containing charcoal and charred bones, the earliest of which gives a ^{14}C date of 16,000 B.P. At Koonalda Cave in South Australia, a series of human implements have been found, the oldest of which also gives ^{14}C dates of about 16,000 B.P., but the tools continue downwards and suggest a total age of about 30,000 years. Radiocarbon dates on charcoal from hearths from this cave now show that it was occupied at least as long ago as 22,000 B.P. (Frank, 1971).

The latest evidence on Australian occupation comprises a number of skeletons from Kow Swamp in northern Victoria which are late Pleistocene in age and which show definite similarities with the skulls of Solo, Java, more than 100,000 years earlier. Another recent find is a cranium from Lake Mungo, in western New South Wales, which has been accurately dated at 26,000 $\pm$ 1000 years B.P. and appears to be essentially modern in form (Thorne, 1971). In Arnhem Land, in the extreme north of Australia, two successive tool-making episodes can be recognized, the first of which, on radiocarbon dates, lasted from about 22,000 to 6500 B.P. The second is from 6500 B.P. until the present. It would seem, therefore, that Australia has been inhabited for at least 30,000 years and possibly longer.

For long periods during the Pleistocene, Australia and New Guinea were certainly united as one continent by a land bridge where today exists the Torres Strait. There are unfortunately no radiocarbon dates from New Guinea, but there is definite evidence that some Highland areas were settled by at least 16,000–25,000 B.P., although unfortunately no skeletal material has been recovered to throw any light on the nature of the people living in New Guinea at this time. Certainly when the Wadjak skulls of Java are included (although the date for these cannot be accurately established), it seems that we have a reasonable succession from Australia, through Java, via the Niah skull of Borneo, back to South-East Asia, both in geographical and temporal terms.

The only possible route for large-scale migration into the American continent is from Asia via the Bering Strait into Alaska and then southwards. Ice barriers would have probably closed this route from about 21,000 to 11,000 B.P. The most favourable period for human migration would seem to be about 25,000 years ago, although it now seems probable that much migration did, in fact, take place well before this. The isthmus between North and South America probably only came into existence about 15,000 years ago and this, for a long time, was thought to be the probable earliest date for the entry of man into the South American continent. In this case, too, it now seems likely that man entered South America rather earlier than previously thought. Unfortunately firm dates and fossiliferous material to provide details in this outline of man's colonization of the Americas are at present lacking. The degree of genetic divergence in the Indian races now living in the continents, and the few suggestive bits of archaeological evidence certainly indicate an earlier rather than a later date for colonization, but much remains to be discovered.

Recently R. S. MacNeish (1971) has announced the finding in the Ayacucho valley of Central Peru of a series of remains of cultures which apparently date from 20,000 B.C. to A.D. 1500. In one cave the bone from a sloth *Megatherium*, associated with stone tools, has a radiocarbon date of 14,150 (±180) B.P. Other dates of bone fragments in these tool assemblages are successively 14,700 (±1400), 16,500 (±1200), and 19,600 (±3000). These dates, and others compatible with them must, if substantiated, mean a significant rethinking of our ideas on the occupation of North and South America. There are, for instance, bone tools from a site in the Canadian Yukon with radiocarbon dates of between 23,000 and 28,000 years B.P. In Texas there are some flake tools estimated to be 38,000 years old. These dates now assume more importance in suggesting that man first entered North America perhaps 40,000 years ago, and South America 5–10,000 years later.

Knowledge of the time scale of human evolution is the framework on which all else must hang. We have seen the appearance of two or possibly three species of *Australopithecus* nearly two million years ago, and then the evolution of *Homo erectus* in S.E. Asia about 700,000 years ago, together with his appearance in mainland China, Africa, and Europe around about 500,000 years ago. We have then seen the first appearance of *Homo sapiens* possibly 400,000 years ago in Europe and then his apparent disappearance or alteration in different parts of the world from between 150,000 to about 35–40,000 years ago, after which he reappears almost everywhere. Within this interval a rather different type, Neanderthal man, appears, although there are several straws of evidence to suggest that he too contributed to the gene pool of modern man, albeit differentially in different parts of the world. Given this time scale, our next task is to determine how it corresponds with the biological levels reached by man and his forerunners. What, in other words, is the grade of each of the fossils we know?

3.3. The Biological Grade of Human Forerunners

We have so far mentioned four evolutionary stages of the development of man, the *Australopithecine* stage, the *Homo habilis* stage, the *Homo erectus* stage, and the *Homo sapiens* stage. We must now consider each of these a little more closely, and in particular the times of transition between these grades in different parts of the world. One of the potentially most important hypotheses which we must consider is that of Carleton S. Coon (1963), who suggested, following an earlier idea by Weidenreich, that in fact the major races actually pre-date the species *Homo sapiens*. He believes that they are continuations of races or separate lineages of *Homo erectus*, and that the time of transition of these lines from the *erectus* grade to the *sapiens* grade varied in different parts of the world. Clearly this has implications for the status and ages of the races of man living today and therefore we must trace man's development from this point of view as carefully as we are able.

The earliest genus of the Hominidae is *Australopithecus*. Although, as we have seen, it comprises at least two rather different species, we can say quite a lot regarding the grade reached by this genus. The cranial capacity is relatively small, ranging from about 450 to 700 c.c. and it appears likely that the limits of cranial capacity are about the same as the present gorilla, although the body size was much smaller than that of the gorilla, perhaps 60–130 lb. in weight depending on the species and perhaps ranging from 4 ft. 6 in. to more than 5 ft. in height. Although the jaws are relatively large the dentition is essentially hominid, the main differences between the species being that *A. africanus* has small teeth with distinct breaks between canines and premolars, and between premolars and molars. In *A. robustus* the teeth are larger and there is but one break between the incisor/canine field and the molar/premolar teeth. It is probable that these differences in dentition are closely related to the increased body size. From the postcranial skeletal remains which we know of *A. africanus*, particularly the pelvic girdle, it is certain that the gait was bipedal and upright, and it is probable that *A. robustus* was also bipedal. There is clear-cut evidence from Swartkrans and Makapansgat in South Africa, and from Olduvai in East Africa, of the toolmaking abilities of *Australopithecus*. At Makapansgat, about 1·5–2 million years ago, there is a collection of animal bones showing clear evidence of hunting, and many of the bones from these killed animals have been used as tools. There is also some evidence of the use of stone tools in the

Makapansgat cave which was inhabited over a period of possibly 50,000 years exclusively by *A. africanus*. At Swartkrans a number of simple stone tools have been found which are believed to be between 0·75 and 1 million years old. We have already mentioned that at Swartkrans it is likely that both *A. robustus* and either *A. africanus*, or a form closely related to it, lived together and we cannot tell, therefore, which of these species was responsible for the toolmaking. At Olduvai there has been found an astonishing variety of stone tools, some of considerable sophistication, and these have been found associated with the earliest finds of hominid remains in Bed I dated as about 1·75 million years old. Since, as we have seen, there were two and possibly three lineages living at Olduvai, it is impossible to assign the responsibility for the toolmaking with any certainty. Tobias (1967) suggests that *Australopithecus* lacked the ability to use a tool to make a tool, which is an important technological shortcoming.

Many authorities suggest that the robust lineages of *Australopithecus* (including *A. Boisei*) show so many specializations of skull, jaws, teeth, and post-cranial skeleton, and possibly behaviour and ecology, that they contribute few, if any, genes to later species. It is the single lineage of *A. africanus*, it is thought therefore, which gave rise to *Homo habilis* and thence to *Homo erectus*. *Australopithecus africanus* is widespread in Africa during the early Pleistocene and almost certainly was more widely distributed by the middle Pleistocene. *Homo erectus* is a species which was also widely distributed but *Homo habilis* is, at the moment, known only from the lower Pleistocene of Olduvai gorge. But however these fossils and this lineage is classified, there is little doubt that it gave rise to *H. erectus* and probably within the continent of Africa, a conclusion reinforced by the intermediate form found at the top of the Olduvai Gorge sequence in Kenya.

One of the objections to this idea is the fact that the earliest specimens of *Homo erectus* are almost entirely confined to Asia. However their discovery there is probably only a chance effect and *Homo erectus* was even at this time probably a fairly wide-ranging species. Specimens which can be attributed to *H. erectus* are certainly found throughout the old world by the Middle Pleistocene, and show a number of features of evolutionary advance.

Fig. 3.4 shows some of the subspecies of *H. erectus* which are recognized by various scholars (Washburn, 1964; Howells, 1966; Tobias, 1971), although we have already noted that *Homo erectus habilis* and *H. erectus capensis* are placed by some authorities in the genus *Australopithecus* and others recognize a separate species *H. habilis*. The important thing to discover, though, is what are the biological facts which persuade anthropologists to place these rather varied fossils within the same genus, thereby implying a similar grade of biological organization? What is the significance of their varied distribution in time? The lowest Java skulls from the Djetis levels are very primitive; they have very thick walls and their cranial capacity is less than 900 c.c. The Lantien skull is similar, with a capacity of about 780 c.c. The Pekin fossils are slightly more

Grade	Europe	North Africa	East Africa	South Africa	East Asia	South East Asia	Australia
5	Homo sapiens (vertesszollos) 200,000 B.P.			H. sapiens rhodesiensis (Broken Hill)			H. sapiens (N. Victoria)
4				Some affinities with H. erectus leakeyi 40,000 B.P.		Homo erectus soloensis 150,000 B.P.	Some affinities with H. erectus soloensis 10,000 B.P.
3	Homo erectus heidelbergensis 400,000 B.P.	Homo erectus mauritanicus 500,000 B.P.	Homo erectus leakeyi 500,000 B.P.		Homo erectus pekinensis 500,000 B.P.		
2						Homo erectus erectus 700,000 B.P.	
1			Homo erectus habilis	Homo erectus capensis	Homo erectus lantienensis 1,000,000 B.P.	Homo erectus modjokertensis 1,000,000 B.P.	

Fig. 3.4.—Some representatives of the species *Homo erectus*. These are given subspecific status by Campbell, Howell, Tobias, and others, but the most important thing is to note the widespread distribution of the species, its division into four evolutionary grades, and the fact that a fifth grade which is really across the species border into *Homo sapiens* nevertheless shows relationship with *Homo erectus* rather earlier in time.

advanced than those from Java and Lantien, with a higher braincase, shorter face, and a cranial capacity of nearly 1100 c.c. Nevertheless, the general construction of skull and jaw is similar and the dentition of both is somewhat larger and more primitive than that of modern man. It is in Java that the nearest approximation to a stratigraphical sequence exists and with finds from the Djetis beds (1 million years old), Trinil beds (700,000–500,000 years old), and the Solo River (150,000 years old) some evolutionary progress can be seen, particularly if the finds from Lantien (probably 700,000 years old) and Choukoutien (500,000 years old) can be interpolated into this sequence. The postulated existence of *Homo erectus* in this region only 150,000 years ago is very late compared with the time of the other geographical subspecies, but nevertheless seems certain because of the large, primitive tooth size, the flatness of the skull, and the cranial capacity at about 1100 c.c., which is almost identical to Pekin Man and very much less than the average 1400 c.c. of modern man. The existence of *Homo erectus* over this span of time in Asia therefore appears certain and the break between *H. erectus soloensis* of 150,000 years ago and *H. sapiens* from Niah in Borneo of 40,000 years ago is quite profound. Although we have mentioned some similarities between *H. erectus soloensis* and much later Australian skulls it should be emphasized that this involves only a small number of cranial characters and there is no doubt that the latter are *Homo sapiens* (Thorne and Macumber, 1972).

When we turn to the fossil sequence in Africa, the persistence of relatively primitive *H. erectus* to even more recent times is perhaps even more puzzling. We have mentioned that there is some doubt as to whether the earliest postulated members of *H. erectus*, *H. e. habilis*, and *H. e. capensis* should in fact be assigned to this grade rather than *Australopithecus*. It would appear, therefore, that these must represent a population very close to the transition between the *Australopithecus* level and the *Homo* level, rather more than a million years ago. After this transitional period later representatives of *Homo erectus* are very rare anywhere in Africa. One cranium has been found in upper Bed II (dated at 500,000 years ago) at Olduvai with a cranial volume of about 1000 c.c. Although this skull is basically similar to the Asian *Homo erectus*, it has a number of features much more similar to later African hominids. As already mentioned, three mandibles and a parietal bone were found at Ternefine which resemble those from Choukoutien, the parietals being similar right down to the markings made by the arteries supplying the blood to the outside of the brain. There are a number of other later finds in North Africa, principally jaws and teeth, which all show a reduction in size and certain morphological changes. In South Africa, as we have already mentioned, it appears that this grade of evolution, represented by the Makapansgat mandible and the Saldanha and Rhodesian skulls, persisted through to a time only 40,000–50,000 years ago. These skulls have brain volumes of about 1250–1300 c.c., which fall within the range of modern man and yet exhibit characteristic *erectus* features of long skulls and large faces, with inflated nasal regions and thick projecting brow ridges. They are certainly very similar to *H. erectus* from Bed II at Olduvai and are probably part of the single subspecific lineage which must have evolved slowly in Southern Africa for more than half a million years. The Ternefine finds show similarities to Choukoutien, and the Makapansgat mandible has at least one feature in common with the Lantien jaw—congenital lack of a third molar. Too much cannot be made of these similarities, but they are at least suggestive of a very widespread, or at least intercommunicating, population of essentially *H. erectus* grade, varying both spatially and temporally on a world-wide basis. In addition, as we have mentioned, these late African fossils show some characteristics of Neanderthal man. The fate of this group is equally as obscure as Neanderthal Man elsewhere. Some believe they became extinct without contributing in any way to the gene pool of modern man, others that they hybridized to some extent with other populations before themselves becoming extinct. Professor Tobias believes they may have evolved into Bushmen (Tobias, 1959), while Carleton Coon suggests that they gave rise to the Negroes.

The crossing of the *Homo sapiens* border appears to be earlier in Europe than anywhere else, at least so far as we can at present tell from fossil material. The fossils from the inter-Mindel site at Vertesszöllös in Hungary are as close to any boundary as can be drawn. These are 400,000 years old and yet while the teeth show resemblances to Asian *H. erectus*, the skulls are advanced. The occipital bone found comes from a large skull (a volume over 1400 c.c.) and are more evenly rounded than *H. erectus*. The next skulls to be found in Europe from Steinheim in Germany and Swanscombe in England are definitely *H. sapiens*, with brain volumes of about 1200 c.c., although with rather large faces and prominent brow ridges. Certainly they are rather primitive, but equally certainly they have crossed the *erectus/sapiens* threshold.

Just what is this threshold? If we examine the grade reached by *H. erectus* at different periods of prehistoric time, or indeed the grade reached by different major races at the present time, does it tell us anything of the origin of these races? Following C. S. Coon, there appear to be three principal criteria by which we might distinguish between *H. erectus* and *H. sapiens*—brain size, cranial form, and tooth size. We shall consider each of these in turn, and also return to them later when considering possible lineages within these two species.

Eighteen skulls of *H. erectus* have been found which are clearly differentiated from those of living persons and these have cranial capacities ranging from 750 c.c. to 1225 c.c. Skulls of modern man range from 1200 c.c. to 1800 c.c., with an average of about 1400 c.c. The three series of *erectus* skulls, those from China and Java, have means of 848 c.c., 1043 c.c., and 1091 c.c. The largest skull regularly classified as *H. erectus* is that of Broken Hill, Rhodesia, with 1280 c.c. This is now normally classified as *H. sapiens rhodesiensis* (see p. 107). The smallest fossil skull excluded from *H. erectus*, and included within *H. sapiens*, is that from

Steinheim, which has a capacity between 1170 c.c. and 1290 c.c. (it is a badly crushed skull) and this is the skull from a small woman. It is at any rate within the range of modern females. The threshold between the cranial capacity of *Homo erectus* and *Homo sapiens* can therefore be set at approximately 1250–1300 c.c. with the proviso that a skull will be named as *erectus* or *sapiens* on the total range of characters, not on cranial capacity alone. These data on cranial capacities *of Homo erectus* are summarized in *Table* 3.3.

Skull size, of course, is less important than shape. In general we can say that *Homo erectus* crania are long and broad at the base and converge toward the top, both lengthways and sideways. The skulls of *Homo sapiens* bulge more to the front and to the back and also to the sides, because a larger brain is set on a proportionately smaller base. The principal measurements which illustrate this are measurements of individual bones on the skulls of both species, particularly the frontal, parietal, and occipital bones. Two principal measurements are taken on each bone, its length measured along the centre line of the skull, and its chord, i.e., the distance in a straight line between its two ends. A saggital

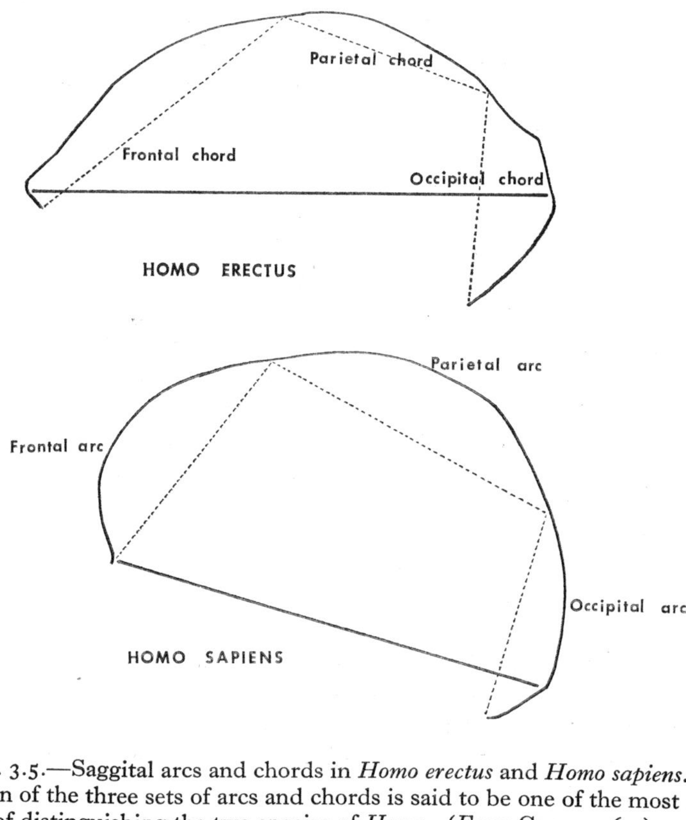

Fig. 3.5.—Saggital arcs and chords in *Homo erectus* and *Homo sapiens*. Comparison of the three sets of arcs and chords is said to be one of the most reliable ways of distinguishing the two species of *Homo*. (*From* Coon, 1963.)

curvature index can then be calculated by expressing the length of the chord as a percentage of the length of the arc (*see Fig.* 3.5). In the *H. erectus* skulls the parietal chord is 93–97 per cent of the length of the arc, whereas in modern skulls it runs from about 89 per cent to 92 per cent. There is very little overlap in this ratio between the successive species and this represents the best distinguishing feature, although the other two are also useful. It should be emphasized, though, that while these indices are useful at delineating grade, they are insensitive with regard to differentiating between different lineages.

Tooth size, as we shall see, is very variable, both within lineages and also with regard to its rate of change throughout hominid evolution. Nevertheless teeth have, on average, become smaller, while brains have become larger, and any index that mirrors these movements is more sensitive to change than either component. Two such indices were used by Coon (1963): a brain–palate index which he derived from an earlier index constructed by Keith (1925); and a brain–molar index which he developed himself. The brain–palate index is the cube root of cranial capacity divided by the square root of palate area, while the brain–molar index is the cube root of cranial capacity divided by 100 times the square root of the sum of the crown areas of the 6 upper molar teeth.

$$I = \sqrt[3]{CC} \div \sqrt{\sum 1 \times b(M^1, M^2, M^3)} \times 100.$$

These two indices divide up the hominid fossil sequence as shown in *Table* 3.2. While the first two grades are comprehensive, the *H. sapiens* section contains mostly European and Near-Eastern material. They are listed to show that evolution proceeds at different rates in different lines.

Table 3.2.—INDICES OF THE SKULL OF MAN (*From* Coon, 1963.)

BRAIN–PALATE INDEX			BRAIN–MOLAR INDEX		
Australopithecus					
Age Rank	*Subject*	*Index*	*Age Rank*	*Subject*	*Index*
(3)	Zinjanthropus	1·1	(3)	Zinjanthropus	22
(2)	Swartkrans	1·2	(2)	Swartkrans	22
(1)	Sterkfontein	1·4	(1)	Sterkfontein	24
H. erectus					
(4)	Pithecanthropus (Java)	1·4	(4)	Pithecanthropus	30
(5)	Sinanthropus (Pekin)	1·7	(5)	Sinanthropus	34
(9)	Broken Hill (Rhodesia)	1·7	(10)	Broken Hill	37
H. sapiens					
(11)	Tasmanian (Mod)	1·8	(6)	Steinheim	39
(7)	Gibraltar	1·9	(8)	Skhul V	40
(8)	La Chapelle	1·9	(9)	Combe Capelle	40
(6)	Steinheim	2·0	(7)	La Fergaissie	43
(10)	Combe Capelle	2·2	(12)	Cro-Magnon	43
(11)	English (mod)	2·3	(11)	Grimaldi	43

Table 3.3.—Average Cranial Capacities in Hominids

a. Fossil Material (Coon, 1963; Tobias, 1967; Wolpoff, 1969)				
Australopithecus	$n = 7$			502·4 c.c.
Homo erectus of Java (Trinil and Sangiran, >500,000 B.P.)				
I Trinil		850 c.c.		
II Sangiran		775 c.c.		
III Sangiran		*c.* 880 c.c.		
IV Sangiran		750 c.c.		
V Sangiran		875 c.c.		
			mean $n = 5$	848 c.c.
Homo erectus of Choukoutien (>400,000 B.P.)				
II		1030 c.c.		
III		915 c.c.		
X		1225 c.c.		
XI		1015 c.c.		
XII		1030 c.c.		
			mean $n = 5$	1043 c.c.
Homo erectus of Java (Solo River 150,000 B.P.)				
I		1035 c.c.		
V		1225 c.c.		
VI		1035 c.c.		
IX		1135 c.c.		
X		1055 c.c.		
XI		1060 c.c.		
			mean $n = 6$	1091 c.c.
Homo erectus of Lantien				780 c.c.
Homo erectus of Olduvai				1000 c.c.
Total *Homo erectus*	$n = 18$			990 c.c.
Homo sapiens neanderthalis	$n = 20$			1342 c.c.

b. From Extant Populations (Montague, 1960)		
Australoid	Australian Aborigines Northern Territory	1256 c.c.
	Tasmanians	1264 c.c.
	Australian Aborigines South Australia	1278 c.c.
	New Guineans	1280 c.c.
	Australian Aborigines Victoria	1338 c.c.
Negroid	Bushmen	1329 c.c.
	African Negroes	1346 c.c.
	American Negroes	1350 c.c.
	Hottentots	1408 c.c.
Caucasoid	British	1370 c.c.
	Czechs	1438 c.c.
	Modern Europeans	1446 c.c.
	Swiss	1468 c.c.
Pacific People	Polynesians	1451 c.c.
	Maoris	1476 c.c.
	Tahitians	1487 c.c.
Mongoloid	South American Indians	1442 c.c.
	Western Eskimos	1473 c.c.
	East Central Amerindians	1495 c.c.
	Mongols	1573 c.c.

It is often stated that *H. erectus* is slowly evolving and, for instance, changed but little during the 500,000 years that he is known from Southern Africa. This is not altogether true since a number of characters of the skull and dentition show fairly marked evolutionary changes and the cranial capacities also show noticeable increases, as mentioned in the preceding pages. It is true, though, that similar grades of *H. erectus* existed at different times in different parts of the world. We have *H. erectus* from Europe at about 500,000 years, *erectus* from the Solo River, Java, at 150,000 years, and from Broken Hill, Rhodesia, at 50,000 years and Australia at 10,000 years B.P. similar skulls from the *erectus/sapiens* boundary. Have we any evidence that such differences in grade are perpetuated within *H. sapiens*?

Table 3.3 lists some of the average cranial capacities of different groups of man, both living and fossil. In a sense these are rather false comparisons, because the most important statistic regarding cranial capacity is the range of values which any particular group takes, not individual values. In very few

Table 3.4.—THE RATE OF EVOLUTION OF THE CRANIAL CAPACITY OF THE HOMINID SKULL (*From* Campbell, 1963.)

FOSSILS	CRANIAL CAPACITY	RATE IN MILLIDARWINS
Australopithecus 1·2 m. yr.	500	
to		280
H. erectus erectus 0.5 m. yr.	900	
H. erectus erectus 0·5 m. yr.	900	
to		351
H. erectus pekinensis 0·4 m. yr.	1000	
H. erectus pekinensis 0·4 m. yr.	1000	
to		375
H. sapiens steinheimensis 0·15 m. yr.	1325	
H. sapiens steinheimensis 0·15 m. yr.	1325	
to		6
H. sapiens sapiens 0·0 m. yr.	say 1400	

cases, so far as the fossil material is concerned, have we more than a very few skulls available to give us an idea of this range. It is also often impossible in the case of fossil material to measure directly the cranial capacity due to the incomplete or imperfect nature of the skull. In these cases recourse must be made to estimates and calculation. The figures in *Table* 3.3 for living groups are mostly those which derive from actual measurement of a number of skulls. This table shows certain differences between the different races so far as cranial capacity is concerned, but the differences can in part be related to unequal body size, and in any case it seems reasonably clear that evolution of the size of the skull was largely complete by the time of Swanscombe or Steinheim man and what we now observe is merely perfectly normal polytypic variation. Campbell (1963) gives some figures for the changing rate of evolution of the cranial capacity which indicates that its evolution has been very slow since the advent of *H. sapiens*. This is shown in *Table* 3.4. It seems unlikely that future discoveries will upset this general conclusion, even though the figures themselves may be changed somewhat.

Tooth size is not of primary consequence in determining evolutionary grades. It is of interest, though, to investigate the average rate of evolution of teeth in hominid evolution. Two important conclusions emerge from the various

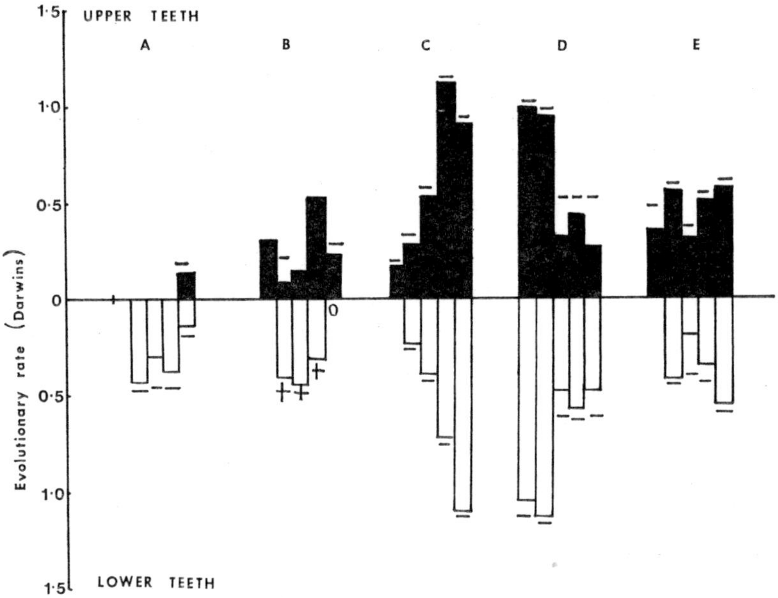

Fig. 3.6.—Rates of change of total size (crown area) of premolar and molar teeth within the hominid lineage. A, Olduvai Bed I–Olduvai Bed II. B, Olduvai Bed II–*H. erectus* (Java). C, *H. erectus* (Java)–*H. erectus* (Pekin). D. *H. erectus* (Pekin)–*H. sapiens* (Modern). E, Olduvai Bed I–*H. sapiens* (Modern). For each segment of the lineage the histogram bars represent from left to right P3; P4; M1; M2; M3. (*From* Bilsborough, 1969.)

studies which have been undertaken. Firstly, evolutionary rates vary markedly for different parts of the dentition, and, secondly, it seems that the teeth, like the cranium, underwent their most rapid evolution during the later Middle Pleistocene and that during the Lower Pleistocene the rate of change of both dentition and cranial capacity was very low. *Fig.* 3.6 from Bilsborough (1969) shows the mosaic of the rate of evolution of the total size (crown area) of premolar and molar teeth within the hominid lineages. A similar mosaic of evolutionary rates in canine teeth is demonstrated by Kinzey (1970).

The evolutionary rate which is used here is a proportional change per million years calculated from Haldane's (1949) formula:

$$\text{rate} = \frac{\log_e x_2 - \log_e x_1}{t_2 - t_1}$$

where x_1 and x_2 are the values for a parameter at times t_1 and t_2. The rates are expressed in millidarwins (1 darwin is a change by a factor e in 1 million years). *Fig.* 3.7 from Brace (1967) shows the absolute changes in cranium and

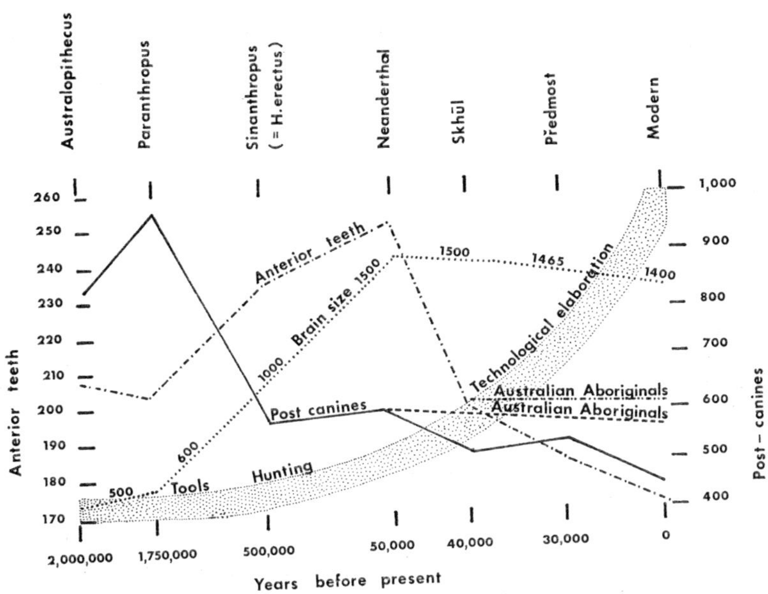

Fig. 3.7.—The major hominid cranial changes plotted with the major adaptive changes that have occurred in the cultural dimension throughout the course of human evolution. (*From* Brace, 1967.)

dentition plotted together. This variation in evolutionary rates sheds some light on the present variation in teeth size in *Homo sapiens*.

The most often used classification of human dentition is Flower's index, which compares the length of the premolar–molar row to a saggital dimension on the skull, the basion–nasion chord. Flower divided the range of variation he obtained into three parts, microdont, mesodont, and megadont, as shown in *Table* 3.5. The polytypic variation which is shown here undoubtedly makes many of the comparisons of evolution between different grades rather inaccurate in detail, firstly because fossils were probably not typical and secondly because the mean of a microdont and a macrodont population was used by both Bilsborough and Kinzey to represent modern *Homo sapiens*, which of course obscures the true state of affairs. For instance, the means of the Australian Aborigines dental parameters fall well within the range of the sample of teeth for Pekin man and some of the teeth, particularly the upper molars, exceed the Pekin means, so that apparently several of the teeth increased in size during the past 370,000 years. The microdont Bushmen, on the other hand, have a mean dental dimension which, with a single exception, lie outside the Pekin range,

Table 3.5.—FLOWER'S INDEX AND RACIAL CLASSIFICATION.
(*From* Coon, 1963.)

	MALES	FEMALES	MALES AND FEMALES	
Polynesians	40·1			
Non-British Europeans	40·5	41·6	41·3	
Ancient Egyptians	40·8	41·2	41·0	Microdont
British	41·0	41·6	41·3	
Central and South Indians	41·4			
Bushmen, South Africans			42·4	
Chinese	42·6			
American Indians	42·8			Mesodont
African Negroes	43·2	44·6	43·9	
Javanese and Sumatrans	43·3			
Melanesians	44·2			
Andamanese	44·4	46·5	45·5	
Australian Aborigines	44·8	46·1	45·5	Megadont
Tasmanians	47·5	48·7	48·1	

and it turns out in fact that the fastest rates of evolution in the Australian line only just exceed the slowest rates in the Bushmen.

The differential rate of evolution needed to give rise to either Australian Aborigines or Europeans is in fact relatively low in comparison with evolutionary rates early in hominid ancestry. We conclude, then, that there is no evidence for the existence of different grades within the many populations or races of *Homo sapiens* living today.

3.4. A GEOGRAPHICAL HISTORY OF MAN

3.4.a. *The Europeans*

The first definitive remains of *Homo sapiens* were found, as we have already mentioned, at Vertesszöllöss in Hungary dating from the second interglacial, perhaps 400,000 years ago. The jawbone of Heidelberg man, the Steinheim skull, the bones from Swanscombe, and the fragments from Fontechevade in central France and other small less accurately dated finds, provide conclusive proof that *Homo sapiens*, and judging by the later finds, a modern form of *H. sapiens*, lived in Europe for maybe 200,000 years prior to the last glacial period. The time scale of evolution of *H. sapiens* in Europe set in the framework of climatic change and cultural stage is illustrated in *Fig.* 3.8.

During this last glacial period in Europe, the middle Paleolithic period, these modern-looking forms were replaced by what were apparently the much more primitive Neanderthals. Nevertheless, as we have seen, characteristic Neanderthal skeletal traits were probably specialized rather than regressive.

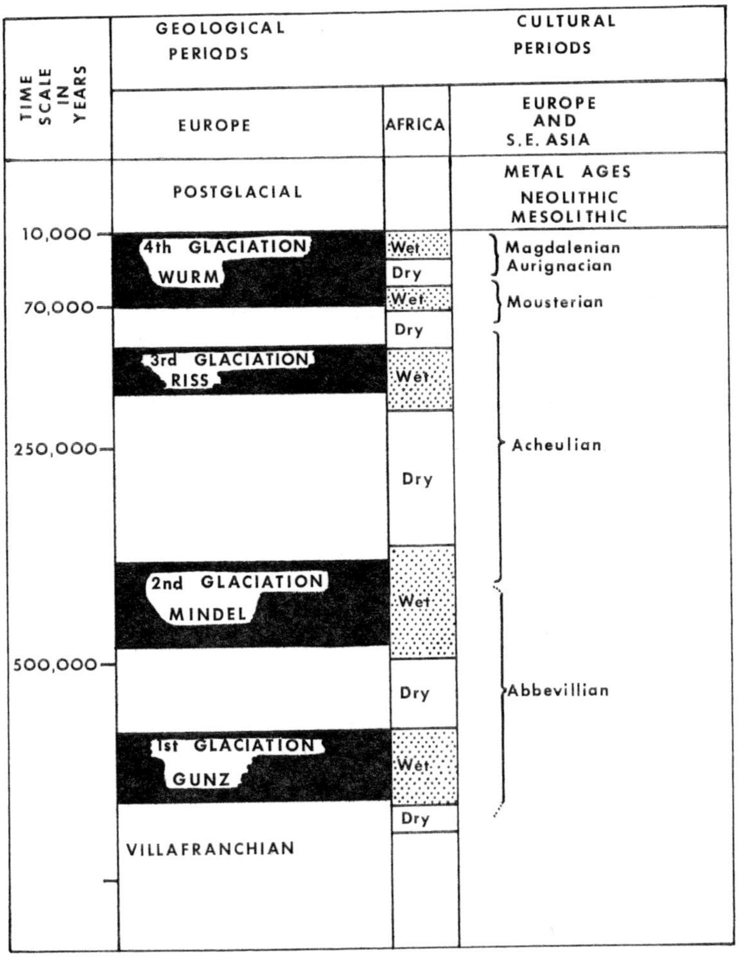

Fig. 3.8.—The Pleistocene succession in Europe and Africa and the cultural sequence associated with it. (*After* Harrison and others, 1964.)

Certainly their brains were somewhat larger on average than ours, and their way of life reasonably well advanced. Many authorities now believe, from both archaeological and anthropological evidence, that the Neanderthalers contributed significantly to the gene pools of later European people.

About 40,000 years ago we have already noted the appearance in Europe of types essentially modern in appearance, the Cro-Magnon types, skeletally identical in every respect with many of the living Europeans. The Cro-Magnards were generally taller than the Neanderthalers and had very different skull morphologies. There are a number of skeletons from Czechoslovakia exhibiting a combination of Neanderthal and *sapiens* characters which strongly suggest hybridization between the two races. As well as a number of physical differences between these two types, the people who replaced the Neanderthalers in western Europe possessed superior technology, the so-called Aurignacian, and all their weapons and tools were superior to those of the indigenous Mousterian flint industry. Archaeological evidence from eastern Europe and southern Russia shows that flint making of undoubted Mousterian inspiration continued to exert a strong influence on certain of the later Palaeolithic industries and this contributes to the impressions of intermixture between the Neanderthalers and the newcomers in this area.

It is generally thought that these new peoples entered Europe from the East, but from where is not known. Nor is anything known about their racial relationships. We might remember, however, that at Mount Carmel in Israel a number of skeletons have been found which show some features suggesting hybridization between immigrants and established Neanderthalers.

In summary, then, we can say that the beginnings of *Homo sapiens* can be seen faintly in eastern Europe much more than 300,000 years ago, but that the crucial stages of evolution of truly modern man are still unplaced. Since Neanderthal Man came to occupy much of Europe, and he is almost certainly not the main contributor to the gene pool of modern man, many of these important evolutionary changes must have taken place outside Europe. This could be in Africa, central Asia, or eastern Asia. We will be in a better position to comment on these possibilities when we have examined the history of the other main racial groups. But when the modern types do reappear in Europe it can be fairly confidently argued that they were members of the White, or Caucasian, race. This is mainly because their skulls and facial skeletons suggest this after biometrical analysis. There is, in addition, a small painting at Angles-sur-Anglin in France dated from about this period which is of a white-skinned man with black beard and black hair.

The people of Europe today show great variety of physical features and genetic constitutions. In the light of modern genetic knowledge it is a misinterpretation to believe that they can be divided with any real accuracy into different races. The archaeology and history of Europe dispels this notion as much as genetics. The races of Europe, as we shall see, are partly differing mixtures of successive immigrant waves, partly locally differentiated stocks, presumably each responding slightly to special local environmental conditions and genetic background.

The Palaeolithic immigrants who replaced the Neanderthals all had large long skulls and they occupied southern Europe and northern Africa while northern Europe was still under glaciation. The Palaeolithic gave way to the

Mesolithic, which extended past 4000 B.C. without any evidence of further immigration, although there are well-marked tendencies for a reduction in size of the skull and the appearance of rounder heads among the long ones—a tendency which gradually became more pronounced.

More or less simultaneous with the Mesolithic in Europe, but in the Near East, north and east from Palestine, farming developed and this led to a marked increase in population numbers. Coupled with this were slight reductions in body size and skull size. After about 4000 B.C. there were probably a number of migrations into Europe from the East and Near East which succeeded in changing still further the physical types of the indigenous Europeans. The result was that by the end of the Neolithic in Europe there were people with long slender skulls in southern Europe, larger skulls in the west and parts of the north, possibly brought in by the Battle-Axe people, and rounder skulls in the north and elsewhere remaining from the Mesolithic. Various migrations continued in the Bronze Age and Iron Age, including the Bell Beaker people, the Urnfield people, and, later, the great invaders from the east, Goths, Huns, Avars, Mongols, and others. It may be that some of these invaders introduced the trend to round-headedness which become so prominent in the Alpine peoples of central and western Europe, but this is only part of the story. Southern Europe remained long-headed and in the north they were also long-headed. The change to rounder heads cannot be ascribed simply to the migration of a new race into the area. We have seen that this change arose in the Mesolithic in different places and it also arose in other parts of the world. Round-headedness is simply a more economic way to accommodate a larger brain in the rather smaller skull of modern man. It is a general trend not a racial characteristic of much significance, and it is a trend, moreover, which has resulted in very rapid change in some European populations (Bielicki and Welon, 1966). Blond hair and blue eyes are other features widely distributed in northern Europe and not just restricted to the 'Nordics' of racial classifiers. These characters often appear, at least in partial expression, all over Europe and therefore the genes responsible for them must be very widespread, even though concentrated in the 'Nordics' and 'East-Balks'.

We have talked of the Whites or Europeans inhabiting the Near and Middle East and migrating westwards from the time of Cro-Magnon onwards. But it is clear that they have also migrated eastwards, and of course we have still not specified their main centre of evolution, which might well be deep in Asia.

In India, for instance, there have been two European migrations. The first, probably from the Middle East, established the Indian civilization with agriculture and settled cities along Middle East lines. The people had darker skins than those from the Middle East, but then we have already seen that this is a trait which might well have evolved very rapidly. In about 1800 B.C. there was a further European or 'Aryan' migration (possibly from 'Iran' since these two words derive from the same word root) and these peoples established their own civilization. The original inhabitants of India are still represented by small numbers of dark-skinned people speaking non-Indo-European languages, and it is probable that hybridizing with these dark natives has contributed more than a little to the darkness of the Indians' skin today.

It is probable that much of Russia and Central Asia are original territories of the White peoples. Indeed the racial name 'Caucasoid' is a recognition of this fact. The Cro-Magnon type persisted throughout the Neolithic at least as far as Lake Baikal in Central Siberia and then gradually changed, probably partly by mingling after eastward migration, into the smaller types of northern Europe. These types were probably continuous across the whole of Eurasia in Iron Age times and Indo-European languages were spoken in Chinese Turkestan until about a thousand years ago.

In Japan there exists on the northern islands of Hokkaido and Sakhalin a small isolated group, the Ainus. These people are almost certainly of Caucasoid stock, both from their phenotypic appearance and genetic similarities. They apparently date back to about 7000 B.C. and may be remnants of a number of varied primitive tribes in the region, indicating that the Caucasoids then stretched across the whole Eurasian continent. Some people have suggested that the Ainu are related to the Aborigines of Australia. This now appears unlikely (Yamaguchi, 1967).

3.4.b. *Pacific Peoples*

The peoples living in and around the Pacific are very varied and yet it seems likely that they derive from one founding population, albeit a long time ago. Some of these relationships can indeed be reconstructed from archaeological, geographical, and historical evidence. Most suggested relationships are unfortunately still speculation.

It is tantalizing that within this area has been found a relatively complete succession of skulls which are undoubtedly of *H. erectus* but that almost nothing has been found for early *H. sapiens*. *Table* 3.6 summarizes our knowledge in this area by mentioning some of the more important fossil finds which have been fairly well dated.

This geological succession encompasses geographical areas where today three racial groups live: the Asiatics or true Mongoloids of China, the Melanesians of S.E. Asia and the Pacific Islands, and the Australoids of Australia. What evidence is there, if any, of racial origins in this and other fossil material?

Despite assertions and suggestions by a number of workers (e.g., Weidenreich, 1945; Coon, 1963) it must be recognized that no very definite conclusion on racial origins can be drawn from these fossils. This fact in itself is perhaps of some significance, since it means that any of their supposed and oft-postulated relationships with one or another racial group extant today must be rather slight or else they would generally be established.

Coon and others (1950) have argued that many of the characteristic features of the Mongoloid face are purely secondary adaptations against the cold. Thus the face has become flattened to decrease its area of exposure to frostbite, the nose is low, and the cheekbones have been brought forward and widened so that the nose and the eyes become recessed into the face. It is

Table 3.6.—HOMINID SUCCESSION IN EAST AND SOUTH-EAST ASIA

DATE	PLACE	DESCRIPTION OF FIND
1,000,000 B.P.	Java, Djetis beds	Two skulls and mandible of *H. erectus*. Some similarities with Australopithecus from Bed II Olduvai
700,000 B.P.	Lantien, N.W. China	Skull and jaw of *H. erectus*
700,000–500,000 B.P.	Java, Trinil beds (Sangiran)	Several skulls and limb bones
500,000–400,000 B.P.	Choukoutien near Pekin, China	Several skulls of *H. erectus*. Similarity of some detail with finds from Ternefine, N. Africa
150,000 B.P.	Java, Solo River	Eleven skulls and two limb bones. Some features in common with European Neanderthaloids
40,000 B.P.	Borneo, Niah Cave	Juvenile skull. Different from Solo skull. Similar to modern S. Australian
25,000–30,000 B.P.	Australia, Lake Mungo	Largely modern Australian Aborigines
13,000 B.P.	Australia, Keilor Terrace	Skull essentially modern
10,000 B.P.	China, Choukoutien Upper Cave	A number of skulls. Unspecialized Mongoloids
10,000 B.P.	Australia, Kow Swamp, Northern Victoria	A number of skeletons. Skulls show definite relationships with Solo River fossils

possible to reconstruct the origin and migration of these specially adapted Mongoloids from archaeological and historical evidence. Although, of course, in line with the degree of sheer speculation involved, a good deal of uncertainty must still remain even concerning these peoples, let alone the other Mongoloids. Coon and others (1950) think that these special facial features arose during the last glacial period in Siberia 20,000–100,000 years B.P., where the peoples lived between the glaciers to the north and the mountains to the south. Here they lived, it is suggested, subjected to intense cold, with pneumonia and infected sinuses acting as the selective agents.

When the glaciers retreated these specialized Mongoloids migrated outwards, almost to western Europe and northern Europe and Scandinavia in one direction, to Korea and parts of Japan in the other, and southwards to Tibet, tropical South-East Asia, Indonesia, the Philippines, and Formosa. Many of the details of these successive migrations can be determined archaeologically. We know particularly that in their southward migrations into South-East Asia and the islands they mingled, mixed with, and finally largely superseded the Australoids who were already living there.

But this does not really solve the problem of the origin of most of the Pacific peoples. We have implied that the extreme manifestations of the Mongoloids arose from a pre-existing generalized stock, and it is to these people, therefore, that we should look for further information on racial origins.

In many ways the Indians of America are suitable candidates for the generalized racial ancestors, even though they are fairly variable themselves. Their variability, though, with the exception of the Eskimos, who have probably only arrived within the last 3,000 years, is fairly superficial. Three broad groups of Indians can be recognized as we go backwards successively in time, those of the present, those of around 3000 B.C.—late Mesolithic types, and earlier still the 'Otamid' type. These last had a large long skull and a long face with a high nose as opposed to the small-skulled, medium-length Mesolithic people.

There are few good fossils to take the story much farther. A skeleton from Tepexpan, Mexico, is dated at about 10,000 B.P. and this skull shows the individual to be similar to the Indians of today, with a rather round head, not a long one. Another skeleton from Texas has been given a date of about 20,000 B.P. The skull is extremely long and high with a light brow ridge and small teeth and jaws. It seems as if this is a representative of the pre-Mesolithic Otamid type.

Some of the most interesting and potentially most informative sites in North and South America have so far not yielded any skeletal material, and we have already indicated some of the radiocarbon dates of these sites on p. 104. MacNeish (1971) describes four distinct stages of man's occupation of South America, based on the characteristic assemblages of tools that are found. MacNeish claims to be able to recognize similar assemblages in North America which are between 5,000 and 10,000 years earlier than the South American sites. He then suggests an even more tentative association of three of these cultural traditions with sites from China or the U.S.S.R. The overall very tentative conclusions from these recent finds are that man arrived in the Western Hemisphere more than 40,000 years ago and conceivably as long as 100,000 years ago. This therefore pushes back the time of separation of the non-specialized Mongoloid lines into Americans and Asians to at least this time. Diagrammatic representation of MacNeish's suggestions are shown in *Fig.* 3.9.

Fairbridge (1960) suggests that crossing the land bridge across the Bering Strait would have been possible only during the peak of the

Riss-Illinoisian and all of the Wurm-Wisconsian glaciations; that is from about 125,000 to about 100,000, and from about 70,000 to 10,000 years ago, although during the later part of this second period it might well have been ice covered. Haag (1962) agrees, but suggests that it was also available during much of the

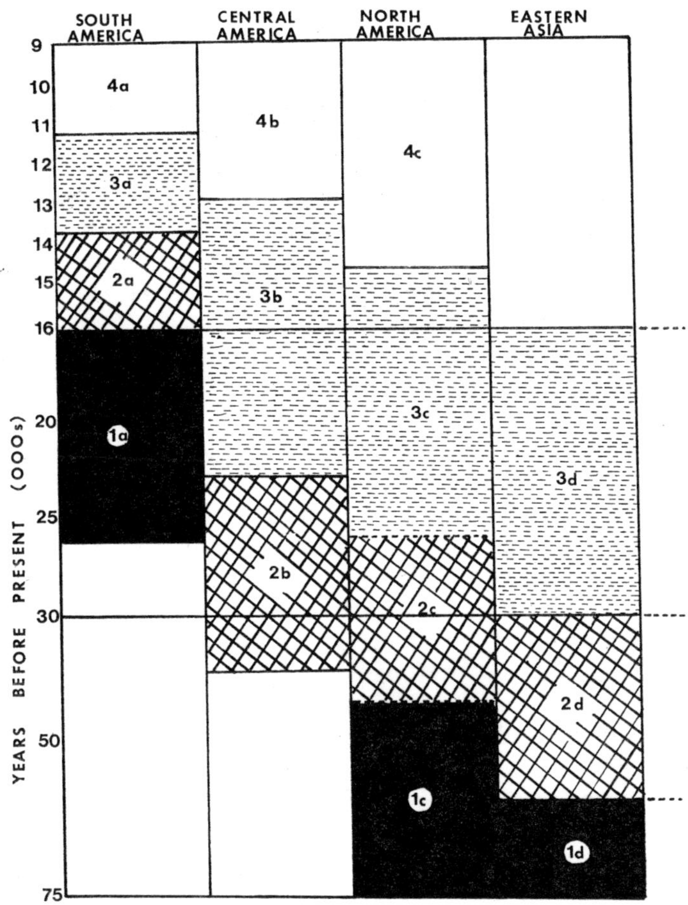

Fig. 3.9.—Some suggested correlations between new world and old world prehistoric cultures

1, Core Tool Tradition

2, Flake Bone Tool Tradition

3, Blade, Burin, Leaf Point Tradition

4, Specialized Point Tradition.

Suggested sites include: **1a**, Paccaicasa 19,600 B.P.; **1c**, Calico; **1d**, Fenho, China. **2a**, Ayacucho 12,500 B.P.; **2c**, Lewisville 38,000 B.P.; **2d**, Shuitung-Kio, China. **3a**, Tagua-Tagua 11,300 B.P.; **3b**, Tlapacoya 23,150 B.P.; **3c**, Wilson Butte 15,000 B.P.; **3d**, Malt'a-Buret, Russia. (*From* MacNeish, 1971.)

interglacial preceding the Wisconsian. So there is no problem here regarding our possible revised interpretations of the occupation of the Americas.

We return then to Asia to seek the origin of these peoples. In the upper cave of Choukoutien, the same locality as that where Pekin man was discovered, a number of skulls have been found which are probably about 10,000 years old, and these are certainly unspecialized Mongoloid types and in a number of details resemble the Otamid Indians of North America.

These skulls are very late Upper or Early Post-Pleistocene. From lower down in the Upper Pleistocene come two skulls, a female skull from Tze-Yang Szechian, which is certainly *H. sapiens* but really shows no conclusive racial resemblances, and a male skull from Liukiang Karangsi province. This has shorter incisors and lacks a third molar, both of which are Mongoloid tendencies. In addition the various indices which can be calculated from the skull, such as the frontal index of facial flatness, are all within the ranges of modern Mongoloid peoples. None of these features are conclusive, yet taken together they do indicate Mongoloid affinities.

This is really as far as we can go with the fossil material. The next important fossils in time are Middle Pleistocene, those of lower level Choukoutien in China and Solo in Java, and they are generally accepted as being *H. erectus* and therefore as being across the species border. Coon's (1963) suggestions that these can respectively be described as Mongoloid and Australoid ancestors have not generally found favour and the reader is referred to Coon's book and Birdsell's review (Birdsell, 1963) for the detailed arguments. Nevertheless it should be mentioned that there are certain similarities between the Solo River fossils and the very recent finds at Kow Swamp in northern Victoria, Australia, as we have already pointed out in earlier sections.

The most that can be said of Mongoloid origins is that they arose in Mainland Asia possibly 100,000 years or more ago. Migrants from this stock then spread southwards to Burma and Tibet and northwards and eastwards into America. During the height of the glaciation some of their number evolved separately and these then came to replace some of their unspecialized relatives in much of mainland and S.E. Asia.

Let us return, then, to the other Pacific peoples, the Melanesians and Australoids. These are a group of peoples occupying parts of South-East Asia and the islands of the South-West Pacific. As they exist today they show great variety; being Negroid types, Negritos or pygmy types, and Australoids. The Negroid peoples are largely inhabitants of the Melanesian Islands. The Negritos show a very discontinuous distribution, being found in the New Guinea Highlands, in central Malaysia (the Semang Negritos), on some of the Philippine Islands, and in the interior of North-Central India. Earlier reports of their presence in Australia (Gates, 1941, 1960a, b) are incorrect. The Australoids are found in Australia, of course, coastal New Guinea, Ceylon (the Veddas), and northern India (the Dravidians). The problem is to relate these groups to one another and to decide on their original region of origin (if indeed it is a single place of origin). It should also be mentioned that the various divisions just listed are by no means today held to be as rigid and meaningful as they once were. The uncertainty about whether this variety is merely of a minor geographical nature, or indicative of important genetic differences, in turn puts doubt upon previous interpretations of racial origins. To try and solve some of these puzzles we can examine the fossil evidence successively from Australia, Melanesia, Indonesia, Indo-China, and India.

There is hardly any fossil material known from anywhere in the Indian sub-continent. Although most of the population today are Caucasoids, there are a number of Australoid peoples, some Negritos, and many millions of people who are a mixture of Mongoloid and Australoid. Skeletal material from the Indus Valley indicates that Caucasoids, Mongoloids, and Australoids were all present in northwest India as early as 2400 B.C. Before then probably only the Caucasoids and Australoids were in India. We have already mentioned the civilization with Middle East connexions dating from a very early Caucasoid invasion, but the very discontinuous nature of the distribution of the Australoias suggests they have long inhabited India and may well have been largely displaced by the first Caucasoid immigration, whenever that was.

Indo-China has not been very productive of fossil remains either. Some early Neolithic skulls have been found at a number of localities and these have been assigned to Mongoloids and also to Negritos. In fact it is suggested that a few skeletons are of hybrids between these peoples (Coon, 1963). Indo-China lies right in the path of the southward expansion of the Mongoloids that we have already referred to, and again, therefore, it would seem that the Australoids were the pre-existing peoples who, apart from a few remnants in the Malay Peninsula and offshore islands, were almost completely replaced.

We have already seen that man was living in Australia at least 25,000 years ago. Nevertheless the racial components of the people who have lived in Australia then and now are by no means well known. A number of authorities, in particular J. B. Birdsell (1950, 1967), have suggested a trihybrid origin for Australian Aborigines in which there were successive waves first of Oceanic Negritos, who are represented, it is said, by the now extinct Tasmanians and the Barrineans of Queensland. Secondly came the Murrayians who, it is alleged, are basically an archaic Caucasoid group, and thirdly the Carpentarians, said to have arisen in India. Few people now believe this literally and neither the Tasmanians (Thorne, 1971) nor the Queensland Negritos (Larnach and Macintosh, 1970) are today believed to represent separate racial groups.

The fossil material which has been found in Australia does seem to be of two rather different types. Some are large and rugged with large brow ridges, large palates, and with marked frontal retreats and postorbital constrictions. On the other hand many are almost the reverse, orthnographic, with well-rounded frontal areas, minimal postorbital constrictions, and horseshoe-shaped palates

(Thorne, 1971). These morphological variations may correspond to two cultural traditions, an early one from 20,000 B.P. to 6,000–7,000 B.P. and a later one subsequent to this.

Morphologically, genetically, and linguistically, the Aborigines are similar to some of the inhabitants of New Guinea—mainly those of the coastal areas. In fact there has been considerable contact between parts of northern Australia and its northern neighbours for the last few thousand years, a situation which is reflected in the social changes, particularly in Cape York and Arnhem Land, apparent during this period. There is also a striking temporal correlation between the probable beginnings of agriculture in Melanesia and Indonesia and the commencement of the second of Australia's cultural/technological traditions. Furthermore Steinberg and Kirk (1970) have shown, using the gamma globulin (Gm) variants, that up to 30 per cent of the genotype of the northern Aborigines came from New Guinea, whereas little if any admixture from this source is seen in western and central Aborigines (*see Table* 3.7). On

Table 3.7.—FREQUENCY OF Gm ALLELES IN FOUR POPULATIONS OF ABORIGINES (*From* Steinberg and Kirk, 1970.)

POPULATION	NO. IN SAMPLE	Gm^{1}	$Gm^{1,2}$	$Gm^{1,5,13,14}$
Western Desert	289	0·73	0·27	0
Central Region	84	0·79	0·21	0
Total	373	0·74	0·26	0
Kimberleys	268	0·58	0·25	0·17
Northern Region	96	0·64	0·26	0·10
Total	364	0·60	0·25	0·15

the other hand calculations using this estimate of 30 per cent admixture for the Group Specific Component, the Gc^{Ab} gene, gives a figure close to that which is actually observed for Cape York, but very different to that found in the Kimberleys. This suggests that the northern Aboriginal populations themselves are not genetically homogeneous, a suggestion confirmed by other analyses (Kirk, Blake, Moodie, and Tibbs, 1971; Sanghvi, Kirk, and Balakrishnan, 1971).

Now in Melanesia there is an important distinction between those peoples speaking Melanesian languages (MN) and those speaking non-Austranesian languages (NAN), and it is found that this distinction is quite marked in the distribution of the gammaglobulin (Gm) factors, particularly in New Guinea. These NAN peoples of the highlands and western coast of New Guinea could be clearly differentiated from the MN peoples of the rest of New Guinea and New Britain using this genetic factor (Curtain, Van Loghem, Baumgarten, Golab, Gorman, Rutgers, and Kidson, 1971).

The archaeological evidence is that there were at least two successive migrations into this area from the north. The first was possibly 30,000 years ago (Mulvaney, 1971) and the migrants came south from Asia, mostly on foot. This migration may have been long continued, even after the end of the Pleistocene. The later immigrants, the MN speakers, derive from eastern Asia, from whence they migrated across the Melanesian Islands to New Guinea, arriving there between 2000 and 700 B.C.

It seems possible from this that there might well have been successive waves of migrants into Australia and elsewhere. From the present distribution of the Negritos, from the physical distinctions within the Australian Aborigines, and from the genetic evidence of the distribution of the Gm factors, some of which are absent in western and central Aborigines whereas others reach high frequencies there, some conclusions do emerge. It seems probable that it is the Negritos who first made this outward migration from Asia, both westwards into India and southwards and eastwards to Australasia and Melanesia. The Australoids *sensu strictu* undertook similar migrations subsequently, although whether from the same geographical area or the same stock we cannot tell. The great complexities of later migrations and minglings throughout S.E. Asia, Melanesia, and Polynesia are superimposed on all of this.

Kellock and Parsons (1970) compared the incidence of minor non-metrical cranial variants in Australian Aborigines with the incidence of the same variants in crania from Melanesia and Polynesia. They found that the differences between the Australian and Polynesian samples were greater than the differences between samples from three areas of Australia, broadly speaking Northern Territory and Northern Western Australia, Central Australia, and South-Eastern Australia. Comparing the samples from these three areas of Australia separately, they found that the northern area had more association with Melanesia than Polynesia, while the more southern regions of Australia had more association with Polynesia than Melanesia, perhaps, they suggest, by migration from Asia. The samples which Kellock and Parsons examined were not large and their findings are probably best regarded as indicating general relationships rather than certain origins.

There is one very important piece of fossil evidence bearing on these varied interpretations which it is now appropriate to consider—the Niah skull from Borneo. This was discovered in 1959 and dated at 39,000 ± 1,000 years. It has very little brow-ridge development and a vertical forehead. The palate and teeth are smaller than those of the fossil Australian skulls, but Brothwell (1960) concluded that it is related to the Tasmanians. If this postulated relationship is correct then it indicates a time when presumably Indonesia was occupied by the Australoids. Some authorities believe that the Tasmanians were a Melanesian stock who were never present on the mainland of Australia. More recently (Thorne, 1971) it has been suggested that they are merely a southern variant of the Aborigines. Whichever of these interpretations is correct, the presence of Tasmanian or south Australian ancestors in Borneo 40,000 years ago is an important landmark in reconstructing S.E. Asian pre-history.

Finally we might briefly mention the fossil record of the dog. This is an animal which appears comparatively late in the Australian fossil record. It is not found in Tasmania and therefore presumably appeared after the land bridge between Tasmania and the mainland disappeared. However, nowhere in South-East Asia until very recently are dogs found anywhere in the fossil sequence. The nearest place where dog fossils are found is India and in India, too, are found assemblages of artefacts similar to those of the earlier cultural tradition of northern Australia.

The place of origin of these Australoid peoples is then almost certainly in Asia, but just where is not known nor how many different places of origin there might be. What are their relationships with the Mongoloids? Firstly it is clear over much of this geographical area that there has been considerable intermixture between Mongoloids and Australoids. Secondly every new investigation on the archaeology and physical anthropology of this region shows how complex the situation is. Genetic drift, gene flow, and natural selection have all been shown to have played important parts in the prehistory of these peoples. If there is any common origin postulated for Mongoloids and Australoids, then it must be of some antiquity, rather more than 50,000 years ago at least. It does seem, nevertheless, in our present state of ignorance, that the Pacific peoples do have a common origin somewhere on the continent of Asia, and that this common origin does not include very much contribution from the other great group of people on this earth, the Africans.

3.4.c. *The Africans*

The Negroid peoples of Africa can be divided into a number of racial groups and over much of northern and eastern Africa there has been considerable Caucasoid immigration. More recently, within the last three hundred years that is, there has been considerable White settlement in southern Africa together with smaller but important importations of people from India and South Asia. The situation in Africa is therefore as complex as anywhere in the world.

Most of the north of Africa is a Caucasoid realm of relatively recent occupation, but southwards, down the east coast, the Caucasoid migration was considerably earlier and penetrated much farther into the continent. They have become mixed to a greater or lesser extent with the indigenous peoples. Thus the Somalis and Ethiopians show clearly their European origin, or at least the fact that their principal component is European, and many Negro tribes must also have absorbed a good deal of European admixture. So far as can be ascertained at the moment it is the Europeans who entered eastern Africa first, possibly during the Palaeolithic, and the Negro admixture which came later, as the result of internal migrations by various Bantu tribes.

The true Negroes occupy much of central and southern Africa today, although it is certain that much of their southward migration is very recent and their arrival in what is now the Republic of South Africa probably only just pre-dates the Whites. The Negroes themselves are very variable, encompassing the Pygmies of very short stature and the Nilotic Negroes who are exceedingly tall. Many of the inhabitants of East Africa, the Hamitic Negroes, show the affects of the varying degrees of White admixture (from Hamitic-Semitic peoples) that is present in them, as we have mentioned. Then there are the normal Forest Negro or Bantu, the characteristic inhabitants of West and Central Africa. The last group are the Bushmen and Hottentots, who are now restricted to small areas of southern Africa but who, it is believed by some, were once much more extensively distributed over the whole of Africa and may, at least according to Coon, have had their origin in the extreme north of Africa.

As we did for the other races of man, we will first consider the fossil evidence to see what light this throws on African origins. *Homo erectus* first appears in Africa about 0·5 million years ago, represented today by the three Ternefine mandibles and the skull of Hominid 9 from the top of Bed II at Olduvai. Although there is a time gap of at least 400,000 years before the next group of fossils from South Africa, the morphological similarities are sufficiently clear for them to be recognized by most authorities as a single racial line. The question is which racial line (if any) in terms of the people living today, and to what extent has it remained isolated from other racial lineages.

As we have already mentioned, Tobias (1971) has shown that the Cave of Hearths skull, dated at about 55,000 B.P., has features in common with the much earlier north-west African remains including the Ternefine mandibles from Algeria, a mandible from Temara, Morocco, and also a mandible fragment from Dire Dawa in Ethiopia. Both the first and second molars are 5-cusped, which are characteristic Negro features, and the teeth are all fairly large, which is another Negro characteristic. It seems likely, therefore, that the widespread population which these fossil finds presumably represent, contributed at least some genetic material to modern Negro populations. The skull from Broken Hill in Rhodesia is similar in many respects to the earlier forms we have mentioned and also in some features to modern Negroes. It certainly has a number of specialized features including enormously thick and projecting brow ridges and a very unusual premaxilla index of facial flatness. Its teeth are megadont, and again most of these features are those found to some extent in modern Negroes. Other finds of about this period in time, say 50,000 B.C. to 30,000 B.C., include a skull cap from Saldanha Bay, 90 miles north of Capetown, a skull from Florisbad near Bloemfontein, and some very modern-looking material from Kanyera on the shore of Lake Victoria.

All this material brings us very close to the modern African Negro. None of it, though, tells us much of the place of origin of the different Negro types in

Africa today, or of the Bushmen and Hottentot groups. Anthropology and archaeology can only help us a little here.

It is generally agreed the Negro presence in southern Africa is recent and that they have migrated there probably from West Africa within the last 2,000–3,000 years. Some of the consequences of the meeting of these immigrant Bantu with the indigenous Bushmen are shown in *Fig.* 3.10. This shows the estimated

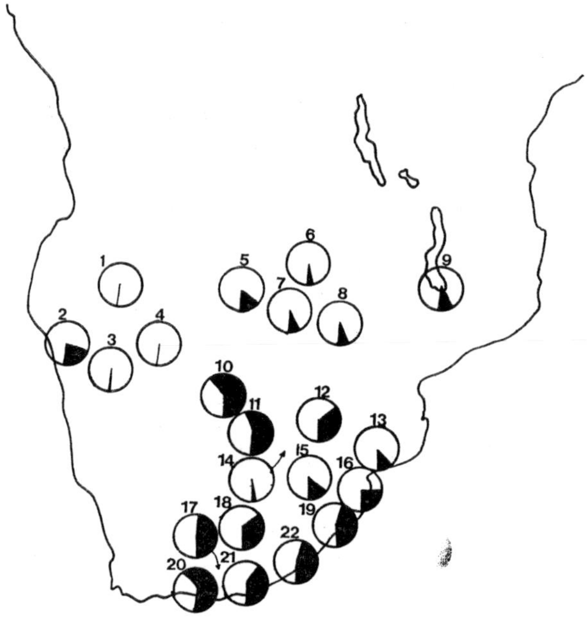

Fig. 3.10.—Map showing the percentage estimated amount of Bushmen admixture in various Negro tribes in South Africa. (*After* Steinberg, 1969.)

1, Angolans 0; **2**, Ovahimba 22; **3**, Ovambo 3; **4**, Okavango 0; **5**, Mlozi 13; **6**, Lenje 6; **7**, Plateau Tonga 11; **8**, Valley Tonga 12; **9**, Malawians 9; **10**, Ngalagawi 60; **11**, Bechuana 53; **12**, Venda 38; **13**, Shangaan Tonga 15; **14**, Hoebele 5; **15**, Pedi 17; **16**, Swazi 28; **17**, Baga 49; **18**, Basuto 29; **19**, Zulu 45; **20**, Xhosa 60; **21**, Hlubi 37; **22**, Pondo 45. The shaded segment is amount of Bushmen admixture.

amount of Bushmen admixture in various Negro tribes in southern Africa based on the gamma globulin antigens. Before the Bantu Negroes arrived the Bushmen and Hottentots were widespread in southern Africa. On linguistic grounds it seems probable that the Bushmen lived in East Africa, and it is suggested that Hottentots obtained their cattle via a route southwards through East Africa.

Into East Africa, as we have mentioned, Caucasoids migrated at the end of the Pleistocene, either during the last glacial or after. Various East African Negro groups have since become absorbed into them to produce Hamitic-speaking peoples. It is possible, too, that a number of Bushmen have also become absorbed into these peoples. The Masai tribes are probably of mixed Negro–White parentage, while others, such as the Watussi, are probably more directly related to these ancient Whites. There are a number of lines of evidence, including genetical, linguistic, and geographical, which put the immediate home of the Negro into West Africa south of the Sahara. Livingstone (1958) has reviewed much of the evidence, particularly that concerning the rise of malaria and the spread of genes giving protection against malaria. We cannot say where the Negro stock came from originally or when the Bushmen separated from them, primarily because of a lack of fossil evidence. It seems almost certain, though, that the widely spread species *H. sapiens rhodesiensis*, which in Africa had slowly evolved from an even more widespread *H. erectus* over the course of 400,000 years or so, contributed a significant proportion to the African people (Tobias, 1959). Within the past 6,000–8,000 years there has obviously been a considerable influx of peoples and their genes into Africa and its indigenous populations from Europe and Asia by way of the Middle East. The resulting populations, often genetically rather mixed, have spread right across the north of Africa and southwards down the east coast. What is really unknown to us is how long this process has carried on. In other words, did this influx from Eastern Europe or South-West Asia begin 8000 years ago or had migration been carried on from these areas for a long time before that? Could it be that modern *H. sapiens*, carrying with them many of what we now regard to be Negro characteristics, arrived in Africa and provided the speed up of evolution that otherwise perhaps had lagged in Africa? Perhaps these same peoples, which spread eastwards into Africa, also migrated westwards, carrying the same genes and achieving the same speed up of evolution in the indigenous populations of South-East Asia and Melanesia. This would then be an explanation for the similarities of the Negroes on the one hand and the Negritoes on the other. The ancestral home of these people and the time they left it are of course utterly unknown.

3.5. A Genetic History of Man

As we have seen in Chapter 1, the classification of man into different races long pre-dates the development of genetics, and even more so the subject of Human Genetics. The ABO blood group was discovered in 1900 and it was fairly soon realized that it might be useful in determining racial differences. The MN and P blood groups were discovered in 1927, but it was not really until the discovery of the Rhesus system in 1940 that blood group differences could really be used to supplement the sort of classifications built up on physical differences. Since that time an enormous amount of information has been gathered on an ever increasing number of useful genetic loci. Cavalli-Sforza and Bodmer (1971) list 27 polymorphic blood groups and enzyme systems about which useful information is known in at least two racial groups and Harris (1969) gives data on another nine. This number excludes the Gm polymorphism and the HL-A and other white-cell and platelet antigen polymorphisms which are now being actively investigated in many different racial groups.

We have already presented in Chapter 2 some of the analysis of this variation, indicating the amount of genetic difference between the races and the light it throws on racial relationships. It is appropriate to consider this in a little more detail here, particularly with a view to reconstructing man's racial history.

There are basically two ways of going about the calculation of genetic distances between human races. Firstly one can consider racial groups at a fairly detailed level, aiming to choose those which might be representative of the major geographical races. Of course one can only choose those local races which have in fact been analysed for the genes thought to be useful, and this may bias the results. The second approach is to amalgamate the smaller local races and take the average for them. This makes the analysis easier and gaps in the information on race markers are more easily filled.

The results of using these two approaches are given in *Fig*. 2.33. Cavalli-Sforza and Edwards, who carried out this analysis, point out that the number of possible evolutionary trees that can be constructed with fifteen racial groups is enormous and that many of the alternatives would have an almost equally good fit to the data. However, they also say that all trees that do give a good fit to the data retain the characteristic racial groupings shown in *Fig*. 2.33. That is, the three African peoples stay together, the two peoples from South Asia, the Veddahs (from Ceylon) and the Gurkas (from Nepal and possibly of mixed Indian and Oriental ancestry) are intermediate between Europeans and other Asians. The American Indians stay together, as do the Australians and Melanesians.

One of these alternative trees is shown in *Fig*. 3.11 projected on to a world map. The relative lengths of the branches do not, in this case, show the amount of genetic difference between the races but the sequence of racial splits is maintained. It is very apparent that this tree fits very cloesly the probable routes of man's migrations and therefore strongly reinforces the interpretations of man's racial history that we have put forward in the previous sections of this chapter.

These conclusions were reinforced by the second approach in which some of the local races were amalgamated to allow a larger number of gene loci to be incorporated into the analysis. This emphasized the closeness of European and extra-European Caucasoids (the Arabs, Jews, Indians, and others) which we have already noted, suggested that all the Pacific populations could be included with the Mongoloid peoples, and indicated how the known intermixture between the extra-European Caucasoids and Mongoloids has decreased the genetic distance between them.

Having then confirmed the reality of the world's major racial groupings, we can now examine the time at which they are thought to have diverged and the more detailed evolutionary relationships of some of the better known local races around the world. It is possible to predict the evolutionary rate of divergence of human races if the differentiation is due to genetic drift, and although, as we have seen in Chapter 2, this is often a plausible mechanism at the micro-geographic level it cannot really be accepted on a larger scale. The measurement which is used in this case is *f*, the coefficient of kinship. If indeed genetic change is due mainly to genetic drift, *f* values for individual loci should be equal. However, if selection is operative *f* values might be lower than under drift alone (if heterosis is maintaining the polymorphism), or higher if differential selection

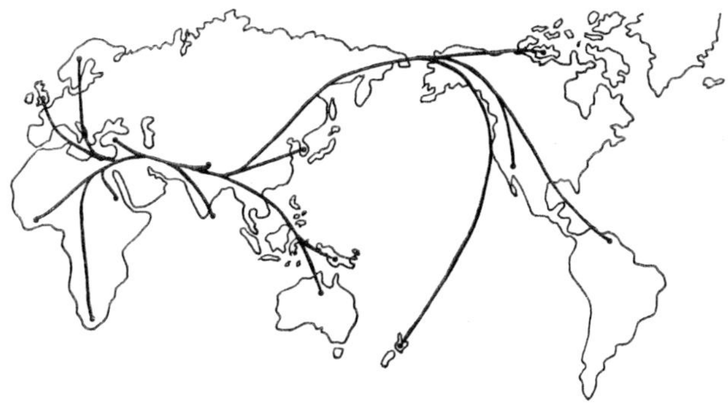

Fig. 3.11.—A phylogenetic tree of man projected on a world map to show correspondence with possible routes of migration. (*From* Edwards and Cavalli-Sforza, 1964.)

exists in the various environments occupied by the main racial groups of man. It is therefore difficult to know which, if any, of the observed values for different genes are due to drift or selection. A simple empirical approach, followed by Cavalli-Sforza and other workers in this field, is to estimate the average observed *f* value between two groups that had diverged from a single group at a known point in time. The value so obtained can then be compared with that for two other groups whose time of separation is not known, and this can then be assessed on the ratio of the *f* values.

Among the most useful events in human evolution are those of the first penetration of North and South America. The ancestors of all the present Indian tribes crossed the Bering land bridge from Asia not less than 25,000 years ago and it is likely that there was a greater homogeneity in the founding stock than is true for any other large population. In the case of the South American Indian it was for long believed that the first availability of the central American isthmus about 15,000 years ago would be the date at which their divergence began. Although doubts on both these dates now exist, it seems probable that the occupation of South America followed that of North America by a few thousand years. In any event, in the last few years people have realized the enormous importance of genetically investigating these South American tribes, and in fact we now have estimates of genetic distance between 12 tribes and also between some of the villages within 2 of these tribes (*see* pp. 89–91). The kinship coefficient of these 12 Indian tribes is 0·082 and we can therefore use this as a reasonably firm baseline for determining genetic differentiation per unit time.

There are few other dates about which we can even be so roughly accurate as this. Most of the recent prehistories of both Africa and India are very poorly known and in Australia, where it might be hoped we could obtain a reasonably accurate date of penetration of the Aborigines, there is still great uncertainty even over whether there have been one, two, or three major migrations let alone when they might have occurred. In line with the evidence discussed in section 3.4.*b* it is thought that the earliest Aborigines arrived in Australia at least 30,000 years ago. Australia's neighbour New Guinea has been occupied for at least 20,000 years (Golson, 1968) and possibly up to 50,000 years (Bulmer, 1969). In New Guinea, both on linguistic and genetic grounds, the inhabitants can be divided into two groups of people, pre-Austranesians and much more recently arrived Malayo-Polynesians, although now of course the distinction is blurred a good deal by gene flow between these peoples (Friedlander and Steinberg, 1970), so this makes any use of New Guinea dates very suspect. Furthermore, it is now fairly certain on the basis of the distribution of gamma globulin (Gm) factors that the northern Aborigines have 30 per cent genetic admixture from New Guinea. This admixture differentiates them markedly from Aboriginal populations in the centre and Western Desert areas. The overall estimates of genetic differentiation between Australian Aborigines and New Guineans must therefore be an underestimate, given this evidence of admixture.

Within New Guinea it has been calculated that for some of the non-Austranesian languages of the New Guinea Highlands, up to 4,000 years would be required for the degree of differentiation existing today (Wurm, 1964, 1969) and we have some evidence on the genetic differentiation of these tribes (Simmons, Graydon, Zigas, Baker, and Gajdusek, 1961). The two groups of tribes examined in these two studies do not entirely coincide and therefore the degree of genetic differentiation must have taken place within a time span greater than 4,000 years, the minimum for this language requirement, but less than 50,000 years, the likely maximum for occupation of New Guinea.

On the basis of published information (Cavalli-Sforza, 1969a, b; Cavalli-Sforza and others, 1969) it is possible to construct *Table* 3.8, which incorporates the above time intervals and, taking into account all the genetical problems and variation, and despite all the archaeological uncertainties, the internal consistency of this table is quite encouraging.

The data in *Table* 3.8 are based on a number of blood groups, enzyme and protein variants, and the exact ones which have been used vary from one estimate to another. It should be emphasized that the f values are greatly influenced by the nature and force of the selection operating on the polymorphism concerned. Bodmer and Bodmer (1970) present alternative f values between different races which are in general much lower than these given here. These are derived from the HL-A (Human Leucocyte Antigen) polymorphism, and Bodmer and Bodmer suggest that the low values for racial differentiation with respect to this trait indicate that relatively uniform stabilizing selection is acting on it.

In *Table* 3.8 the three crucial dates for the times of origin of the three major racial groups of the world are very uncertain. Indeed accepting the ideas put

Table 3.8.—SOME STEPS IN RECENT HUMAN EVOLUTION

	POPULATIONS	TIME AVAILABLE (yr.)	f
Differentiation within	Villages in Parma Valley, Italy	>1000	0·036
	Some New Guinea Tribes	>4000	0·045
	Australian Aborigines	~30,000	0·040
	African Main Group (Pygmies, Bushmen, Bantus, West Africans, and Southern Nilotes)	?	0·042
	S. American Indian Tribes	~25,000	0·082
Differentiation between	Australian versus New Guinean	15,000 (an overestimate because of gene flow)	0·103
	(Australian + New Guinean) versus S. American Indian	>50,000 (could be much more than this)	0·286
	Caucasoid versus Mongoloid	? 100,000	0·242
	Negroid versus Caucasoid	? 150,000	0·352
	Negroid versus Mongoloid	? 200,000	0·416

forward in s. 3.*4* means that to talk of dates of origin in this way for each of the races is to ignore their subsequent mixings and the fresh evolutionary starts because of these mixings. Nevertheless it is possible, in fact, to make some reasoned estimates of when these times of original divergence might be, following the arguments, though not always the dates, set out by Cavalli-Sforza (1969a) and Cavalli-Sforza and Bodmer (1971). The evidence we have reviewed so far certainly suggests that the tripartite division of the peoples of the world is the fundamental one, although the Mongoloid, or Eastern group, including all populations living around the Pacific, may be more heterogeneous than the others. The f values given in *Table* 3.8 for differentiation between the three main racial groups are compiled on information from 16 genetic loci. If we believe that the f values are a function of the line of differentiation of the groups concerned we might estimate the time of divergence of these three main racial groups as between 100,000 and 200,000 years, and these, therefore, are the dates which have been entered in *Table* 3.8.

Using this genetic evidence, three possible modes of origin for African, American, and Australian racial groups can be constructed, as indicated in *Fig.* 3.12. If the genetic differentiation is assumed to be due entirely to drift or

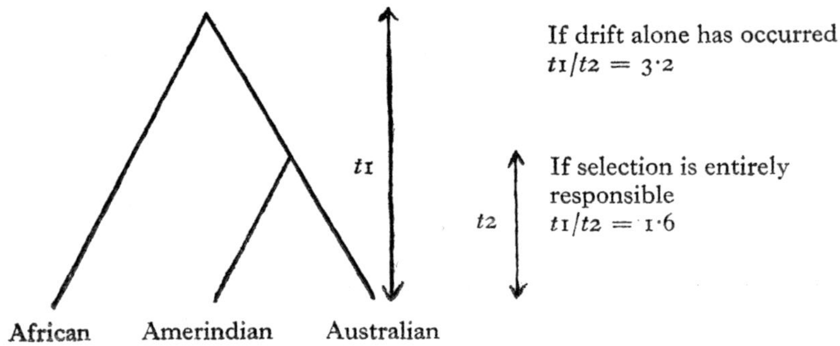

	Drift		Selection
If t_2 is 25,000 years t_1 is	80,000	or	40,000
50,000 ,,	160,000	or	80,000
75,000 ,,	240,000	or	120,000
100,000 ,,	320,000	or	160,000

Fig. 3.12.—Some possibilities regarding the time of divergence of different racial groups. (*After* Cavalli-Sforza, 1969a.)

entirely to selection we get what can be regarded as maximum and minimum estimates respectively of the time spans involved. On the basis of the information available to him, Cavalli-Sforza (1969a) obtained ratios of $t1/t2$ of 1·75 and 1·42 respectively with these two alternative hypotheses. With regard to the time of separation of the Africans from the Easterners using 14 genetic loci, and then the time of separation of the American Indians from the Australians, Cavalli-Sforza finds a ratio $t1/t2$ of 3·2 on the basis of genetic drift and 1·6 on the assumption of selection, where $t1$ is the time since separation of Africans and Australians and $t2$ the time of separation of Australians and Amerindians. Now we have already mentioned that the first American Indians entered North America at least 25,000 years ago. One would suppose therefore that the absolute minimum time of separation of these people from the Australian group must be 30,000 B.P. It is likely, of course, to be rather more than this. The various possibilities are therefore as shown in *Fig.* 3.12.

There are some other lines of evidence we can review which will help us determine which of these estimates is most likely, and how the sort of time scale which we have carried out on genetic grounds will compare with that derived from archaeological and palaeontological evidence.

Certain alleles of some genetic loci are restricted to certain races or certain populations in one or a few geographical regions of the world. An example of such an allele is that controlling the Di (a) antigen of the Diego blood group, which is found in the majority of all Mongoloid populations but only rarely in other racial groups. Another variant only present in Mongoloid populations is a transferrin variant D_{chi} which is widely distributed among South American Indians, throughout East Asia, among the Lapps in Finland, and also in some tribal populations in northeast India. This widespread pattern of distribution of D_{chi} suggests that it arose before the migrations from Asia to America took place. Now it can be assumed that since the time of origin of this mutant it has been gradually increasing in frequency at an approximately linear rate, since there is no evidence that this transferrin polymorphism is of the balanced type. On the basis of evidence for the rate of complete amino-acid substitution in protein molecules in various species of mammal, it can be calculated that the length of time required for the D_{chi} mutant to achieve its present observed frequency would be in the range 30,000 in 80,000 years.

There is another transferrin variant D_1 which is widely distributed in Australian, Melanesian, and black African populations. Assuming that this mutant form derives back to a common ancestral population, then it can be calculated in a similar way that the D_1 variant arose in a population which was ancestral to Australians and Melanesians and black Africans but was already distinct from populations ancestral to other racial groups approximately 100,000 years ago.

The other possible check on this postulated time scale is to study the extent of variation in gene frequency among all the races of the world. One can then calculate how long it would take for this variation to arise under various assumptions of population size and on the single assumption of genetic drift and not selection being responsible for observed variation. *Fig.* 2.24 shows the curves derived by Cavalli-Sforza (1966), and they indicate that the observed variation for even the most variable genes would be compatible with an isolate size of 10,000 and a separation time of 300,000–400,000 years—the duration of existence of our species *Homo sapiens*. We have already mentioned that the drift hypothesis is unlikely to be completely true, but selection will in some cases increase the rate of accumulation of variance and in others decrease it, so the figure for drift alone is probably not an unreasonable estimate. A second important point is that the normal isolate size for man is unlikely to have been as high as 10,000—indeed in many cases will normally have been closer to 1000. This would mean that the total separation time must be rather less than 300,000 years.

3.6. Synthesis

If the foregoing brief and fragmentary racial history of man has shown anything, it has shown how mixed up are the races of man and how difficult it is to piece together their history.

The origin of *Homo sapiens* from *Homo erectus* can be dated at between 300,000 and 400,000 years ago, probably earlier rather than later, but in exactly which part of the world it is not easy to say. Fossils close to the transition are found in Eastern Europe and we may therefore suppose that the active area of evolution was either here or in western and central Asia. Neanderthal man may have been an evolutionary dead end in Europe to be largely replaced by true *H. sapiens* coming in from the east. These newcomers can be called Caucasoid or European, and as well as spreading into Europe they penetrated into Africa via the Middle East and also moved eastwards to the extremities of Asia.

Now *Homo erectus* was a widespread species and was evolving, albeit at different rates, wherever it existed. It seems possible that as it evolved rapidly into *H. sapiens* in Asia the southernmost margin of this evolving population should incorporate a number of features appropriate for tropical areas. We have seen in Chapter 2 how skin colour at least may evolve rapidly and there is every reason to suppose that *H. erectus* would itself show some tropical adaptation ready for incorporation into *H. sapiens*. We know that the products of this rapid evolution, the Caucasoid *H. sapiens*, migrated eastwards and westwards since we have noted their fairly sudden appearance as Cro-Magnon man in Europe about 40,000 B.P. and their representation among the Ainu in the extreme east. In all probability they also migrated southwards, incorporating these genes for tropical adaptation, and then south-west and south-east to Africa and S.E. Asia respectively, becoming more and more differentiated as they encountered and absorbed other groups in the *H. erectus–H. sapiens* continuum on their way. Some of the specific tropical adaptations have a common origin in South Central Asia, others are peculiar to Africa or S.E. Asia and Oceania where more slowly evolving groups were absorbed.

Just as in the southern areas the rapidly evolving *H. sapiens* incorporated tropical adaptations, so in the north it incorporated temperate and cold climate adaptations—those of the Mongoloids. These then spread mainly westwards and then south-westwards as the general unspecialized Mongoloid types right through to the New World. The specialized Mongoloid types then arose in response to an increase in severity of the climate in a small area, and then when conditions became appropriate followed in the migratory footsteps of their Mongoloid cousins, both of them overruning much of the area already occupied by the western tropical group who were diverging into the Negritos, Australoids, and Melanesians.

The complicated secondary migrations and invasions can then be followed, at least in outline, through archaeological evidence. The Caucasoids move back westwards and southwards into India and later into Africa. They find preexisting dark-skinned peoples in India just as they do in Africa. The division of the Negroids in Africa into the Negroes and Bushmen probably took place in Africa, with perhaps differential admixture from the peoples already living there (*H. sapiens rhodesiensis*) being an important contributory factor. The migrations within Africa of these two peoples and the entry of the Caucasoids can then be followed fairly well.

In Asia we have seen that the Pacific peoples, the Mongoloids and Australoids broadly defined, have a good deal in common genetically. This may be explained partly through genetic exchange throughout their long history in contact with one another and their migrations through one another's territory. Secondly it might reflect the incorporation of similar genes from the eastern peoples of the *erectus/sapiens* line that they both met migrating eastwards. And thirdly it might reflect a similarity of origin anyway on the eastern segment of the centre of rapid evolution in Asia.

The sort of scheme suggested here is shown diagrammatically in *Fig*. 3.13. The important features of this proposal are: (1) The principal, i.e., the fastest, line of evolution from *Homo erectus* to *Homo sapiens* occurred in one relatively

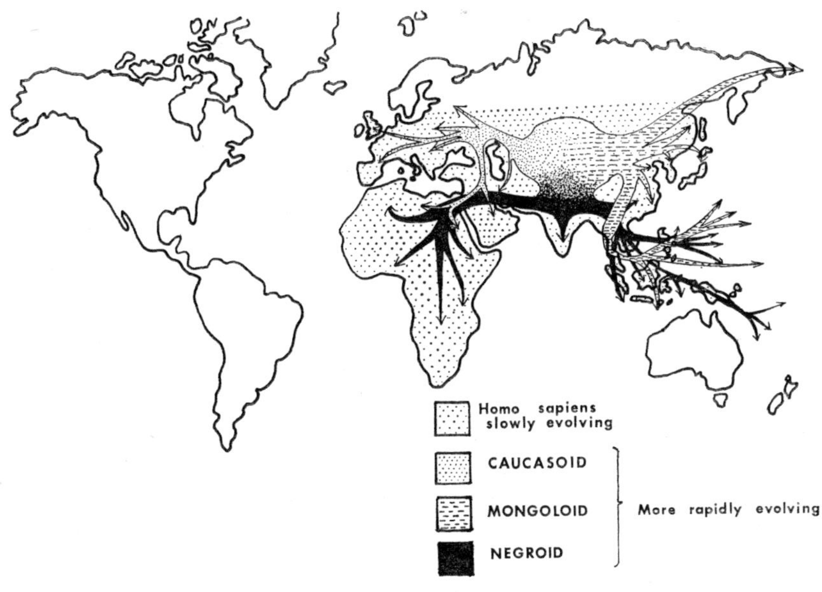

Fig. 3.13.—A highly schematic suggestion for what might have been the stages in the racial evolution of man. We can distinguish five possible stages:

1. Early *Homo sapiens* widespread and slowly evolving but nevertheless differentiating into *H. s. rhodesiensis*, *H. s. neanderthalensis*, and one or more eastern representatives (*H. s. australiensis?*).
2. Rapid evolution somewhere in Asia of modern types rather uniform in appearance but with some peripheral local differentiation.
3. Outward migration between about 150,000 and 50,000 B.P. or a little later of a few major population groups. It is from early in this period that the major racial groups date.
4. The partial mixing of each migrating group with the men in the areas into which they arrive.
5. Subsequent racial differentiation but still confused by large-scale migrations and integrations right up to the present day.

circumscribed geographical area. (2) Radial spreading from this area of divergent populations gave rise to the main racial groups of man. (3) Varying degrees of admixture with less rapidly evolving lines became incorporated into these racial groups as they spread into new environments. (4) Considerable degrees of secondary admixture within and between these main groups has characterized the rest of man's history.

With particular reference to this last point we shall consider in some detail recent hybridization in man, that is in the last 500 years or so, in Chapter 8. But it is no new phenomenon; we may remind ourselves that in this chapter we have seen the Caucasoids enter both India and Africa as a second wave of migrants, so that today in these continents many millions of people are 'hybrids' between the major races. Within Africa we have seen extensive Bantu/Bushmen hybridization and also mixing between western and eastern Negroes. In South-East Asia, Melanesia, and Polynesia the Mongoloids and Australoids are inextricably mixed and successive migrations and invasions between the Mongoloids and Caucasoids have moved back and forth across Asia since, so far as we can see, the time of the origin of these races up till the present.

It is therefore obvious that it is a meaningless question to ask 'How old are the races of man?'. In part the major races appear to owe some of their differences to the very early differentiation of a rapidly evolving population of *Homo sapiens* somewhere in central or western Asia from about 250,000 years ago. Geographical isolation of more slowly evolving African and East Asian populations laid a foundation for racial distinctness. However, it also seems probable that the races share a considerable degree of genetic identity dating from much more recently. The rapidly evolving line of *H. sapiens* seems to have migrated outwards and caught up the local populations in so doing. Thus they no doubt incorporated a good deal of the gene pool of *H. sapiens neanderthalensis* on their way into Europe, of *H. sapiens rhodesiensis* on their way into Africa, and the equivalent of their migration south-eastwards and eastwards. This eastern population appears to be just as real and as widespread as the other two, judging by the similarities which are apparent between recent finds in Java and Australia spanning more than 100,000 years. It might perhaps be termed *H. sapiens australiensis*. From this point of view, racial origins date back between 50,000 and 100,000 years. This is certainly in line with the molecular and genetic evidence and we remember that the maximum genetic difference between races is of the order of 20 per cent.

Throughout the whole of this chapter has run the thread of incorporation of varying populations of already dispersed peoples of the genus *Homo* to different extents in different places in the old world into a population which perhaps evolved a little faster than they, or at least a little differently. The history of man is the history of racial integration. This is therefore the reason we have examined the history of man in some detail in a book concerned primarily with racial integration.

CHAPTER 4

THE BIOLOGY OF RACE CROSSES

4.*1*. Heterosis and Hybrid Vigour

The crossing of individuals from two different races is often referred to as 'hybridization' and the offspring of such unions as 'hybrids'. Like the word 'race' itself, these two terms are used in rather imprecise ways. In general biology, as we shall see, they are used to describe a number of distinct, if related, situations, and one particular aspect of their meaning has been adopted for human usage, often with a faintly pejorative air.

Ernst Mayr (1963) defines hybridization as 'the crossing of individuals belonging to two unlike natural populations that have secondarily come into contact', and yet the term is also used by plant and animal breeders (and was so used by Mendel) to denote crossing between organisms which differ at one or a few genetic loci. Some workers would restrict use of the term 'hybridization' to crosses between distinct species, but for the wide variety of biological situations which must be described in some terms, Mayr's definition seems appropriate.

We are in this book primarily concerned with crossing between populations which have been isolated for varying periods of time and which differ to varying extents, but which are never members of different species. The degree of genetic difference between human populations is often uncertain, but for old-established races may reach a maximum of 25 per cent of the genome (*see* Chapter 2, s. 2.5) and some of these races are considered as subspecies by Coon (1963) and others. More recently established races, such as the American Negro, are clearly much less distinct, and much of the evidence we have on the consequences of race crossing in man has come by extrapolation from studying the offspring of marriages between people of different countries, tribes, or geographical locations; in other words, people who can hardly be said to represent different races at all. It is therefore highly relevant and very instructive for the present study, to examine hybridization in the broad sense as well as in the narrow. That is, to consider everything from interspecies crosses down to crosses between strains or populations with very minor genetic differences. The whole will contribute to our understanding of race crossing in man.

It has long been known that the cross-breeding of two different strains will usually produce more vigorous hybrid offspring than either of the parental strains separately. Examples which are readily thought of include the mule, a horse–donkey hybrid which has been preferred by man for more than 2,000 years for certain tasks, and hybrid corn—derived from the crosses of 2 or 4 inbred strains—which has helped to double the average corn yield in the United States from about 25 bushels per acre (1923–32) to about 50 bushels per acre (1953–62). More recent developments include hybrid poultry and hybrid wheat. The superiority of the hybrid can, in fact, be shown to be due to two slightly different phenomena—different in so far as their consequences are concerned, that is, since the molecular bases of the phenomena are not yet completely understood. The nomenclature describing these situations is rather confused. The terms 'heterosis' and 'hybrid vigour' are often used interchangeably, although there is a tendency for them to be used differentially for the two rather different phenomena with which we are concerned.

The first of these phenomena, true hybrid vigour, includes improvement in general fitness characters such as longevity, resistance to disease, more rapid growth, and greater fertility. These all indicate improvement in those characters contributing to Darwinian or evolutionary fitness. Improvement in other aspects, such as simple increase in size, does not necessarily mean an increase in evolutionary fitness and may, as in the case of the mule for instance, be associated with organisms which have effectively zero fitness since they are sterile. The most dramatic expression of this hybrid vigour is normally restricted to the first generation offspring of crosses between two distinct populations, after which it is gradually lost. In many instances it appears that second generation hybrids lose most of their advantages and the expression of many of their characters falls below the mean value of the two parental populations, or even below the value of either of these contributing populations.

In direct contrast to this situation are the many instances of increased biological fitness, or heterosis, which can be attributed simply to the fact that one or a few genetic loci are heterozygous—a situation which is obviously not related to the number of generations by which the hybrids are separated from the original interpopulation crosses. The expression of heterosis in this case is largely independent of population structure, since it is manifest whenever the different alleles of the gene or genes concerned are present in an individual, no matter what circumstances produce that individual.

The first of these situations, hybrid vigour produced in interpopulation crosses, has been termed 'luxuriance' by Dobzhansky (1964). As he says, it is really an accidental condition brought about by the complementary action of genes found in the original parental populations that are crossed. The second situation, that produced by the heterozygosity of one or a few loci, is true heterosis, termed by Dobzhansky '*euheterosis*' and divided by him into *mutational* and *balanced heterosis*, which are due primarily to the sheltering of deleterious recessive mutants by dominant alleles and to overdominance respectively.

Our first task, therefore, is to review what is known of the molecular bases of heterosis and hybrid vigour and to see if the magnitude of expression of heterosis can be related in any way, either to the amount of genetic difference between the parental populations, or to the number of generations which have elapsed since the initial hybridization. We can then consider the possibilities of heterozygote superiority in more general terms. What follows is a brief treatment of heterosis. For a really comprehensive review consult J. W. Gowen's volume *Heterosis* published in 1964.

4.2. The Genetic Basis of Heterosis

4.2.a. *Dominance and Overdominance*

The first problem which emerges is whether heterosis is or is not manifest in some individual or by some population. Some geneticists insist that, for the particular characteristic concerned, the hybrids must exceed the values shown by both parents; others regard it as sufficient that the hybrid exceed the average of the two parents.

Given that we have decided whether heterosis is or is not shown in some situation, there are two different genetic situations which can produce it, although in neither can we specify in most cases the molecular processes involved. These two situations are respectively dominance and overdominance of the individual genes concerned. Part of the problems surrounding these terms and their use is, firstly that they each apply to relationships between two alleles at one locus whereas, of course, most hybridizations involve heterozygosity at many loci, and secondly there is no mention of the environment in this sort of discussion and the individuals and the populations concerned may successively inhabit a whole series of different environments, in only some of which might dominance and overdominance be expressed.

The difference between dominance and overdominance can be explained in terms of a diagram (*Fig.* 4.1), where the linear scale represents the expression of some character, height, weight, fertility, etc. We shall first examine overdominance. A number of examples are known where overdominance appears to exist and which illustrate its distinctively important features. One of the ways in which the heterozygotes may exceed either of the homozygotes in the expression of some character, is that they produce a unique 'hybrid substance' different from either of the homozygous products. In maize, for instance, Schwartz (1960, 1964) demonstrated electrophoretically the presence of hybrid esterase enzymes in strains of maize derived from highly inbred lines, and hybrid alkaline phosphatases have been found in an intergeneric bacterial

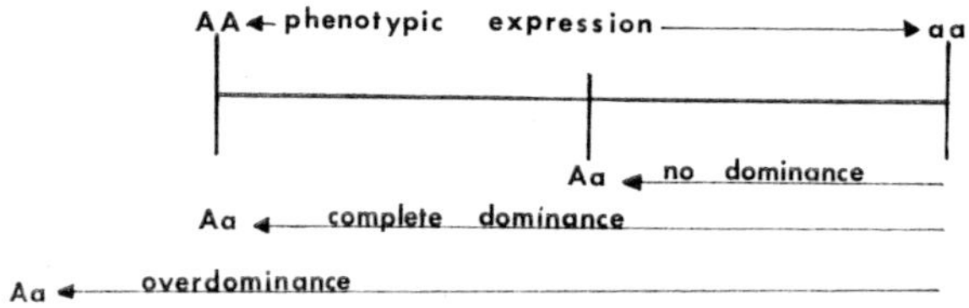

Fig. 4.1.—Overdominance, expressed in terms of differential phenotypic expression.

hybrid (Levinthal, Fan, Higa, and Zimmerman, 1963). The AB antigen of human blood is another example of a compound which is different from that produced by either of the homozygotes for the alleles concerned. One of the most interesting examples of this situation is that for the formation of hybrid haemoglobins in some centrarchid sunfish hybrids (Manwell, Baker, and Childers, 1963). In most cases no hybrid molecules are formed by the hybrid fish, but in two cases a new haemoglobin was detected, as shown in *Fig.* 4.2.

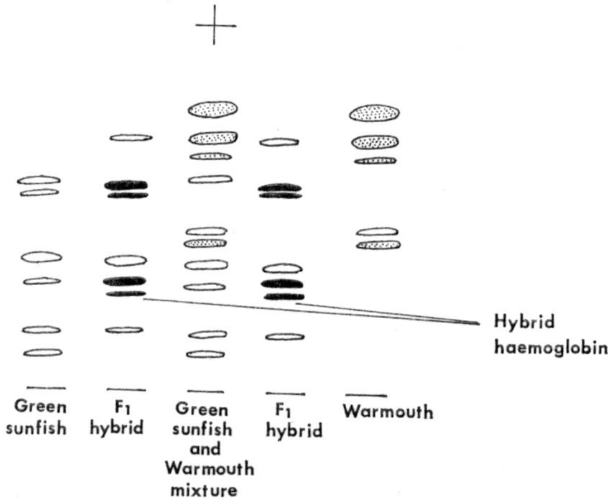

Fig. 4.2.—Diagrammatic representation of the starch gel electrophoresis of the major haemoglobins from the green sunfish, the warmouth, the F_1 hybrid between the two species, and a test-tube mixture. (*From* Manwell and Baker, 1970.)

The interesting point was that haemoglobin obtained from the hybrid green sunfish/warmouth sunfish showed consistently stronger haem–haem interactions than those of the haemoglobins in the parent species (*Fig.* 4.3), and as

Manwell, Baker, and Childers point out, the 'hybrid haemoglobin' is better suited for providing a higher rate of metabolism in certain hybrid sunfish than is the haemoglobin of either of the parent species. There is no question, they state, 'but that hybrid sunfishes are more energetic and aggressive than the parent species'.

Sick, Frydenberg, and Nielsen (1963) found that the electrophoretic pattern of haemoglobin from hybrids between the plaice, *Pleuronectes platessa* and the flounder *Pleuronectes flesus*, is different from that of a mixture of haemoglobin

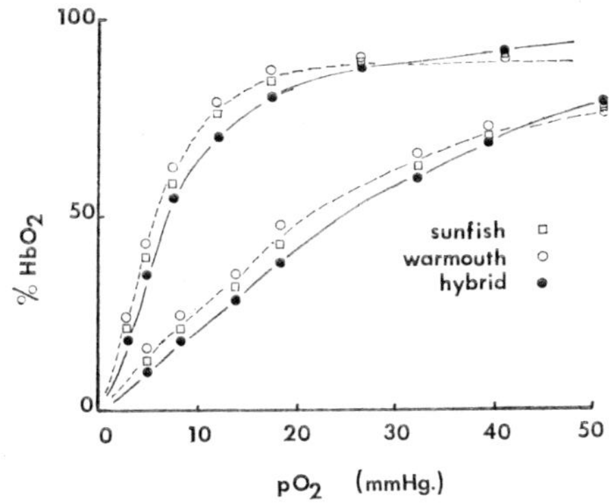

Fig. 4.3.—Oxygen equilibria from the green sunfish *Leponius cycinellus*, the warmouth *Chaenobrythus gulosus*, and the F_1 hybrid. Note that at either *p*H, which cover a range of those expected *in vivo*, haemoglobin from the hybrid has a more sigmoidal oxygen equilibrium curve (stronger haem–haem interactions) and a slightly lower oxygen affinity than haemoglobin from either parental species. (*From* Manwell and Baker, 1970.)

from the two species. Hybrid vigour there certainly seems to be in crosses between these species, because Sick and his co-workers report that hybrids account for 10 per cent of the total flounder catch in the Danish Belt Sea, and yet there is little or no introgression of genes from one species into the other.

It is sometimes the case that, although no hybrid molecules are formed, another kind of hybrid metabolism still occurs, simply because the heterozygote produces an intermediate amount of the same substance that each parent produces. For example, in the fungus *Neurospora* a heterokaryon (a culture containing two sets of different nuclei) containing *pab* and *pab*+ produces a more optimum amount of *p*-aminobenzoic acid for growth than does either homokaryon (Emerson, 1948).

Yet another situation in which the heterozygote has an advantage in molecular terms over either homozygote is exemplified by the sickle-cell situation in man. As already mentioned, the heterozygote here produces both forms of the haemoglobin for which it carries genetic information, Hb^A and Hb^S, and thereby escapes suffering from anaemia and also receives protection from endemic malaria.

Sometimes non-nuclear or non-chromosomal structures may be involved in hybrid vigour, for example polymorphism of mitochondria associated with heterosis in certain maize crosses. Sarkissian and McDaniel (1967) investigated the activity of cytochrome c oxidase in mitochondria of inbred and hybrid maize. The strain 'Ohio 45' has a total activity of 0·303 units while the strain 'Wf 9' has a total activity of 0·754 unit; the hybrid 'Ohio 45 × Wf 9', however, has a total activity of 1·176 units. The hybrid plants also show a region of hybrid mitochondria, when extracts are sedimented in a sucrose density gradient, which is not present in similarly treated extracts from the parental lines. Some mitochondrial proteins are specified by nuclear genes and others are specified by mitochondrial genes, so there exists the possibility of complementation between nuclear and cytoplasmic inheritance.

Whatever the cause in molecular terms, it seems clear that in a variety of cases the heterozygote is better protected or 'buffered' so that it can produce a normal phenotype under a wider variety of environmental stresses. Some data collected together in *Table* 4.1 exemplify the fact that hybrids are quite frequently

Table 4.1.—VARIANCE COMPARISONS BETWEEN INBREDS AND HYBRIDS FOR DIFFERENT CHARACTERS* (*From* Strickberger, 1968.)

		INBREDS	HYBRIDS
Drosophila melanogaster			
Lifetime egg production	CV	48	31
Wing length	CV^2	2·35	1·24
Abdominal bristles	M.S.	0·1466	0·0857
CHICKENS			
Egg shape	V.C.	4·235	3·585
Egg weight	V.C.	8·347	4·610
Shank length	V.C.	0·0589	0·0351
MICE			
Birth weight	CV^2	119	59
Weight at 3 weeks	CV^2	98	47
Weight at 60 days	CV^2	24	19

* The units used are measures of variation, either mean squares (M.S.) or variance components calculated from mean squares (V.C.), or coefficients of variation (C.V.).

superior in their ability to produce a uniform phenotype under a variety of environmental conditions, whereas homozygotes, in this case inbred lines, exhibit much greater phenotypic variability.

The usually stated alternative to overdominance, the dominance hypothesis, is based on the superiority of dominant alleles when the recessive alleles are deleterious. This hypothesis suggests that as a population or strain tends to become inbred, the individuals within that population are much more likely to become homozygous for one or more deleterious genes (*see* Chapter 2, s. *2.1* and Chapter 5, s. *5.3*), with the populations differing in the array of deleterious genes for which they become homozygous. When crosses between such inbred populations occur, the deleterious recessives of one line will become hidden by the dominants of the other and vice versa, and the hybrid offspring, since they are thereby homozygous for no deleterious recessives, therefore appear to be heterotic.

As already mentioned, we do not yet know which of these two mechanisms is normally operative, or whether both might occur in average population crosses. One important point, however, might be made regarding overdominance strictly defined. As Wallace (1968) points out, overdominance requires that the heterozygotes fall outside the range of the two homozygotes, whatever the genetic background or the environment in which the alleles are found. Since this requires that alleles at one locus act independently from those at all other loci, a situation which we know not to occur, overdominance strictly speaking cannot exist. Wallace suggests the term 'marginal overdominance' to overcome this objection, and *Table* 4.2 is taken from his book *Topics in Population Genetics*

Table 4.2.—General Scheme illustrating how Heterozygous Individuals may come to exhibit Marginal Overdominance even though Overdominance, strictly defined, is lacking under each Individual Condition (*From* Wallace, 1968.)

Condition	Genotype		
	AA	Aa	aa
1	1	1	$1-s_1$
2	$1-s_2$	1	1
3	1	1	$1-s_3$
.	.	.	.
.	.	.	.
n	$1-s_n$	1	1
Margin	$1-S_{AA}$	1	$1-S_{aa}$

to illustrate this idea. In the example illustrated, overdominance does not exist in any one single case, but overall the heterozygotes are clearly superior to both homozygotes. The conditions under which performance is measured, Wallace points out, can represent different time periods, geographical areas, background genotypes, developmental stages, etc. Thus the idea of marginal overdominance includes most situations where heterosis might occur. It is probably better to consider the causative factors of heterosis in this rather non-specific way, in view of our general ignorance of the matter at the present time.

4.2.b. *Population Experiments on Heterosis*

Our further understanding of the genetic basis of heterosis or hybrid vigour can best be advanced by considering a number of experimental situations in which it has been investigated. Hampton L. Carson (1958) studied two laboratory populations of *Drosophila melanogaster* which had been kept for sufficient periods of time until they had reached equilibrium with regard to size and productivity. Into one of these populations Carson introduced a single haploid set of chromosomes. He did this by putting into the population a single male fly which was a hybrid between a wild type male and a female extracted from the experimental population. The experimental population then showed a threefold increase in size and productivity, an increase which was maintained for more than ten generations after the introduction of the new genome. The changes occurred without any environmental change and took place over the short span of only three generations. The nature of the response of Carson's populations to the addition of a single haploid genome is illustrated in *Fig.* 4.4.

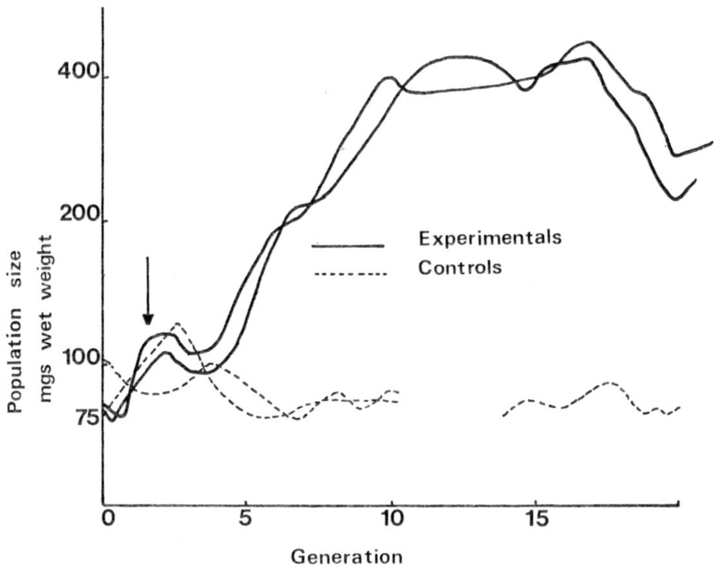

Fig. 4.4.—The response of stable closed laboratory populations of *Drosophila melanogaster* to the introduction of a single haploid set of autosomes of 'foreign' genetic material.

But the most frequent and most thoroughly analysed situations in which heterosis occurs concerns the crossing of inbred strains, in the laboratory or on the farm, of animals such as *Drosophila*, mice, chickens, pigs, and cattle (*see*, for example, the review of Manwell and Baker, 1970).

One of the most frequent experimental modes of demonstrating heterosis is from a diallele cross which is the cross between all inbred lines derived from the

Table 4.3.—HETEROSIS IN CHICKENS. THE RESULTS OF A DIALLELE CROSS (*From* Nordskog and Ghostley, 1964.)

CHARACTER	AVERAGE PERFORMANCE		
	Purebreds	Line Crosses	Breed Crosses
Chick Mortality (per cent)	15	11	10
Growth Mortality (per cent)	22	19	18
Adult Mortality (per cent)	33	30	31
8-week weight (g.)	600	625	641
Egg-production rate (per cent)	41	45	46
Egg-weight (g.)	60·0	59·3	59·3

same base population in all possible combinations. Nordskog and Ghostley (1964) report results from such a cross involving eight strains from four breeds of chicken using two strains from each breed. *Table* 4.3 shows that the performance of breed crosses was better than line crosses within breeds, which in turn were better than pure strains for all these different characters, with the single exception of egg weight.

In *Drosophila* a large number of traits are known to show hybrid vigour and *Table* 4.4 from Lints and Lints (1968) shows the response of various quantitative characters to inbreeding and subsequent hybridization. *Fig.* 4.5 shows that the superiority of egg laying is manifest, not only in the total number of eggs laid, but in the number of days over which egg laying is continued. It is apparent that the greatest degree of hybrid vigour is manifest in those characters such as fertility and longevity, and to a lesser extent oxygen consumption, which are of prime evolutionary and general fitness significance.

Hybrid vigour in maize has been one of the most used manifestations of this phenomenon and has been carefully studied by Moll, Lonnquist, and co-workers (Wellhausen, Roberts, Hernandez, and Mangelsdorf, 1952; Moll, Salhvana, and Robinson, 1962; Paterniani and Lonnquist, 1963; Moll, Lonnquist, Velez-Fortuno, and Johnson, 1966). Wellhausen recorded the performance of F_1 hybrids for a large number of Mexican races of maize and found that the heterotic responses ranged from large positive effects down to performance below either parent variety. Moll, Salhvana, and Robinson examined F_1 hybrid yields from 66 crosses between 12 parental strains from the cornbelt of the U.S.A. and found that 61 gave yields greater than the midparent value and 5 below it, a result essentially confirmed by Paterniani and Lonnquist with 63 F_1 crosses among South American races of maize. Moll, Salhvana, and Robinson also studied the results of 15 crosses of 6 varieties representing three widely dispersed geographical regions. In each of these studies there was a positive correlation between genetic diversity, geographical race or plant type, and the magnitude of heterotic response. However, each of them had a number

Table 4.4.—Rates of Oxygen Consumption and Other Quantitative Characters in Wild, Inbred, and Hybrid Genotypes of *Drosophila* (*From* Lints and Lints, 1968.)

Genotype	Oxygen Consumption ($\mu l.O_2$(mg./hr.)	Thoracic Size (arbitrary units)	Weight (mg.)	Fecundity* (number of eggs)	Fertility* (number of eggs hatching)	Longevity† (days)
Gabarros wild	2·84 ± 0·10	41·75 ± 0·08	1·37 ± 0·02	1490 ± 73	617 ± 71	35·3 ± 1·5
Gabarros inbred	3·04 ± 0·10	40·01 ± 0·07	1·18 ± 0·02	668 ± 155	323 ± 77	29·8 ± 4·0
Gabarros inbred ♀ × Abeele inbred ♂	3·69 ± 0·11	41·10 ± 0·10	1·31 ± 0·02	2227 ± 100	1846 ± 102	46·4 ± 2·5

* Means for 10 individuals.
† Mean imaginal lifespan of 20 individuals.

of individual crosses that did not perform as well as expected when judged on the basis of parental yields. The story was taken a little farther by Moll and

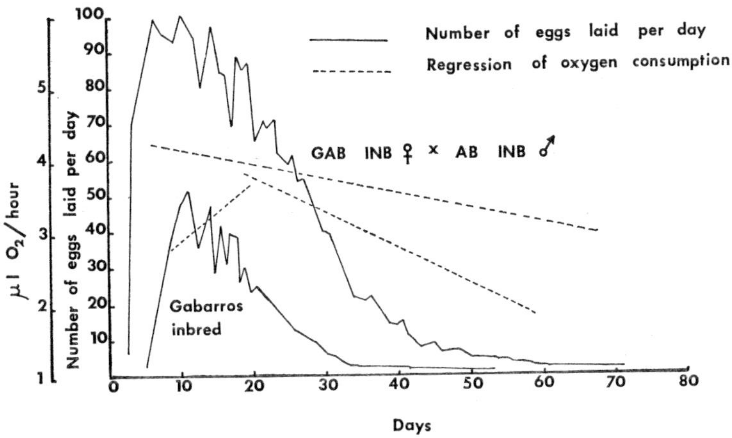

Fig. 4.5.—The fecundity and rate of oxygen consumption of inbred and hybrid *Drosophila* throughout life. (*From* Lints and Lints, 1968.)

others (1966) who took two populations from each of four geographical regions, south-eastern United States, Midwestern United States, Puerto Rico, and southern Mexico. They intercrossed these populations in all possible combinations, and from the resulting F_1 populations an F_2 generation was obtained. The parental F_1 and F_2 populations were grown in the four regions of origin. The results indicated that heterosis increased with increased divergence within a restricted range of divergence, but extremely divergent crosses resulted in a decrease of heterosis. They implication of all these results is that genetic difference, or heterozygosity, is essential for the manifestation of heterosis, but that simple genetic difference need not necessarily give rise to heterosis. Indeed too great a difference may be positively counterproductive.

All this suggests that the nature of the genetic background is of considerable importance in determining the expression of heterosis, and this is confirmed by some laboratory experiments with heavily irradiated populations of *Drosophila* (Dyer, 1971a, b). In these experiments two populations of *Drosophila melanogaster*, visibly distinct but genetically very similar, were allowed to compete together for 25 discrete generations in small population cages. Hybrids between these populations were removed at each generation and therefore, by irradiating one population and not the other, the cumulative effects of mutation induced by radiation could be studied.

The fundamental experimental design is shown in *Fig.* 4.6. In several replicates of this experiment the following conclusions were arrived at (Dyer, 1969c, d, e; 1971a, b):

1. The irradiated population showed little in the way of detrimental effects measured in terms of competitive ability, a result which was thought to be due to the favourable overdominant effects of several induced mutations in this small closed population.

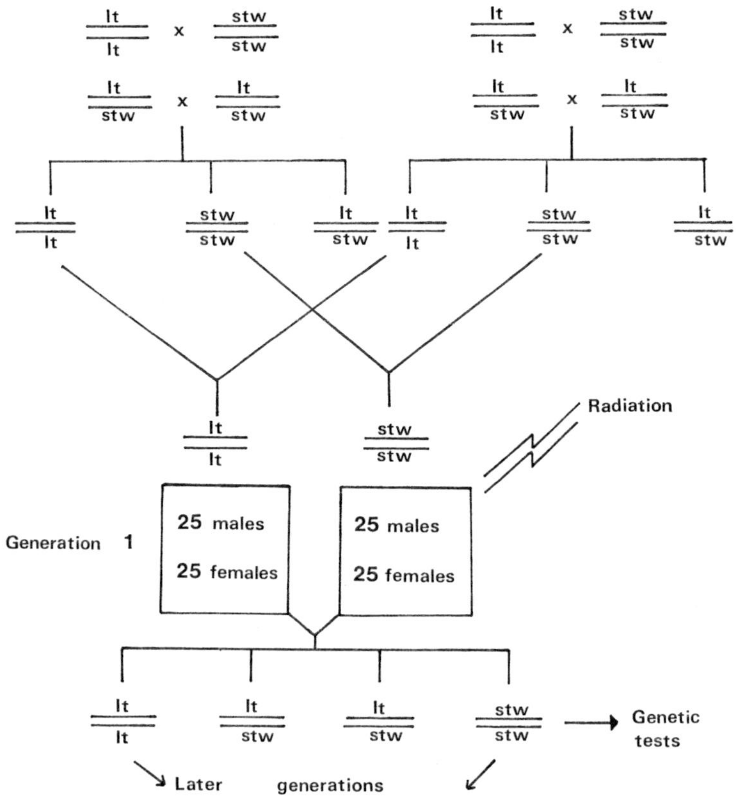

Fig. 4.6.—The fundamental experimental design of Dyer's experiments on the cumulative effects of radiation on *Drosophila* populations.

2. Hybrids between irradiated and non-irradiated populations showed much greater detrimental effects, despite the fact only 50 per cent of their chromosomes had a history of irradiation.
3. The provision of just a small permanent gene pool of the inter-population hybrids resulted in a 100 per cent increase in numbers emerging; presumably, therefore, evolution of the genetic background occurred, allowing the expression of heterotic effects.
4. Three hybrid irradiated populations constructed afresh at each generation by mating males and females from each of these continuous irradiated populations showed drastically reduced competitive ability (*Fig.* 4.7).

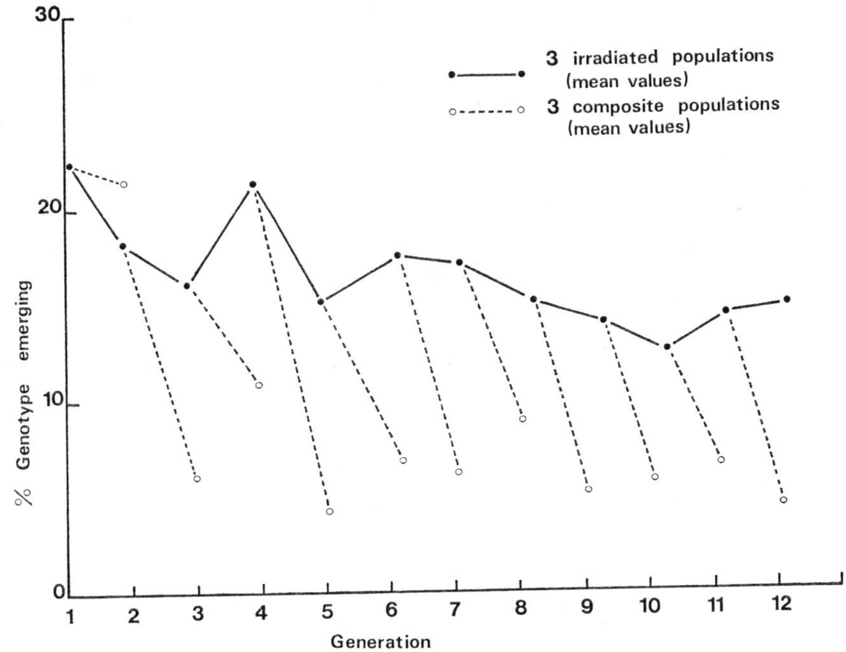

Fig. 4.7.—Percentage of an irradiated genotype emerging in competition with a non-irradiated genotype. The figures plotted are the mean of three replicates of this experiment, together with the mean percentage hatching of three composite genotypes derived from the first three and subjected to the same competition.

The same mutants, in other words, merely rearranged into a discontinuous genetic background, confer a fitness and competitive ability which is about 20 per cent of the normal levels.

Such results, although from rather special experiments, are clearly significant in suggesting what could occur in different, more normal situations. Their relevance for what in fact does occur in natural populations or in man himself may, of course, be questioned. Indeed Mukai (1969a, b), discussing a theory of optimal heterozygosity based on extensive experiments with laboratory populations of *Drosophila*, concludes that new mutants are overdominant only under restricted conditions and that they do not ordinarily manifest overdominance in natural populations. We might, therefore, conclude this brief investigation of the genetic basis of heterosis with a survey of some of the demonstrated cases of heterosis occurring after interpopulation crosses, particularly those in man himself.

Prior to extensive investigations on this problem, it might have been supposed that interpopulation crosses would normally lead to a drastic loss of fitness, since the genes in one population are the product of a long history of selection for optimal interaction. However, in *Drosophila* interpopulation crosses, whether in *D. Hydei*, a world-wide species building up very large populations, or in *D. virilis*, with a more scattered distribution, first generation hybrids were superior both in fertility and fecundity. This result was confirmed by Wallace and Vetukhiv (1954, 1955, 1956) for *D. pseudoobscura* and *D. Willistoni*, and some of their results are shown in *Table* 4.5. This introduces one of the most

Table 4.5.—COMPARISON OF THE VIABILITY, FECUNDITY, AND LONGEVITY OF INTRA- AND INTERPOPULATION HYBRIDS OF *D. pseudoobscura* AND *D. willistoni*. (*From* Wallace, 1968.)

	F_1	F_2
SURVIVAL		
D. pseudoobscura		
Intrapopulation	1·00	1·00
Interpopulation	1·18	0·83
D. willistoni		
Intrapopulation	1·00	1·00
Interpopulation	1·14	0·90
EGGS/♀/DAY		
D. pseudoobscura		
Intrapopulation	1·00	1·00
Interpopulation	1·27	0·94
LONGEVITY		
D. pseudoobscura		
Intrapopulation	1·00	1·00
Interpopulation (16°)	1·25	0·94
Interpopulation (25°)	1·13	0·78–0·95

important topics in the biology of race crosses in man, that of reproductive performance. We can examine the available evidence on the fertility of race crosses in man within the general context of possible expression of heterosis.

4.3. HETEROSIS IN MAN

Evidence for the existence of hybrid vigour in man comes from both direct and indirect measurements. The examples from direct measurement are of two main types: The first type is that due to single gene heterozygote advantage. Into this category fall the advantage of sickle-cell heterozygotes in malarial areas (*see* s. 2.3.*a*), the possible increases in fertility for AB mothers (Penrose, 1955), the postulated reproductive advantage of up to 5 per cent for heterozygotes for Tay-Sachs disease (Shaw and Smith, 1969), the even higher heterozygote selective advantages suggested for carriers of cystic fibrosis (Danks and others, 1965; Knudson and others, 1967), and the evidence for increased family size in siblings containing cases of hereditary colobema (a dominant defect) and phenylketonuria (a recessive defect) which is mentioned by Penrose (1955). Many of these examples, and others of a similar nature, which may or may not of themselves be statistically significant, suggest that heterosis in man may be shown not only by metrical overdominance of the heterozygote but by increases in fertility connected with mechanisms for potential resistance to disease. Evidence for such a phenomenon must be carefully looked for when we consider the results of various race crosses, but it is worth pointing out in this context the survival of hybrid South American Fuegians and the death of pure bred natives after contracting measles when they were first exposed to this disease (Bridges, 1948). Such examples of heterosis, as are shown in the next chapter, are different in each case and their full consequences for racial hybrids depend very much on the two races concerned, the individual genes, and their frequencies in the populations. Detailed consideration is therefore deferred until the next chapter.

The second type of heterosis known in man from direct measurement is analogous to that which we have already seen in other organisms—namely,

Table 4.6.—Arithmetic Means of Stature, Quetelet Index, and Chest Circumference in Children 4, 8, and 16 Years old from Stetin in relation to Distance between Mother's and Father's Birthplace. (*From* Wolanski and others, 1970. Copyright, University of Chicago Press.)

Distance between Parents' Birthplaces (Km.)	4-Year-old		8-Year-old		16-Year-old	
	Boys	Girls	Boys	Girls	Boys	Girls
	Stature (body height) in cm.					
0–10	103·8	102·2	128·7	126·7	170·6	160·4
11–50	104·2	104·8	125·8	126·0	173·1	161·0
51–100	104·6	103·3	127·5	122·6	173·5	161·2
101–300	103·9	103·6	127·5	126·2	173·2	161·1
301+	104·3	103·7	128·6	126·6	171·1	161·5
	Quetelet index					
0–10	165·29	162·39	201·09	194·99	347·39	332·39
11–50	164·49	164·49	198·19	198·19	354·49	345·99
51–100	166·19	158·39	202·39	185·69	371·19	342·09
101–300	166·09	163·09	203·59	200·39	351·79	346·49
301+	169·09	163·29	205·49	197·58	361·49	343·59
	Chest circumference in cm.					
0–10	54·03	52·70	62·36	58·99	82·66	75·08
11–50	53·69	53·12	60·71	59·12	84·61	75·78
51–100	54·39	52·27	60·87	56·60	86·32	75·25
101–300	54·28	52·89	61·16	59·24	83·46	75·17
301+	54·41	53·24	61·46	59·29	84·17	75·55

the increase in absolute value or manifestation of some continuously varying character controlled by a number of genes, no matter whether this character is a physical feature, or a trait such as fertility or 'fitness'. Manifestation of this sort of heterosis is, of course, one of the most important things we must look for in any offspring of interracial matings, but there are certain indications from various surveys of those sorts of characters most likely to show heterosis in man, although not necessarily in interracial hybrids, and it will be as well to examine this evidence first.

A recent study in Poland (Wolanski, Jarosz, and Pyzuk, 1970) related heterosis in a number of traits in children to the distance between the birthplace of the parents. This problem is particularly relevant in Poland, where after World War II about 12 million people were resettled, and internal migrations resulted in towns like Szczecin (Stetin), for instance, having 97 per cent of their population exchanged. *Tables* 4.6 and 4.7 show some of the basic data obtained by

Table 4.7.—ARITHMETIC MEANS OF HEAD CIRCUMFERENCE AND CEPHALIC INDEX IN CHILDREN 4, 8, AND 16 YEARS OLD FROM STETIN IN RELATION TO DISTANCE BETWEEN MOTHER'S AND FATHER'S BIRTHPLACES. (*From* Wolanski and others, 1970. Copyright, University of Chicago Press.)

DISTANCE BETWEEN PARENT'S BIRTHPLACES (Km.)	4-YEAR-OLD		8-YEAR-OLD		16-YEAR-OLD	
	Boys	Girls	Boys	Girls	Boys	Girls
	Head circumference in cm.*					
0–10	51·08	50·32	53·10	52·06	56·77	55·41
11–50	51·83	50·29	53·26	52·06	56·97	55·61
51–100	51·26	49·95	53·18	51·94	56·66	55·47
101–300	51·50	50·55	53·11	52·17	56·37	54·78
301+	51·62	50·56	53·17	52·33	56·99	55·26
	Cephalic index					
0–10	86·49	87·32	86·33	85·83	86·23	85·51
11–50	86·68	87·27	87·32	84·45	84·43	85·56
51–100	86·99	87·32	85·52	85·43	83·70	84·61
101–300	86·67	86·43	86·84	86·13	84·41	85·05
301+	88·88	85·40	86·61	86·26	84·70	85·00

* By metopion and opisthocranion.

Wolanski and others. They give the magnitudes of five investigated traits in a total of 1896 children: stature, Quetelet's index (body-weight in g. in proportion to body height in cm.), chest circumference measured at the height of the xiphoidale, head circumference at metopion and opisthocranium, and the cephalic index (ev-ev: g-op × 100), according to the distance between birth-places of parents. The principle conclusions which these data give rise to are:—

1. An increase in the distance between the birthplaces of the parents gave rise to an increase in the chest circumference, body height, and body-weight (Quetelet's index) of their children.
2. Heterosis was not found for such traits as head shape (cephalic index) or head circumference.
3. Since the head circumference at birth is 62 per cent of that of the adult, compared with 6 per cent for body-weight, 34 per cent for body height, and 41 per cent for chest circumference, it seems that the heterosis effect emerges mainly through an increased rate of development in the postnatal period.

These authors found no heterosis effect in traits developing particularly rapidly during the embryonic stage, and if this is of general validity it must rule out many previous searches for heterosis in man as being irrelevant, since they concentrated on traits which could not actually be expected to show heterosis anyway.

An important study which showed evidence of heterosis among adults is that of Hulse (1957). Hulse made a number of measurements on descendants of villagers from the Swiss district of Ticino. Because the villages in this mountain district are isolated from one another, a very high proportion of marriages in the past have been between individuals within the same village. Hulse was able to compare the offspring of parents from the same village (endogamous marriages) with those of parents from two villages. Each of these two groups of offspring could be further subdivided into (1) those who remained in Switzerland, (2) those who had emigrated to California, and (3) those who were born in California of immigrant parents. In each of these three groups the mean stature of the exogamous group was approximately 2 cm. greater than that of the endogamous one, a difference which overall was statistically significant (*Table* 4.8). It must be admitted, however, that the interpretation of these results is complicated by temporal trends towards increased stature which seems to have been a world-wide phenomenon for much of this century. There were also differences between the endogamous and exogamous matings in religion, family size, and possibly other factors. Furthermore, the decrease in height with increased inbreeding, i.e., the converse of heterotic effects, is in most cases statistically significant but very small in magnitude for both children (Morton, 1958; Schull and Neel, 1965) and adults (Barrai, Cavalli-Sforza, and Mainardi, 1964; Schreider, 1967), far less than the increase in height with outbreeding which Hulse suggested occurred. Morton and others (1967) therefore suggest that Hulse's result may be due to some non-genetic difference between endogamous and exogamous families. On the other hand Strouhal (1971) found important differences between endogamous and exogamous villages in Nubia, Egypt, for a number of somatometric and dynamometric traits, which he ascribed to heterosis. All in all it seems probable that heterosis is a small but genuine accompaniment of intrapopulation crosses in man for certain metrical traits.

There remains one further area in which heterosis might be expressed and that is fertility itself. Here again the evidence from man is conflicting. Bresler (1962) found in the United States an inverse relationship between the fertility

Table 4.8.—MEAN STATURE IN CENTIMETRES OF ADULT MALE OFFSPRING OF EXOGAMOUS AND ENDOGAMOUS MARRIAGES OF SWISS NATIVES OR THEIR DESCENDANTS (*From* Hulse, 1957.)

SUBGROUP	EXOGAMOUS	ENDOGAMOUS
Swiss	168·51	166·21
Emigrants	168·71	166·90
Californian	172·33	170·50

of the F_1 generation and the recorded number of countries of birth in the P_1 generation. He studied intra-religious matings among Catholics, Protestants, and Jews respectively, and had one, two, three, or four countries represented among the P_1 generation. His samples were not large, but in one case his results were just significant in showing an overall fall in fertility the greater the number of countries represented among the parents. One of his examples concerned the circumstance where one or more of the P_1 generation were American-born. He found 106 matings with two countries represented in the P_1 generation, which provided a mean number of 2·58 live-born offspring; 36 matings with three countries in the P_1 generation had an average of 2·25 offspring; and five matings with four countries in the P_1 generation had a mean of 1·83 offspring. It is difficult to assess the degree of genetic isolation of the countries represented in the P_1 generation, and it is also difficult to assess possible sociological factors which may be serving to reduce the family size of migrant family groups. For instance might large family size itself be a bar to migrating to the U.S.A. in the first place? The value of Bresler's evidence may, therefore, be somewhat questionable.

Contrasting evidence is provided by Cavalli-Sforza (1957), who examined the distribution of fertilities over a period of 300 years in a mountain village in Italy. He made comparisons between the number of offspring from marriages according to the origin of mates, i.e., whether they were both born in the same parish or whether one came from another village. Now there are a number of studies, some of which we mentioned in Chapter 2, showing that in pre-industrial communities there is a measure of genetic isolation between all but the closest villages (Hulse, 1957; Cavalli-Sforza and others, 1966; and Boyce and others, 1967). Cavalli-Sforza (1957) found that the fertility of marriages involving individuals from two different villages was 34 per cent greater than that of marriages between two people of the same village. We can therefore assume some heterotic effects on fertility from these results.

A number of older surveys of hybrid populations are also fairly consistent in suggesting that fertility shows evidence of heterotic enhancement. Davenport (1913a), in his study of Negro–White crosses in Jamaica, found a total of 88 children from 21 such marriages, which is an average of 4·4 children per maternity, although this value, in fact, may be little different from average Negro fertility in Jamaica. Fischer (1913) found that the Reheboth bastaards have an average family of 7·7 and he concluded that 'It is clear that race crossing generally results in a greatly increased fertility'. In support of this contention the Kisar Mestizos had 7·3 children per completed family and Boas (1948) and Ramos (1961) mention the high fertility of Caucasian/Amerindian hybrids. Shapiro (1929) published his history of the Norfolk Islanders, the descendants of the Bounty Mutineers and Tahitian women and their fertility is shown in *Table* 4.9.

Table 4.9.—FERTILITY AMONG NORFOLK ISLANDERS.

	AVERAGE NO. OFFSPRING	NO. OF FAMILIES
Original cross	4·17	6
F_1 matings	7·44	9
F_2 matings	9·10	38
F_3 matings	5·39	77
*F_4 matings	2·96	26

* Probably contains many marriages which are very recent and therefore are still producing children.

Clearly there is an heterotic effect with regard to fertility, but it is also rather delayed in that it reaches a peak among F_2 matings and then declines. In the absence of sociological data on these latter cases, though, it would be premature to regard them as completely proving heterosis in man although, in conjunction with the previous evidence, this would seem to be the most likely explanation for the phenomena that we have described.

On balance, therefore, it seems likely that there is some increase in productivity of interpopulation matings in man, although it is not clear just what mechanisms are involved. There appears to be no evidence as to whether the offspring of these interpopulation matings are themselves more or less fertile than their parents or parental populations. The evidence on this point from laboratory animals, which we have already discussed, suggests that they may be more fertile. The demographic trends which we review in Chapter 8 also lend some support to this view.

Note that this problem of heterosis with respect to fertility is distinct from that of general fertility differences between the races, and the resulting general levels of fertility among interracial hybrids.

4.4. Coadaptation of the Gene Pool

Two of the greatest conceptual advances in biology in recent years can be summed up in fairly short descriptive phrases. Firstly, 'genes not only act but interact' (Mayr, 1963) and, secondly, 'it is not individuals but populations which evolve'. For comprehensive discussion of these ideas the reader is referred to Mayr (1963), Dobzhansky (1951, 1970), Simpson (1953), Wallace (1968), and other comprehensive texts on evolution and genetics. In this section we are concerned with the manifestation of high fitness through specific gene interactions within local populations, and the consequences which ensue when these specific interactions are disturbed by interpopulation crosses.

Of the many possible examples of genetic coadaptation which have been investigated, we shall present a few which indicate different facets of the phenomenon: coadaptation at the level of the individual gene and coadaptation as the integration of gene pools.

The platyfish *Platypoecilus maculatus* is polymorphic at a locus controlling the distribution of macromelanophores on the sides of the body and in the tail, and Gordon (1947) showed that the frequencies of individual fish carrying these different patterns were markedly different in different river basins in Mexico. Presumably, therefore, the allele frequencies were also different. Some of these melanic patterns and the results of hybridizations between fish from different river basins are shown in *Fig.* 4.8 (adapted from Gordon and Gordon, 1957). The gene for 'dorsal-spotted', when present in the hybrids formed by crosses of individuals from two different river basins, causes heavy melanic spots to develop at a number of places on the body of the fish. When such individuals are backcrossed to fish from the second river basin, the 'dorsal-spotted' allele of the first causes enormous numbers of pigmented cells to develop over nearly half of the fish's body. Since a polymorphism exists for this gene in each river basin, there must exist some selective advantage associated with each allele. Since the patterns vary, but vary nevertheless within a precise way, the deposition of melanin must be controlled in a detailed way, partly by the particular allele at the locus concerned and partly through the influence of the genetic background in which the allele resides. Nevertheless, when one of these alleles is removed from its usual genetic background and placed in another by appropriate hybridization and backcrossing, the strict control over its functioning is lost and potentially dangerous, apparently uncontrolled, deposition of melanin results. In other words the genes involved, the 'dorsal-spotted' gene and its modifiers on the genetic background, are coadapted.

The second example concerns some experiments on a number of populations of *Drosophila pseudoobscura* collected in western U.S.A. (Brncic, 1954). Brncic put his collected flies through a complicated series of matings with special laboratory stocks, which enabled him to obtain four sorts of wild type flies whose viabilities could be measured against a genetically marked standard. These flies were as follows: (1) carrying two intact chromosomes from one

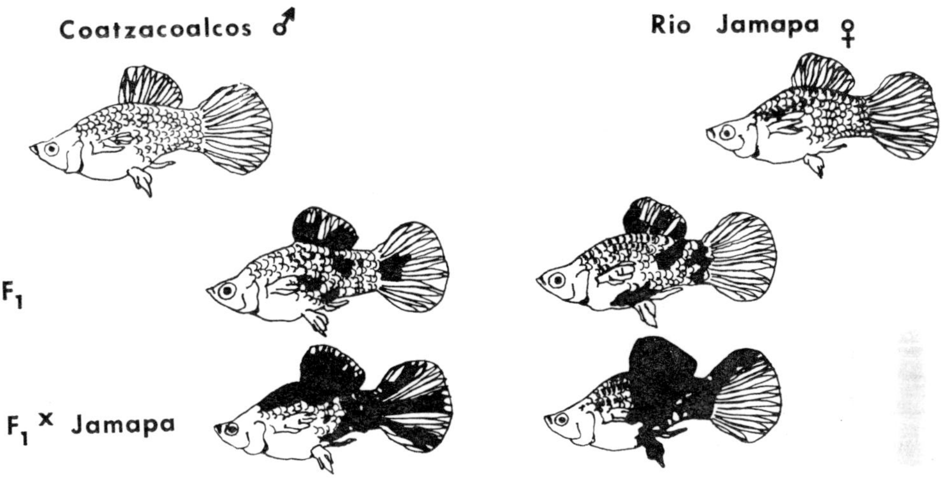

Fig. 4.8.—Parents, F_1 hybrids, and backcross hybrids from two isolated river basins, Rio Tamapa and Rio Coatzacoalcos, Mexico. (*From* Wallace, 1968, *after* Gordon and Gordon, 1957.)

geographic locality; (2) carrying two intact chromosomes from two different localities; (3) carrying one intact chromosome and one containing blocks of genes from two different localities; (4) carrying chromosomes both of which, by recombination, contained blocks of genes derived from two different localities.

Brncic's results are summarized in *Table* 4.10. In all the tests carried out, category 2 flies, the inter-locality hybrids, showed the highest viability and the relative fitness of this class was, in each case, adjusted to 1·00 to provide the standard for comparison. The second highest viabilities were shown by offspring from crosses between flies from a single geographical locality—category 1 or 'parental' flies. Flies with one interpopulation recombinant chromosome (category 3) showed the next highest viability and finally came those flies in which both chromosomes were interlocality recombinants. It seems clear that, at least in this case, the breaking up of gene combinations through interpopulation recombination lowers viability.

A rather different situation regarding the consequences of inter-racial hybridization is shown by the North American frog *Rana pipiens*. This species is unusual in that a number of detrimental effects appear to be demonstrated in F_1 hybrids between different geographical races of this species (Moore, 1950,

1963). In other words, there is a sort of negative heterosis manifest. Moore was able to bring frogs from different environments into the laboratory and obtain a large number of hybrid individuals from many different populations and geographical races. He found that hybrids between individuals from adjacent geographical races were normal in development and morphology, but

Table 4.10.—EFFECT OF RECOMBINATION BETWEEN WILD TYPE CHROMOSOMES OF DIFFERENT GEOGRAPHIC ORIGINS ON VIABILITY IN *D. pseudoobscura* (*From* Brncic, 1954.)

POPULATION	VIABILITY
Parental	0·92–0·93
Hybrid F_1	1·00
Hybrid F_2	0·91
Hybrid F_3A	0·91
Hybrid F_3B	0·86

that the greater the geographical distance between the populations from which the parents came the slower was the rate of development, the greater the morphological defects in the offspring, and the fewer the normal individuals which resulted. As geographical distance between the parents increased, so did the percentage of eggs which failed to develop properly and the earlier in development, overall, did reproductive wastage occur.

In later experiments Moore (1963) produced androgenetic hybrids in which sperm from one race was used to fertilize enucleated eggs from another. These haploids normally died as aberrant tadpoles, but death came sooner and with more conspicuous abnormalities if the eggs and sperm came from different races.

Another example of negative heterosis is that of some inbred mice strain crosses. Although neither CBA nor DBA/2 mice normally show any tendency to develop leukaemia, F_1 hybrids between the two strains have approximately 45 per cent incidence. Furthermore nearly 25 per cent of the hybrids develop plasma-cell metastases with macroglobulinaemia resembling Waldenstrom's macroglobulinaemia in man (Manwell and Baker, 1970).

Tumours are also known to occur in some plant crosses, for example in the hybrids between the lilies *Lillium speciosum* 'Album' × *Lillium auratum* (Emsweller, Asen, and Uhring, 1962). There is apparently a high concentration of the phenolic compound ferulic acid in hybrids with tumours whereas, normally, ferulic acid is detoxified by glycosidation. In some of the hybrids the balance between ferulic acid production and glycosidation has broken down, leading ultimately to the formation of malignant tumours.

There is one other well-known phenomenon concerned with outcrossing from an integrated gene pool which it is worth mentioning briefly. This is the delicate balance of the sex determining mechanism. J. B. S. Haldane long ago predicted that the heterogametic sex (X and Y chromosomes as opposed to two X chromosomes) would show, in general, lower crossing-over frequencies and also possibly some detrimental effects on outcrossing (Haldane, 1922). E. B. Ford (1954) described the occurrence of just such phenomena among crosses of the butterfly *Coenonympha tullia*, which occurs from the extreme north of Scotland southwards to Shropshire and Staffordshire in central England, and into northern and mid-Wales. When males from Wales were crossed with females from Scotland 5 of the 36 female offspring were intersexual, indicating that the balance of the sex genes at different ends of what was once a continuous cline are now differently built-up.

This field of coadaptation is obviously one in which more data are required, but it is one which seems to be curiously neglected by ecological geneticists. The existence of permanent hybrid swarms or centres of introgression are often known and investigated, but the consequences for early generation hybrids are not usually examined in detail. It would seem that statements such as 'hybrids are usually not so well adapted as either of the parental forms' are accepted as describing the situation in adequate detail. A statement by Mayr (1963, p. 376) is typical in this respect. 'The ecological adaptation of the gene complexes meeting in these hybrid zones prevents in all these cases a widening of the zone of intergradation through gene flow.' The mechanisms operative in the early generations of hybridization would seem to warrant much closer study in most animals and plants.

We have made no mention of man in this section so far. Elsewhere we suggest that coadaptation is not likely to be very important in man, nevertheless it is important to understand the principle. It is also important to know that a great deal of fundamental work on the nature and consequences of coadaptation both in man and other organisms still remains to be done. We return to this topic in the next section.

4.5. Race Crosses and Genetic Variability

In the previous sections on heterosis and coadapted gene pools we have failed to discuss the simple amount of genetic or phenotypic variability expressed in interpopulation hybrids as compared with the parental types. Most of the data which we shall present in this section were gathered in the course of investigations into problems connected with genetic isolation through speciation and with the evolution of dominance, most of which are only partially relevant to our present theme.

Several of the conclusions derived from such investigations, however, are of major importance. Perhaps the most fruitful research in this field has been with the mimetic butterfly *Papilio*, principally by Professors C. A. Clarke and P. M. Sheppard from the Nuffield Institute of Medical Genetics in Liverpool (*see*, for example, the following important papers among the many they have published in this field: Clarke and Sheppard, 1960a, b, 1963, 1971; Sheppard, 1961; Clarke, Sheppard, and Thornton, 1968). In *Papilio dardanus* from Africa there are a number of isolated geographical races. The males in all the races are black and yellow, have tails on the hindwings, and are non-mimetic. The females in Madagascar and Grand Comaro and most of those from Abyssinia are like the males, but everywhere else the females are unlike the males. They are tailless and highly polymorphic usually mimicking another, distasteful species. Clarke and Sheppard have shown that within this species there is a multiple allelic system consisting of at least 12 alleles which control the major colour pattern variants, and that there is a second locus, unlinked to the first, which has two alleles controlling the presence and absence of tails on the hind

Fig. 4.9.—Hybridization in the mimetic butterfly *Papilio dardanus*. Three hybrids between the male-like female *meriones* from Madagascar and the mimetic males *hippocoonides*, *cenea*, and *trophonius* from mainland Africa. This black-and-white representation can only hint at the colour differences between the normal mimetic forms and the hybrids which are described by Clarke and Sheppard in 1960.

1, *Meriones* male-like female (yellow); **2**, *Hippocoonides*/yellow heterozygote; **3**, *Cenea*/yellow heterozygote; **4**, *Trophonius*/yellow heterozygote; **5**, *Hippocoonides*; **6**, *Cenea*; **7**, Trophonius. (*From* Clarke and Sheppard, 1960a, b.)

wing. It has been found, though, that when a major gene producing a mimetic pattern is put into an 'unfamiliar' gene complex by hybridization with a race of *P. dardanus* not possessing that particular mimetic form, the pattern produced by the gene is changed and the mimicry becomes less perfect. Some examples of this are illustrated in *Fig.* 4.9. In other words, there must be a number of modifier genes which regulate the function of the genes responsible for the pattern. This breakdown of mimicry occurs in an extreme form in some F_2 and backcross hybrids involving the mimetic South African race and the non-mimetic Madagascan race, in which all the females are male-like. Clarke and Sheppard have also shown that there are usually complete and regular dominance relationships between morphs which are present in the same subspecies of *P. dardanus*. But when races are crossed, so that major genes not normally found together in nature are combined in a heterozygote, dominance tends to break down and intermediate phenotypes are formed. In the Abyssinian race of *P. dardanus* the male-like female form is dominant over the forms *hippocoonides* and *cenea*, but in the F_1 hybrid with the South African race, in which the male-like female form is not found, this dominance relationship breaks down somewhat.

The breakdown was much more pronounced on crossing South African butterflies with those of the Madagascan race, in which *hippocoonides* and *cenea* do not exist. Broadly similar results have been found with *Papilio memnon* from South-East Asia (Clarke and others, 1968). In this species, too, there is an allelic series responsible for the various patterns of the wings, and the alleles concerned do not have the same effect in all gene-complexes. In a number of crosses between different races of *P. memnon* utilizing the same mimetic pattern, the resemblance between model and mimic in the offspring of these inter-racial crosses was better in 0, as good in 10, and poorer in 35 cases, than it was within the races concerned.

In this species the overall mimetic patterns are controlled by three closely linked loci of which one, for the tailed condition, has just two alleles, presence or absence. Examination of the results of crosses between tailed butterflies from Palawan and tailless insects from elsewhere is particularly illuminating with regard to the consequences of interracial crossing. The results of such crosses carried out by Clarke, Sheppard, and Thornton are shown in *Fig.* 4.10. The mean adjusted tail length of pure Palawan *P. memnon* was 11·54 mm. with a variance of 0·9 mm. The tailless races had a mean value of 1·49 mm. and a variance of 0·16 mm. First generation hybrids have a mean of 8·34 mm. and a variance of 5·41 mm. In other words, the F_1 was intermediate in tail length between the two parental stocks, and showed a very great increase in variance. Backcrosses of these F_1 hybrids produced no clear segregation into tailed or tailless individuals, even though normally the pure races are quite distinct. More detailed tests and their analysis showed that the presence of tails is controlled by a single gene with modifiers markedly affecting the actual length. As a result of these modifiers the distinction between the heterozygotes and both homozygotes frequently disappears.

This phenomenon of increased genetic variability is one which is fairly general in interracial crosses, even if often for rather different reasons in different

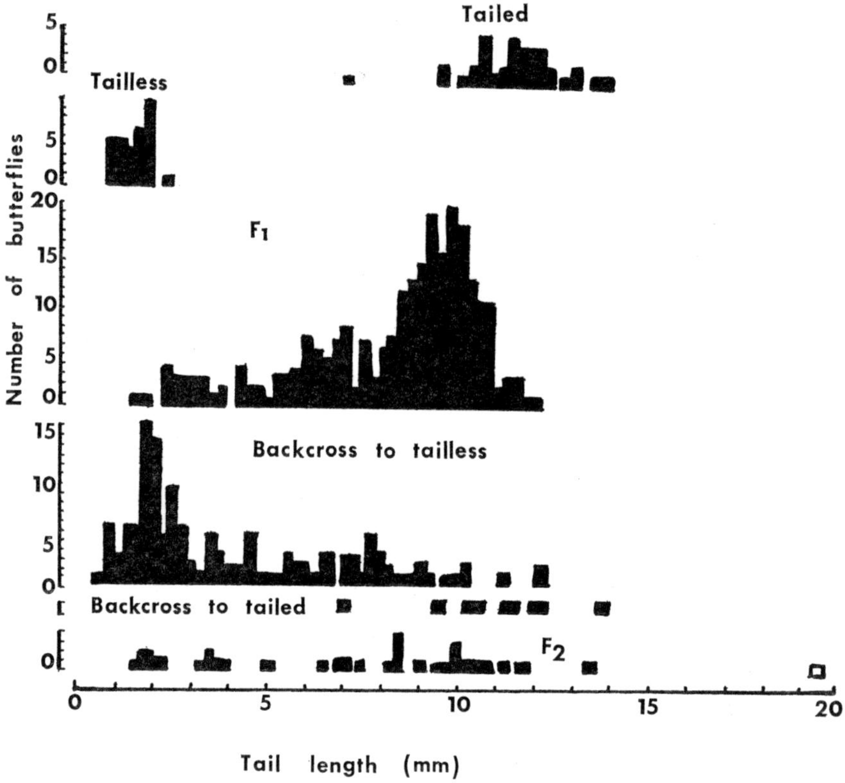

Fig. 4.10.—Histograms showing the lengths of the tails in the 'tailless' races, the tailed Palawan race, the F_1 and F_2 hybrids, and the back-crosses to both the tailed and the tailless races of the mimetic butterfly *Papilio memnon*. (*From* Clarke and others, 1968.)

cases. The extreme example is of course the crossing of two inbred lines which will, among F_2 and subsequent generations, produce enormously more variable hybrid populations. This is because of the segregation and recombination of a variety of genes which were homozygous, but in different combinations in the inbred lines. Clearly every small population in nature will develop homozygosity to some extent, and there is a whole body of theoretical knowledge to show just how much. In other words, depending on effective population size, individual populations will, to a greater or lesser extent, resemble inbred lines. Outcrossing will remove this genetic uniformity, and the increase in phenotypic variety which results depend upon the relative genetic and environmental contributions to the characters concerned and the extent to which homozygosity was attained in the contributing populations. The theoretical details of this topic can be found in Falconer (1960) and are briefly summarized in s. *5.3* in the next chapter. The possible practical implications of all this with regard to human populations we also consider in the next chapter.

We cannot, however, leave this matter here. The work we have so far discussed has been primarily concerned with genetic variability in F_1 hybrids and perhaps the F_2 or backcross generations. But of course in man racial integration is an ongoing process, and biological consequences of relevance to the individuals concerned may continue to be felt for very many generations after racial hybridization. We must, therefore, look more closely at the hybrid zones and hybrid swarms found in nature which are the product of sometimes recently established, sometimes long-continued hybridization between once isolated populations, to see what they can tell us of long-term effects of hyridization.

4.*6*. Long-established Hybrid Populations

Ernst Mayr (1963) in his superb book on animal evolution, makes the following generalizations about hybrid zones: 'The more different the populations coming into contact, the more recent usually is the hybridization; the narrower the hybrid zone, the greater is the probability of imbalance in the recombined genotypes and the greater the probability that the imbalance will express itself as highly increased variability.' The variability is produced, however, only partly by the breakdown of coadapted gene complexes or systems of modifiers of the sort we have already discussed. The principal reason is that there are many generations of hybrids present in such hybrid zones and therefore a wide variety of crosses and backcrosses continually occurring. This ensures complete segregation of all the genetic variability present in the two populations and, as we saw in Chapter 2, this is likely to be considerable.

Nevertheless, this released variation does not affect all characters nor does it necessarily persist, and the forces of natural selection which become operative over longer periods of time can usually be seen controlling it. Thus, for example, Mayr and Rosen (1956) have shown that in some inter-racial hybrid colonies of the Bahama snail *Cerion*, shell sculpture and coloration show increased variability, while size and proportions do not. This, Mayr suggests, may be due to the stabilizing effect of selection which would be probably more pronounced on size than on shell characters; a suggestion supported by the discovery of strong stabilizing selection in the littoral dog whelk *Nucella* (Berry and Crothers, 1968). Mayr has also shown that when two subspecies hybridize in several separate areas—and he has investigated the hybridization of Grey-billed and Black-billed Honeyeaters, *Melidectes*, in New Guinea (Mayr and Gilliard,

1952)—different characters may become stabilized in the various areas of hybridization.

In general, it seems that in these hybrid populations natural selection is continually removing the more extreme phenotypes and the more imbalanced genetic combinations. High variability, which we have seen is so characteristic of early generation hybrids, will normally only be maintained by the continued inflow of parental genotypes, in other words, by the continuing presence of these early generation hybrids.

In a few cases, particularly among birds, investigations have been carried out on hybrid populations which are now isolated from the original contributing populations (Mayr, 1932; Miller, 1941). Thus in the northern Sahara there are a number of isolated sparrow populations which were originally hybrids between *Passer domesticus* and *P. hispaniolensis*. These have now attained phenotypic stability and do not show any evidence of increased phenotypic variability compared with the contributing species.

We may make the generalization, then, that the major biological effects of hybridization become apparent during the first, second, and first backcross generations after the initial event. These effects will be initially to enhance growth rates and productivity and to produce other manifestations of hybrid vigour. Total phenotypic variability will also be much increased. These effects will, on the whole though, disappear fairly rapidly, partly through the loss of luxuriance, partly through the breakup of coadapted gene complexes, and partly through natural selection eliminating the more extreme manifestations of variability. There is some evidence that the relative geographical isolation, and therefore, presumably, the degree of genetic difference, affects the severity of some of the detrimental effects which appear. And certainly, in so far as crosses between very different organisms, that is interspecific crosses, normally produce inviable or sterile offspring, this is to be expected.

There are a few experiments which provide information bearing on this particular situation that are worth discussing. The first concerns the introduction into some isolated populations of *Drosophila willistoni* of foreign genetic material. This species has normally a large number of inversions in its populations, although it is found that the average number of inversions per individual decreases from the centre toward the edge of the species distribution. In particular it is known that in marginal populations the X chromosomes are almost completely free of inversions. The experiment carried out by Cordeiro, Salzano, and Marques (1960), was to introduce a number of hybrids containing X chromosome and autosomal inversions into some isolated marginal populations, and to follow the fate of these introduced inversions during the following 12 months. Six 'new' X chromosome inversions were introduced and the first sample taken after introduction (of some 60,000 flies into each wood) showed a relatively high frequency of these 'new' inversions together, with a significant increase of some introduced autosomal inversions. Samples taken the following year showed that most of these introduced X chromosome inversions had been lost or nearly so, with the exception of two inversions in one of the woods sampled. Similarly the autosomal inversions that had increased in frequency soon after the release of the hybrids had dropped to the control level frequencies after one year. The average number of heterozygous inversions per individual showed a similar increase shortly after the introduction of new flies and then a return to pre-existing values.

The authors draw a number of conclusions from this work but we are here concerned with one important one. Firstly the large-scale introduction of new flies brought about a significant increase in the frequency of heterozygotes. However, the genetic structure of these outbreeding natural populations had strong homeostatic properties that rejected the drastic changes thrust upon them, even though the introduced chromosomes represented a marked increase in heterozygosity. This concept of population homeostasis, or broadly genetic stability, may underlie much of the resistance to marked change shown by hybrid populations of many organisms, including man. There is one very important point we can make with respect to the stability of hybrid zones in animals in comparison with the situation which exists in man—that is the relative stability of the hybrid zones themselves. The hybrid zone between the European crows has apparently existed for about 5,000 years; as have others which were presumably established after separation of the parental populations during the ice age. In the North American butterfly *Basilaechia* the hybrid zone has existed for at least 80 years (Mayr, 1963) and Turner (1971) describes populations of a *Heliconius* butterfly which have hybridized for 2,000 generations. And yet the evidence we present in this book shows that 'hybrid zones' never persist in man unless it be, as in South Africa, where they are supported by a man-made law. This fact alone would suggest that much of the discussion in this chapter of hybrid breakdown and integrated gene pools would have only marginal significance for man himself, a suggestion we will abundantly confirm in later chapters.

It would also appear that most of the racial crosses in man are unlikely to produce the more severe detrimental effects we have observed in this chapter, since we have seen that they do not, on average, approach the amount of genetic differentiation typical of full species. For a few more distantly related races, however, this might be a possible hazard. With regard to the manifestation of heterosis in man the evidence appears somewhat equivocal, but it would be surprising if we could not detect some effects in a few cases. Increased phenotypic variability there seems likely to be in almost all cases of interracial crosses, although often this will be only a temporary manifestation. The full genetic implications of this increased variability, its persistence in time, and its consequences for the hybrid population, can only be determined after more detailed treatment of the genetical dynamics of racial integration. This is the subject of the next chapter.

CHAPTER 5

THE FORMAL GENETICS OF RACIAL INTEGRATION

5.*1*. Gene Diffusion

THROUGHOUT this book we have treated the problem of racial biology in genetic terms. It is appropriate, therefore, to begin our treatment of the formal genetics of race crossing with a consideration of the behaviour of individual genes and their spread from one population to another. This spread of new genes from one population to another and from one geographical locality to another is very dependent on the type of population structure or structures involved. Yet it is a process which is apparently frequently occurring and which often involves the spread of a new gene from one race to another, although probably at lower rates than are characteristic of situations involving any marked degree of racial mixing as such. It is nevertheless necessary to treat this topic in some detail because of its great importance in explaining observed gene frequency distributions.

The gradient of B blood group frequencies from Central Asia to Western Europe is possibly one example of gene diffusion. The map in *Fig.* 5.1 shows this gradient. The highest frequency of the B blood group is in excess of 30 per cent in south central Asia and northern India. From this high point the allele becomes less common westwards until in parts of France and the British Isles (*see* Chapter 1, s. 1.*2*.*a*) the frequency is below 5 per cent. One explanation for this cline is that it is a result of the invasions of the Mongoloids who, from about A.D.500 to A.D.1500, migrated westwards in numbers which decreased with distance from their region of origin. Marriage of the invaders with native Europeans, in whom it was the B group which was absent, led to 'diffusion' of B from Central Asia westwards. The gradient of the B blood group frequency is therefore a reflection of the diminished contact between Mongoloids and populations farther west. It has also been suggested that the B blood group may offer some protection against small-pox (Vogel and Chakravartti, 1966) and possibly other diseases (*see* Chapter 2, s. 2.2). If this is so, then part of the rate and extent of the B allele's diffusion will be determined by the geographic gradient of these various selective forces.

Some interesting information on gene diffusion can be gained by a careful study of gene frequencies among the Aborigines of Australia. One of the most

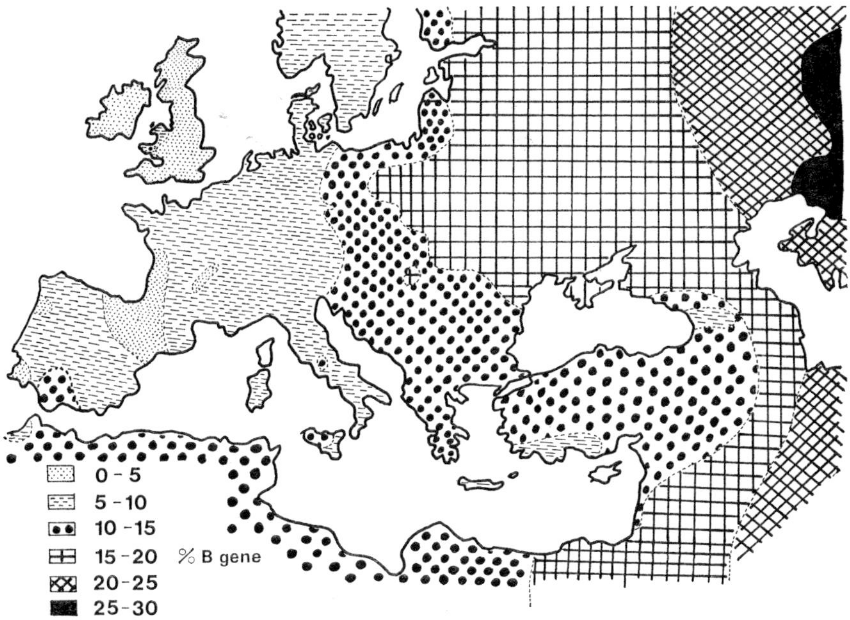

Fig. 5.1.—The frequencies of the I^B allele in Europe showing a general trend from high frequencies in the east to low frequencies in the west.

informative genes which has been examined recently is the antigenic variant E^T in the Rhesus system, which was first detected among Aborigines in the Laverton area of Western Australia (Kirk, 1971). *Fig.* 5.2 shows the varying proportions of Aboriginal people carrying the E^T antigen in various parts of Australia. The pattern of distribution suggests the spread of a mutation which occurred first of all as a deficiency change in the normal E antigen in the Western Desert. From this origin the gene has diffused north and east, and has now reached the Kimberley Ranges, but has not so far reached Queensland. Gene diffusion from the south and centre towards the north coast has therefore taken place.

On the other hand there are a number of genetic markers which indicate that very little gene diffusion in the opposite direction can have taken place, at least for a long period of time. Among the populations of the Western Desert and some of the central tribes are found extreme values of the blood group genes

A, N, and P, and the serum protein genes Hp, Tf, Dl, Gc^1, and $Gm^{1,5}$, for each one of which the population is either approaching or has reached loss or fixation of the gene concerned, suggesting little genetic exchange with the coast. In particular the $Gm^{1,5}$ allele, which is present in low frequency in Cape York and up to 17 per cent in the Kimberleys, is completely absent in the Western

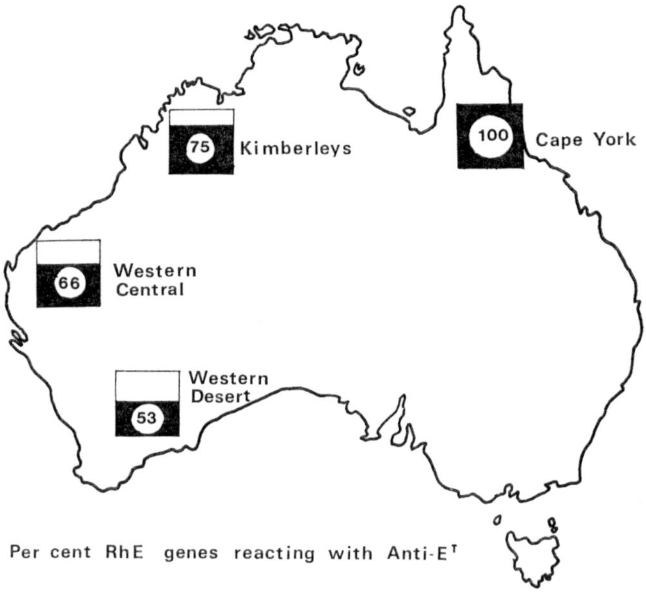

Fig. 5.2.—Area distribution in Australia of RhE positive persons reacting with a naturally occurring anti E^T antibody. (*From* Kirk, 1971.)

Desert. This would suggest a complete absence of gene flow from the coast into the desert regions of the interior.

An interesting variant of the Group Specific Component system shows that gene diffusion has, in some cases, reached the Kimberleys and populations of Central Australia. The allele $Gc^{Aborigine}$ has its highest frequency in Australia in the Cape York area, with lower values in the Kimberleys and central coastal areas of Western Australia. The variant is virtually absent from the Western Desert. The $Gc^{Aborigine}$ gene is widespread in Melanesian populations both in New Guinea and the New Hebrides, although it has not been detected anywhere else in the world. This certainly suggests, therefore, that some genetical material, probably from New Guinea has been introduced into the Aboriginal gene pool via Cape York and has diffused to a limited extent across Australia.

On the other hand the B blood group, which is not present in most Australian Aborigines, is found fairly widespread in the Barkly Tableland area and in Cape York, particularly in the south of this peninsula. Simmons, Graydon, and Tindale (1964) believe that the Kawara tribe has been the main focus for the spread of the B gene to other Cape York tribes. They believe that the introduction of the B group can be explained by separate admixture at different periods with different non-Australian races to the north. A variant of the enzyme 6-phosphogluconate dehydrogenase gives further information on this point. A new allele at the 6PGD locus was discovered (Blake and Kirk, 1969) and named PGD^{Elcho} from its discovery on Elcho Island off the coast of Northern Territory. This allele is present in many populations in Arnhem Land, but has not been detected in Central Australia. More important, perhaps, Kirk has not been able to detect PGD^{Elcho} in several hundred samples from the New Guinea Highlands (Kirk, 1971). A non-Melanesian source for this gene therefore seems indicated, possibly Malayan.

For one of the most informative situations with regard to gene diffusion we turn to the haemoglobin polymorphism involving the three alleles Hb^A, Hb^S, and Hb^C in Africa. We have already discussed in some detail the selective advantages by which the sickle-cell gene Hb^S is maintained in the Negro population. It is believed that Hb^C also confers some protection against falciparum malaria, although the evidence in this case is much less secure (Livingstone, 1971). Now these three alleles can form 6 genotypes, and it is rather a difficult matter to decide on the relative fitnesses of each of these genotypes. Deviations from Hardy-Weinberg equilibria will in some cases, particularly if selection is strong, give estimates of fitness, and Cavalli-Sforza and Bodmer (1971) adopt this approach to try to determine the likely dynamics of these alleles in West Africa.

A survey of Hb^S and Hb^C gene frequencies in 72 West African populations (Livingstone, 1967) showed that there is no population that has C genes but no S genes, and there is a negative correlation between S and C frequencies when both genes are present. By computing Hardy-Weinberg expectations for each population and summing over all populations, Cavalli-Sforza and Bodmer obtain the figures for the fitness of the 6 genotypes shown in *Table* 5.1. There are a number of uncertainties about these figures, although they do in general agree with what is known of the clinical conditions associated with each genotype. We may note two things in particular from this table; firstly, the surprisingly high postulated fitness for CC homozygotes and the apparent equality of fitnesses for AC and AA individuals.

Now we have already seen (*Fig*. 2.25**b**) that Hb^A occurs in high frequencies across a very large belt in Central Africa, but the Hb^C gene is distributed very differently. It shows a maximum frequency in the Upper Volta region and from there decreases fairly regularly in all directions. *Fig*. 5.3 shows the distribution of Hb^C in West Africa, and superimposed on this are some isogenic lines calculated by Cavalli-Sforza and Bodmer on the basis of a diffusion by migration hypothesis explained in their book. The assumption, then, is that a mutant arose, presumably near where Hb^C now shows the highest frequency, and it has

Table 5.1.—OBSERVATIONS OF HAEMOGLOBIN A, S, C GENOTYPES AND HARDY-WEINBERG EXPECTATIONS SUMMED OVER 72 WEST AFRICAN POPULATIONS (*From The Genetics of Human Populations* (L. L. Cavalli-Sforza and W. F. Bodmer) 1971. Copyright, London: W. H. Freeman & Co.)

Genotype	AA	SS	CC	AS	AC	SC
Observed	25,374	67	108	5,482	1,737	130
Expected	25,615·5	306·87	74·69	4,967·2	1,768·6	165·01
Fitness	0·991	0·218	1·446	1·104	0·982	0·788
Standardized Fitnesses ($\overline{AS}$ = 1)	0·89 ± 0·03	0·20 ± 0·11	1·31 ± 0·29	1	0·89 ± 0·035	0·70 ± 0·07

spread and is still spreading outwards from there. As supporting evidence it is found in this area that where Hb^C is highest, Hb^S is lowest, i.e., the negative correlation mentioned above holds good, suggesting that in fact C is gradually supplanting S. The fitness figures in *Table* 5.1 are obviously compatible with this idea. There are some other red-cell defects connected with malaria that

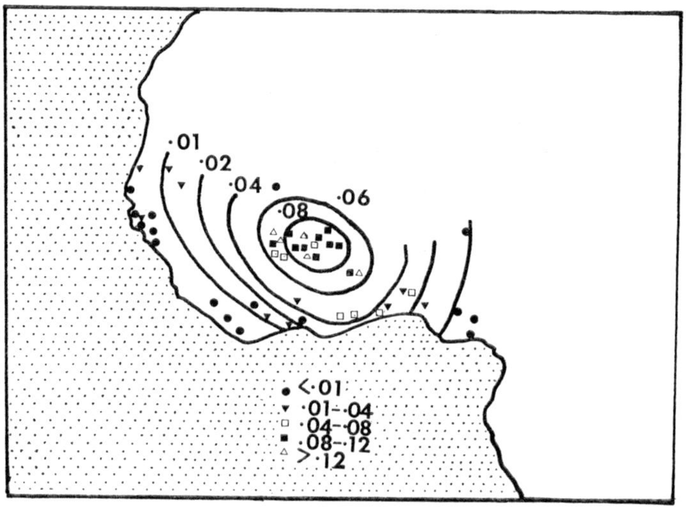

Fig. 5.3.—The spread of the C gene in West Africa. Ellipses indicate expected C gene frequencies on the basis of a simple diffusion model. (*From The Genetics of Human Populations* (L. L. Cavalli-Sforza and W. F. Bodmer) 1971. Copyright, London: W. H. Freeman & Co.)

may represent a similar example of a single origin and diffusion outwards—the 'High F' gene in Africa, for instance, and perhaps the gene for haemoglobin E in South-East Asia.

We see, then, that gene diffusion may come about through two rather different processes: Firstly, the migration of a population into an already inhabited area when the two populations concerned have different gene frequencies; secondly, the appearance of a new gene in a population either by mutation or arrival of one or two migrants and its subsequent spread concentrically outwards at a rate dependent on its selective advantage. In either case a gradient in phenotype or gene frequency results which is called a cline.

Clines may be transient or stable and it is usually impossible to tell which unless we have fossil evidence. However, whenever we observe a cline in a population today it signifies that gene diffusion either has occurred in the past or is occurring at the present time. The existence of many well-documented clines, of both large and small scale, for a variety of different genes among most of the world's populations, is evidence of the importance of gene diffusion in man's evolution (Mourant, 1954).

5.2. Gene Flow

Gene flow is really a special case of diffusion. We are concentrating on the particular circumstance in which the diffusion takes place slowly from one race to another. It therefore implies the existence of some sort of barrier to straightforward gene exchange.

If an admixture is formed between two populations A and B, the first contributing a fraction M and the second $(1 - M)$ to the mixed population, then the gene frequency P_M of an allele in the mixed population is

$$P_M = MP_A + (1 - M)P_B = P_B + M(P_A - P_B).$$

If the three gene frequencies P_A, P_B, and P_M are known then M is given by

$$M = \frac{P_M - P_B}{P_A - P_B}$$

Such an estimate of gene migration M, for a given locus, is equivalent to an estimate of the admixture which has occurred only if: (1) The interracial mating which does occur is random, i.e., there is no assortative or preferential mating between the two populations with respect to the locus considered. (2) The gene migration is entirely from one population into the other. (3) Individuals whose genotype is derived from both populations have no special bias with respect to viability, fertility, social factors, or geographic mobility, which would affect their contribution to the gene pool of the hybrid population.

This last point cannot always be assumed and in fact Workman, Blumberg, and Cooper (1963) showed important effects of selection on some genes in hybrid populations of Negroes in the U.S.A. These workers estimated gene flow from the White into the Negro population in Claxton, Georgia, U.S.A., using seven blood groups and four blood proteins and enzyme variants. They originally suggested that the genes G6PD, Hp_1, Tfd, and Hb^S were affected by selection in favour of the White alleles when Blacks moved into the American environment, in particular because the selective pressure due to malaria was now absent. Cavalli-Sforza and Bodmer (1971) re-analysed these data using only genes for which the African data are most reliable and for which standard errors are small. They concluded that selection probably was implicated in changing the frequency of G6PD and Hb^S among American Negroes. Their figures are given in *Table* 5.2.

Table 5.2.—GENE FLOW AND SELECTION IN AN AMERICAN WHITE AND NEGRO POPULATION (*From The Genetics of Human Populations* (L. L. Cavalli-Sforza and W. F. Bodmer) 1971. Copyright, London: W. H. Freeman & Co.)

Gene	Gene frequencies		West African Gene-frequency Data		Estimates of Gene flow	
	2090 Claxton* Whites	1287 Claxton Blacks	Average	S.D.	No. of Studies	$M \pm \sigma_M$
Fy^a	0·422	0·046	0	0	3	0·101 ± 0·006
R_o	0·037	0·535	0·6051	0·0644	14	0·123 ± 0·0205
A	0·246	0·158	0·1468	0·0336	37	0·113 ± 0·0113
B	0·050	0·129	0·1431	0·0410	37	0·151 ± 0·0110
M	0·508	0·485	0·4644	0·1208	12	0·472 ± 0·2365
G6PD	0	0·118	0·1620	0·0670	25	0·272 ± 0·0133
Hb^S	0	0·043	0·0611	0·0415	72	0·296 ± 0·0066

* Claxton data from Workman and others (1963); West African data from a variety of other sources quoted by Cavalli-Sforza and Bodmer (1971).

If the average estimate of White admixture among Claxton Negroes derived by using these reliable five genes is correct ($M = 0{\cdot}113$), then the gene frequencies of Hb^S and G6PD in Claxton Negroes should be 0·144 instead of 0·118, and 0·054 instead of 0·043 respectively. This gives an idea of the change in gene frequency that may have taken place in the approximately ten generations during which these malarial protective genes have been in a non-malarial environment.

There are thus two principal types of information given by M values: Firstly, on the action of natural selection; and, secondly, on the total amount of gene exchange between two populations. The estimates which are derived using these methods are of course critically dependent on the accuracy of the initial data, and therefore on those genes which provide the most sensitive distinction between the parental populations.

The Fy^a gene of the Duffy blood group may be an almost ideal gene for estimating M between Whites and Negroes in the U.S.A. The mean frequency of Fy^a for all Africans in the areas from which slaves were taken is probably less than 0·03, whereas the frequency for U.S. Caucasians is about 0·43 (Reed, 1969b). Furthermore there is no evidence for natural selection on Duffy blood group phenotypes, and therefore it should give an unbiased estimate of M.

Table 5.3.—ESTIMATES OF M DERIVED FROM Fy^a GENE FREQUENCIES FOR AMERICAN NEGROES FROM VARIOUS AREAS (*From* Reed, 1969b.)

LOCALITY	NEGROES		CAUCASIANS		
	N	*q* ± S.E.	*N*	*q* ± S.E.	M ± S.E.*
Non-Southern					
New York City	179	0·0809 ± 0·0147			0·189 ± 0·034
Detroit	404	0·1114 ± 0·0114			0·260 ± 0·027
Oakland, California	5146	0·0941 ± 0·0038	5046	0·4286 ± 0·0058	0·2195 ± 0·0093
Southern					
Charleston, South Carolina	515	0·0157 ± 0·0039			0·0366 ± 0·0091
Evans and Bullock Counties, Georgia	304	0·0454 ± 0·0086	322	0·422 ± 0·0224	0·106 ± 0·02

* The q for Oakland Caucasians (who are of West European ancestry) was used in all estimates.

Table 5.3 shows estimates of M derived from Fy^a gene frequency in U.S.Whites. The M estimates for the three non-southern regions do not differ significantly and they are all markedly greater than the two southern estimates. They provide irrefutable evidence that there are marked differences between northern and southern U.S. Negroes with respect to the degree of White admixture; a finding which had been suggested by virtually all other comparative surveys in this field.

In addition to knowing the total amount of genetic mixture which has occurred between two races since they first came into contact, we also wish to know just how much gene flow is occurring at each generation, since this of course is a true measure of the rate of biological integration. There are two different approaches which have been adopted to provide an answer to this problem. First, there is an algebraic method (Glass and Li, 1953) which rests on knowledge of the original gene frequencies of the two contributing races. This is, therefore, to some extent an extension of the method we have already examined. The second approach is to measure the difference in gene frequencies between younger and older members of the same race, assuming that they are representative of different generations and that different rates of gene transfer from another race will have prevailed at the time of their conceptions.

The model developed by Glass and Li can be presented as follows: if q_0 is the original allele frequency in one population and Q the frequency of the same allele in another population in which it has remained stable, we have to find m, the fraction of the gene pool of the first population which is derived from the second in each generation. After the first generation of interbreeding, i.e., gene flow, the allele frequencies of the resulting population will have changed from q_o to

$$q_1 = (1 - m)q_0 + mQ.$$

If m remains constant,

$$q_2 = (1 - m)q_1 + mQ,$$

and after k generations

$$q_k = (1 - m)q_k - 1 + mQ.$$

By successive substitution of $q_k - 1$, $q_k - 2$, ... q_2, q_1 we obtain

$$q_k = (1 - m)^k q_0 + mQ + (1 - m)mQ \ldots + (1 - m)^{k-1} mQ.$$

The last k terms on the right-hand side of the equation form a geometric series whose sum is

$$mQ\left[\frac{1 - (1 - m)^k}{1 - (1 - m)}\right] = [1 - (1 - m)^k]Q.$$

Hence the general expression for the allele frequency after k generations of interbreeding is

$$q_k - (1 - m)^k q_0 + [1 - (1 - m)^k]Q.$$

Written in the form

$$(1 - m)^k = \frac{q_k - Q}{q_0 - Q}$$

it enables us to ascertain the value of m under any assigned value of k if the values of q_o, q_k, and Q are substituted. In *Table* 5.2 we have tabulated some of the total amounts of gene flow estimated to have occurred between Whites and Negroes in the U.S.A. Using the above formula it is possible to determine the amount of gene flow per generation. From the data shown in *Table* 5.2 it is found that these genes have undergone gene flow of between 2·5 per cent and 6 per cent per generation. These likely values are confirmed by many other studies of gene flow between Whites and Negroes in North America.

Alternatively the amount of change in the allele frequency of the recipient population in any generation is

$$q_k = q_{k+1} - q_k = -M(q_k - q),$$

the repeated application of which will give the successive allele frequencies in this population.

It is clear that estimates of m, the gene flow, will depend upon the estimate of k, the number of generations of intermixture. This, in turn, will depend upon the length in years of the period throughout which intermixture has occurred and the average length in years of a generation. Furthermore the change of m with changing k is very great when k is low, but becomes less and less as k increases. *Fig.* 5.4 shows this relationship graphically. It illustrates the change in the estimate of gene flow (m) with variation in the number of generations (k)

of intermixture, using as an example gene flow from Whites into Negroes in the United States. On genetic evidence Glass and Li (1953) estimated the total amount of admixture to be about 30·6 per cent (this figure was later revised downwards by Glass (1955)). Now the first shipload of Negro slaves arrived in the American colonies in 1619, although the early build-up of population was slow. It can therefore be estimated that intermixture has been taking place

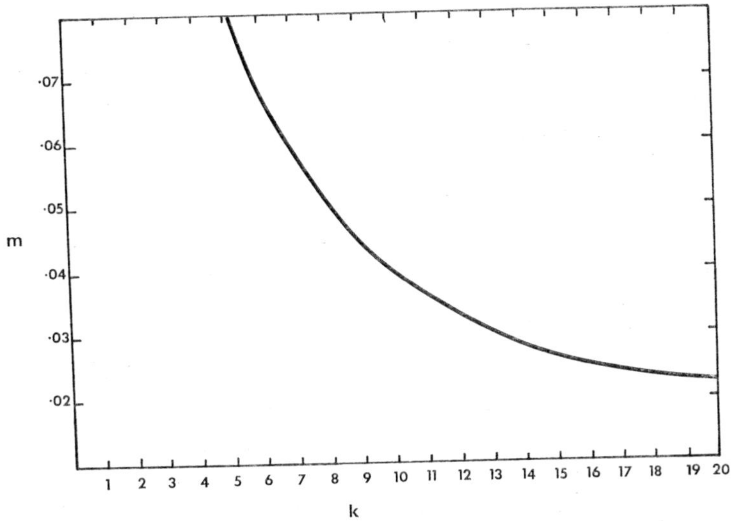

Fig. 5.4.—The change in the estimate of gene flow (m) with variation in the number of generations (k) of intermixture when $(1 - m)^k = 0{\cdot}69435$. (*From* Glass and Li, 1953.)

for about 300 years. The average generation time is between 27 and 30 years. This leads to a minimum estimate of 10 generations admixture and it is obviously important always to use a minimum estimate. This is because if the correct figure is 13 generations our calculation of gene flow per generation is very little wrong. If, on the other hand, we have overestimated the number of generations by 2 or 3 then our calculation of gene flow per generation is seriously in error.

This method, of course, makes no allowances for the fact that gene flow between the populations may have been unequal during the times of possible intermixture, and, as we shall see in Chapter 8, the present rates of admixture may well be increasing in many parts of the world. Roberts (1955) found that for 5 characters: R^0 and R^1 of the Rhesus blood group, PTC tasting, Jk^b of the Kidd blood group, and S of the MNS blood group system, the modal value of gene flow rate occurring at that time was between 0·02 and 0·025 per generation. In southern Brazil, where there has been a considerably greater degree of intermixture between Whites and Negroes, Salzano, da Rocha, and Tondo (1968) calculated that the mean rate of gene flow per generation was 7 per cent.

This brings us to the other major country in the world in which studies of gene flow have been carried out—Brazil. We have seen that it is misleading to generalize about the rate of admixture between Whites and Negroes over the whole of the U.S.A. and there are similar variations of genetic admixture within Brazil. The magnitude of the differences in gene flow is probably greater in Brazil than in the U.S.A., partly because in some areas, as we shall see, there exists almost complete random mating between the races. Saldanha (1957) showed that this variation in gene flow is associated both with the particular tribe of Negroes imported into different areas of Brazil and with the particular Europeans present in any area. The figures are higher, for instance, for those states which have a Bantu ancestry (Sao Paulo and Rio de Janeiro) than for the state in which they have a Sudanese ancestry (Bahia).

The modal values Saldanha found are:

Place	*White Admixture*	*Gene flow* (M)
Bahia	c. 35 per cent	0·035–0·040
Sao Paulo, Rio de Janeiro	c. 55 per cent	0·060–0·065
Brazil	c. 40 per cent	0·045–0·050

It is also stated that Portuguese were more willing to marry Africans than were other European men during the early stages of settlement in Brazil.

There are in fact three races contributing to admixture in Brazil, although Saldanha did not take into account any contribution from the Indians. Roberts

Table 5.4.—A MIGRATION MATRIX BETWEEN THREE NILOTIC PEOPLES OF SUDAN. (*From* Roberts and Hiorns, 1962.)

RECIPIENT POPULATION	DONATING POPULATION		
	Nuer	Dinka	Shilluk
Nuer	0·985	0·0125	0·0025
Dinka	0·0138	0·9775	0·0087
Shilluk	0	0·0098	0·9902

and Hiorns (1962) have shown how the situation of three populations exchanging genes can be analysed by means of the application of migration matrices. We can briefly consider the model they develop by following through their analysis of genetic admixture between three groups of northern Nilotes in southern Sudan—the Nuer, Dinka, and Shilluk.

When mixed marriages are contracted among these peoples the offspring assume the ethnic identity of the village in which they are brought up. Roberts and Hiorns found that in 288 marriages in Dinka villages, 8 spouses were Nuer and 5 were Shilluk, giving admixture rates of 0·0139 and 0·0087 respectively. In Shilluk villages 5 spouses were Dinka and none were Nuer out of 225 marriages, giving admixture rates of 0·0098 and 0. Finally among 200 Nuer marriages

1 was with a Shilluk and 5 were with Dinka, giving admixture rates of 0·0025 and 0·0125. These admixture rates can then be set out as a matrix (*Table* 5.4).

If q_0 and q_1 are vectors of gene frequencies of 3 or more populations at generations 0 and 1 respectively and M the migration matrix then the vector q_1 is given by

$$q_1 = Mq_0$$

and after n generations the gene frequencies of the populations will be

$$q_n = M^n q_0$$

where M^n means the matrix multiplied by itself n times.

The present frequencies of the blood group M gene are 0·5750, 0·567, and 0·5047 in Nuer, Dinka, and Shilluk respectively. We therefore have

$$q_0 \begin{bmatrix} 0{\cdot}5750 \\ 0{\cdot}5670 \\ 0{\cdot}5047 \end{bmatrix} \text{ and } M \begin{bmatrix} 0{\cdot}9850 & 0{\cdot}0125 & 0{\cdot}0025 \\ 0{\cdot}0138 & 0{\cdot}9775 & 0{\cdot}0087 \\ 0 & 0{\cdot}0098 & 0{\cdot}9902 \end{bmatrix}$$

and q_n can be determined by straightforward means. At equilibrium the gene frequencies are expected to be identical and equal to 0·05464, although the time before this equilibrium will be reached is likely to be considerable (*Fig*. 5.5).

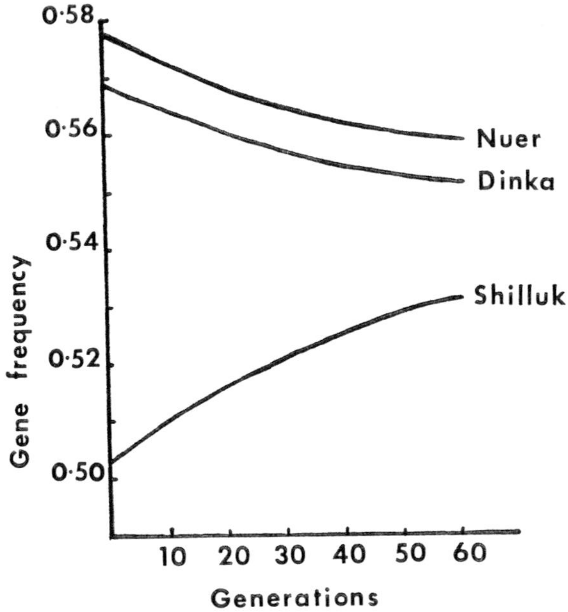

Fig. 5.5.—The predicted approach to equilibrium through intermixture of the gene for blood group M in three tribes of the Northern Nilotes. (*From* Roberts and Hiorns, 1962.)

The same techniques can be used to determine what the gene frequencies were likely to have been in previous generations, and this is obviously often of great value in determining how different gene frequencies have been in the past from what they are today, and therefore, ultimately, how likely it is that the peoples under consideration derive from a common original race.

Roberts and Hiorns (1962) consider this for these three Nilotic groups and extend their analysis back 20 generations in the light of the folklore evidence that the Shilluk entered their present territory about A.D. 1500. They examined 4 alleles at which the Shilluk are today most divergent—M and S of the MNS blood group, d of the Rh system, and B of the ABO group. Their results are shown in *Table* 5.5.

Table 5.5.—RETROSPECTIVE ANALYSIS OF GENE FREQUENCIES IN THREE NILOTIC TRIBES FROM SUDAN (*From* Roberts and Hiorns, 1962.)

	q_0 (*c.* A.D. 1500)				q_n			
	M	S	d	B	M	S	d	B
Nuer	0·5803	0·2559	0·1766	0·1324	0·5750	0·2254	0·1782	0·1270
Dinka	0·5786	0·1197	0·1566	0·1158	0·5670	0·1505	0·1759	0·1155
Shilluk	0·4900	0·1501	0·2709	0·0894	0·5047	0·1477	0·2523	0·0941

From these results the overall effects of intermixture over the last 20 generations have been quite slight. These results also show the importance of the magnitude of the initial distance between gene frequencies for the results of intermixture. For the B and M genes the Dinka were initially intermediate between the other two. Exactly the same population intermixture has moved the Dinka away from the Nuer in M frequency, but has moved the Nuer closer to the Dinka in B frequency. *Fig.* 5.6 shows the changes in the S gene in the three populations, from which it is clear that the positions of populations relative to each other with respect to the frequency of a particular gene are not necessarily constant during the approach to an equilibrium value. Similar techniques were used to confirm beyond any doubt the conclusion of Glass (1955) that the North American Indian has contributed very little to the gene pool of the Negro in the U.S.A.

Many anthropologists, particularly Herskovitz (1928), have asserted that the U.S. Negro has a considerable Indian admixture and this has been generally believed. Although in the past it is possible that there have been quite a number of matings between Indians and Negroes, particularly between runaway slaves seeking refuge among tribal Indians, this fact in itself would not necessarily be apparent today as a high level of admixture. In order for genetic admixture to reach a significant and detectable level, one of three conditions must be fulfilled. Either the hybrids must possess higher fertility than the parental population, or the F_1 hybrids must at some time have formed a significant proportion of the total recipient population; or, what almost amounts to the same thing, hybridization must have continued for a fairly long period of time.

One F_1 hybrid, or heterozygote, will transmit any one allele to only 50 per cent of his progeny on average, and in fact Fisher (1930) long ago showed that most single genes in a population which arise by mutation and which have no selective advantage are inevitably bound for ultimate loss. The length of time which elapses before loss is dependent on family size and the effective population size. In fact if a new allele introduced either by mutation or migration is represented

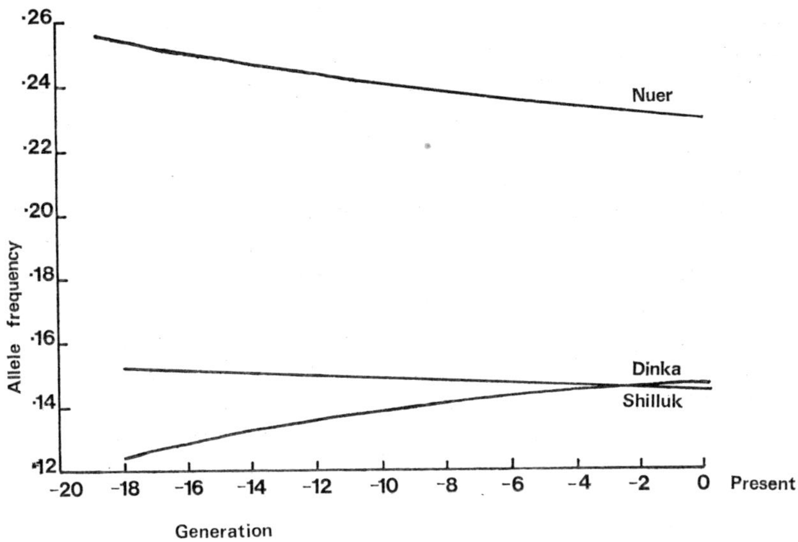

Fig. 5.6.—The changes in frequency of the S gene in Nuer, Dinka, and Shilluk, three tribes of the Northern Nilotes, showing their changing relative positions during the approach to equilibrium. (*From* Roberts and Hiorns, 1962.)

only once in a population of actual size N and effective size Ne, the average number of generations until fixation of such a gene is approximately equal to $4Ne$, whereas the average number of generations until loss is about

$$2\left(\frac{Ne}{N}\right)\log_e(2N), \qquad \text{(Crow and Kimura, 1970.)}$$

In a population of $N = 10^4$ with Ne $0{\cdot}8 \times 10^4$, time to fixation is 32,000 generations, whereas time to loss is 15·9 generations. Most new genes in a population are therefore lost very rapidly, a few increase in frequency very slowly.

All this means that it does not matter how many descendants can be traced back to initial racial crosses, the genes which were initially contributed will not increase proportionately from the time mixing occurred unless they are subjected to favourable selective pressures. Consequently this initial proportion in any racial group must either have been quite large to be detectable by our as yet rather insensitive methods, or the selective advantages of the new genes must be quite marked.

5.3. Racial Amalgamation

It is of course a fundamental axiom of genetics as we presently understand it that the units of inheritance are passed on from generation to generation in an unchanged manner, unaffected by the zygote in which they were active and in which they were replicated. In whatever way they interact with the other genes in the organism in which they are found, they segregate unaffected into the sex cells of diploid organisms, and the only ways in which the units of inheritance can have their activities altered for future generations are by the rather rare processes either of mutation or unequal crossing over within a gene. It is because this is so that we can treat gene frequencies numerically and describe in statistical terms their alteration from one generation to another. It is therefore also the case that we can treat racially mixed populations as simple genetic mixtures of genes and make predictions on this basis as to their current and future genotypic and phenotypic constitutions. We have already considered in Chapter 4 how the biology of racial hybrids might be affected by interactions between genes. We can now consider the straightforward arithmetic consequences of gene mixture in terms of the phenomena of dominance and recessiveness, major genes and minor genes, and continuous and discontinuously varying characters.

5.3.a. *The Hardy-Weinberg Law*

Suppose in a human population under consideration there is one gene which is varying, being represented by two alleles, A and a. Let the proportion of AA individuals in the population be p and the proportion of aa individuals be q. Then $p + q = 1$ and we have:

Marriages		*Offspring*		
Type	Frequency			
AA × AA	p^2	p^2 AA		
AA × aa	$2pq$		$2pq$ Aa	
aa × aa	q^2			q^2 aa

Sum of All Marriages

$$p^2 + 2pq + q^2 = (p + q)^2 = 1^2 = 1$$

Sum of All Offspring

$$p^2\text{AA} + 2pq\ \text{Aa} + q^2\ \text{aa}$$

Thus the first generation produced by random marriage of the two original populations contains the three genotypes AA, Aa, and aa in the proportions $p^2:2pq:q^2$.

Random mating among the individuals of the offspring generation gives:

Marriages		*Offspring*
Type	Frequency	
AA × AA	$p^2\,p^2$	p^4 AA
AA × Aa	$p2^2\;2pq$	$2p^3q$AA + $2p^3q$Aa
AA × aa	$2p^2p^2$	$2p^2q^2$ Aa
Aa × Aa	$2pq\;2pq$	p^2q^2AA + $2p^2q^2$ Aa + p^2q^2aa
Aa × aa	$2\;2pq\;q^2$	$2pq^3$ Aa + $2pq^3$aa
aa × aa	$q^2\,q^2$	q^4aa

$$\begin{aligned} \text{AA} &= p^4 + 2p^3q + p^2q^2 = p^2(p^2 + 2pq + q^2) \\ &= p^2(p + q)^2 \qquad\quad = p^2 \end{aligned}$$

$$\begin{aligned} \text{Aa} &= 2p^3q + 4p^2q^2 + 2pq^3 \\ &= 2pq(p^2 + 2pq + q^2) \\ &= 2pq \end{aligned}$$

$$\begin{aligned} \text{aa} &= p^2q^2 + 2pq^3 + q^4 \\ &= q^2(p^2 + 2pq + q^2) \\ &= q^2 \end{aligned}$$

Thus the proportions of the 3 genotypes have remained the same and clearly will do so for all future generations. A population which is random mating and in which the three genotypes occur in the proportions $p^2:2pq:q^2$ is said to be in equilibrium. One generation of random mating is sufficient to return to this proportion for any one genetic locus, even from a situation of complete negative or positive assortative mating such as the situation outlined above. The law expressing these findings is the Hardy-Weinberg law. It implies, therefore, that in cases of racial mixing, after only one generation of random mating among F_1 hybrids, the racially mixed individuals will show the Hardy-Weinberg proportions for any particular gene at which more than one allele is represented. In fact departure from Hardy-Weinberg expectation is often used as an indication that the population concerned is not breeding at random and that sub-populations or isolates are in the process of formation.

If, at any one genetic locus with two alleles present in a population, one of these alleles is present in very low frequency, then the proportion of homozygous individuals will be very small indeed. For instance if $p = 0{\cdot}9$ and $q = 0{\cdot}1$, the genotypic ratios are

$$p^2 = 0{\cdot}81 : 2pq = 0{\cdot}18 : q^2 = 0{\cdot}01.$$

As the frequency of any one allele decreases the proportion of heterozygotes also falls and the frequency of homozygote individuals falls much more rapidly. In cases of completely recessive genes this means the frequency of those individuals exhibiting the phenotypic condition associated with the gene (homozygotes) markedly decline, and since the carrier (heterozygote) frequency decreases more slowly the ratio between the two increases. Thus, although a rare recessive condition in man, such as alkaptonuria, occurs only in approximately 1 in 10^6 individuals, the gene responsible is carried heterozygously in 1 in 500 individuals—about 2,000 times as many. The relationship between gene frequency, heterozygotes, and homozygotes is given in *Table* 5.6.

Table 5.6.—FREQUENCIES OF AFFECTED AND CARRIER INDIVIDUALS UNDER RANDOM MATING IN VARIOUS CASES OF SIMPLE SINGLE FACTOR RECESSIVE INHERITANCE

GENE FREQUENCY (q)	FREQUENCY OF AFFECTED (q^2)	FREQUENCY OF CARRIERS ($2pq$)	RATIO OF CARRIER TO AFFECTED ($2pq/q^2 = 2p/q$)
0·32	1 in 10	1 in 2·3	4·3 : 1
0·032	1 in 100	1 in 5·6	18 : 1
0·0032	1 in 1000	1 in 16	61 : 1
0·00032	1 in 10,000	1 in 51	198 : 1
0·000032	1 in 100,000	1 in 159	630 : 1
0·0000032	1 in 1,000,000	1 in 501	1998 : 1

Two of the assumptions on which the Hardy-Weinberg law depends are, firstly, effectively infinite population size, and, secondly, random mating among all members of the population. Populations are never in fact infinite and mating, as we have seen in Chapter 2, is rarely random in man. We must therefore consider both of these departures from ideal conditions.

Using the Hardy-Weinberg law we can consider the sort of genetic consequences which result from the division of a very large population into a number of small groups or subpopulations, remembering that in man these subpopulations may well be racially defined.

In any subpopulation of N breeding individuals $2N$ gametes, chosen to a greater or lesser extent at random, unite to form the next generation of N individuals. If this subpopulation is segregating for 2 alleles at a locus, A and a, with frequencies p and q respectively, then variance in the value of q will arise due to sampling and the sampling variance will be that of the binomial $(p + q)^{2N}$. In other words this is the gene frequency in the offspring population of individuals. *Table* 5.7 shows the probability distribution of q in $N = 50$ offspring from a parental population with $p = q = 0{\cdot}5$.

This distribution can be taken as showing the probability distribution for one segregating locus, or the proportion of several varying loci within the specified range; or, if we are considering the subdivision of the species into populations of equal size, then this *Table* 5.7 shows the proportion of such groups with a value of q within the specified range.

Table 5.7.—THE TERMS OF THE BINOMIAL DISTRIBUTION WHEN $p = 0{\cdot}5$, $N = 50$

	q							
	<0·35	0·35–0·40	0·40–0·45	0·45–0·50	0·50–0·55	0·55–0·60	0·60–0·65	>0·65
Prob.	0·002	0·021	0·136	0·341	0·341	0·136	0·021	0·002

The parental gene frequency 0·5 spreads out over a wide range in the next generation and, if this process of spreading continues for a number of generations, the distribution of q will become practically uniform. As the frequencies of q approach and reach the ultimate values of 0 and 1 in these subdivided populations, they become either lost or fixed, and there can then be no further change in gene frequencies. The theoretical rate of fixation and loss of genes each occur at the rate of $\frac{1}{4}N$ per generation and the total overall rate of decay of heterozygosity is $\frac{1}{2}N$. This process is shown diagrammatically in *Fig.* 5.7.

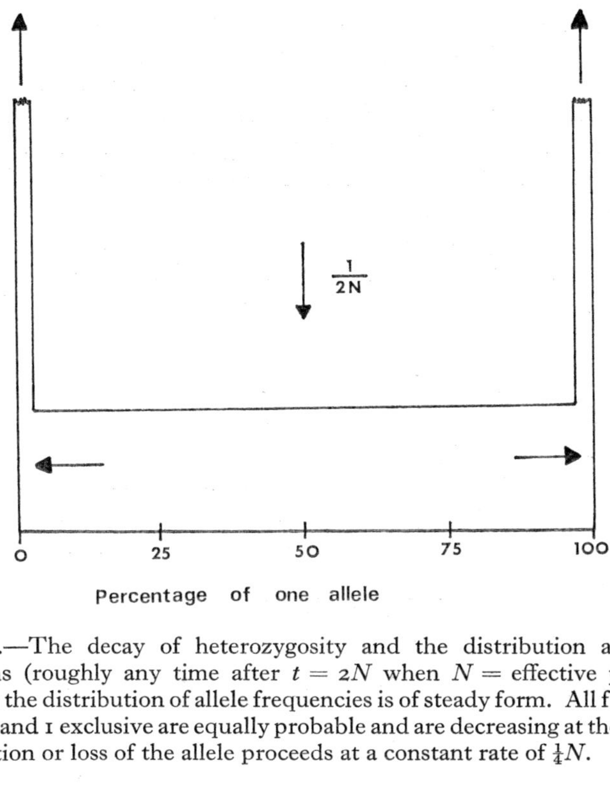

Fig. 5.7.—The decay of heterozygosity and the distribution after many generations (roughly any time after $t = 2N$ when N = effective population size) when the distribution of allele frequencies is of steady form. All frequencies between 0 and 1 exclusive are equally probable and are decreasing at the same rate $\frac{1}{2}N$. Fixation or loss of the allele proceeds at a constant rate of $\frac{1}{4}N$.

Now changes in gene frequency lead to changes in genotype frequency, so the genotype frequencies in small populations follow the changes of gene frequency resulting from the dispersive process we have just discussed. If mating is random within each of these subdivided populations, then the genotype frequencies within any one subpopulation are the Hardy-Weinberg frequencies appropriate to the gene frequencies in the previous generation of that subpopulation. However, genotype frequencies over all subpopulations taken together will not be in Hardy-Weinberg proportions appropriate to the overall gene frequency. Ultimately, if this process is allowed to continue indefinitely and, rather improbably, we suppose selection to be inoperative the whole time, any one allele will become fixed in about half the populations and lost in the other half. Within the whole species the frequency of the two alleles will still be $p = q = 0{\cdot}5$, but there will be a complete absence of heterozygotes.

If these subpopulations become limited in size, then matings between related individuals will become rather more common and therefore, since these are much more likely to carry identical alleles than non-related individuals (*see* s. *2.1*), the number of heterozygotes is likely to be significantly reduced on this account. It can be shown that this decline in heterozygosity will fall by approximately $\frac{1}{2}N$ of its previous value even for moderately large N. The important conclusion is that any finite group of randomly mating diploid individuals will, in the absence of mutations, migration, and opposing selective forces, ultimately reach complete homozygosity. Of course these conditions are unlikely ever to be fully realized in practice, but it does suggest that subdivided populations will always be somewhat deficient in heterozygotes compared with expectations under Hardy-Weinberg equilibria.

The reversal of this process of subdivision into small populations and inbreeding is the breakdown of isolates. From our discussion in this section and in Chapter 2 we can predict the following events accompanying the breakdown of isolates in human populations:

1. A decrease of random inbreeding rates.
2. A decrease of variance of gene frequency.
3. A decrease of homozygote frequencies.
4. An increase of heterozygote frequencies for both rare and common genes, allowing new genotypic combinations of both monogenic and polygenic systems. The effect of this will be particularly marked in heterotic or overdominant loci.
5. A decrease in the probability of random drift as a consequence of the increased population size.

We can now examine some of these processes in more detail.

5.3.b. The Breakdown of Isolates

Thalassaemia is a haemoglobin deficiency of human blood which, in its most severe expression, is caused by homozygosity for the recessive allele Th^T. Neel and Valentine (1947) recorded the occurrence of this disease in a series of consecutive births in one American city between 1928 and 1942. Out of about 100,000 births in this period, 11 babies with the disease were born. The frequency of q, the Th^T allele, is given by $q^2 = 11/100{,}000$, from which it can be calculated that $2pq$ was about 1 in 50. This was therefore assumed to be the frequency of carriers.

The world distribution of thalassaemia is very restricted, however, being largely confined to people living around the shores of the Mediterranean. It was found that the 11 cases of thalassaemia recorded all occurred among people of Italian descent. The total number of Italian births during 1928–42 was estimated at about 26,000 and therefore the frequency of thalassaemia among them was of the order of 1 in 2,400, which in turn suggests that the frequency of carriers ($2pq$) was about 1 in 25.

We therefore have two very different estimates of the frequency of thalassaemia in this city, the first one based on false premises. In fact the overall frequency of carriers can be calculated when we know what proportion of the total population are Italian, and on the assumption that non-Italians do not carry the disease at all. It is (1 in 25) + (0 in 75) = 1 in 100. This is the

Table 5.8.—THE CHANGING PROPORTION OF HETEROZYGOUS CARRIERS AND THE DECREASING FREQUENCIES OF AFFECTED INDIVIDUALS AFTER MIXING OF A POPULATION CONTAINING DIFFERENT FREQUENCIES OF A RECESSIVE GENE WITH A POPULATION WITHOUT THAT GENE.

POPULATION A			MIXES WITH POPULATION B TO FORM VARIOUS FRACTIONS OF HYBRID POPULATION					
q (gene frequency of recessive gene)	$2pq$ (hetero-zygous carriers)	q^2 (homo-zygous individuals)	50 per cent		25 per cent		10 per cent	
			$2pq$	q^2	$2pq$	q^2	$2pq$	q^2
90 per cent	0·18	0·81	0·49	0·20	0·35	0·05	0·16	0·008
80 per cent	0·32	0·64	0·48	0·16	0·32	0·04	0·15	0·006
70 per cent	0·42	0·49	0·45	0·12	0·29	0·03	0·13	0·005
60 per cent	0·48	0·36	0·42	0·09	0·25	0·02	0·11	0·004
50 per cent	0·50	0·25	0·37	0·06	0·22	0·016	0·10	0·002
40 per cent	0·48	0·16	0·32	0·04	0·18	0·01	0·08	0·002
30 per cent	0·42	0·09	0·25	0·02	0·14	0·006	0·06	0·0009
20 per cent	0·32	0·04	0·18	0·01	0·09	0·002	0·04	0·0004
10 per cent	0·18	0·01	0·09	0·002	0·04	0·0006	0·02	0·0001

frequency which will be reached when the Italian isolate is broken down by intermarriage with the rest of the population.

This example illustrates one of the most important points with regard to race mixing—namely, that when the gene frequencies between the two populations involved differ, then on amalgamation considerable change in genotype frequencies will occur. Gene frequencies in such a mixed population will always be, at least initially, the average of the two parental populations, but the genotype frequencies will always show a considerable increase in the proportion of heterozygotes, and therefore a decrease in the number of affected individuals, for traits controlled by fully recessive genes. *Table* 5.8 shows some calculations illustrative of this situation. One population, A, with varying proportions of a recessive gene is assumed to amalgamate with another population, B, which does not possess this recessive gene. It is assumed that A will form 50 per cent, 40 per cent, and lower frequencies of the hybrid population, and the table shows the decreasing proportion of heterozygous carriers and the much more rapidly decreasing frequency of affected homozygous individuals. These genetic consequences of racial amalgamation are illustrated in *Fig.* 5.8.

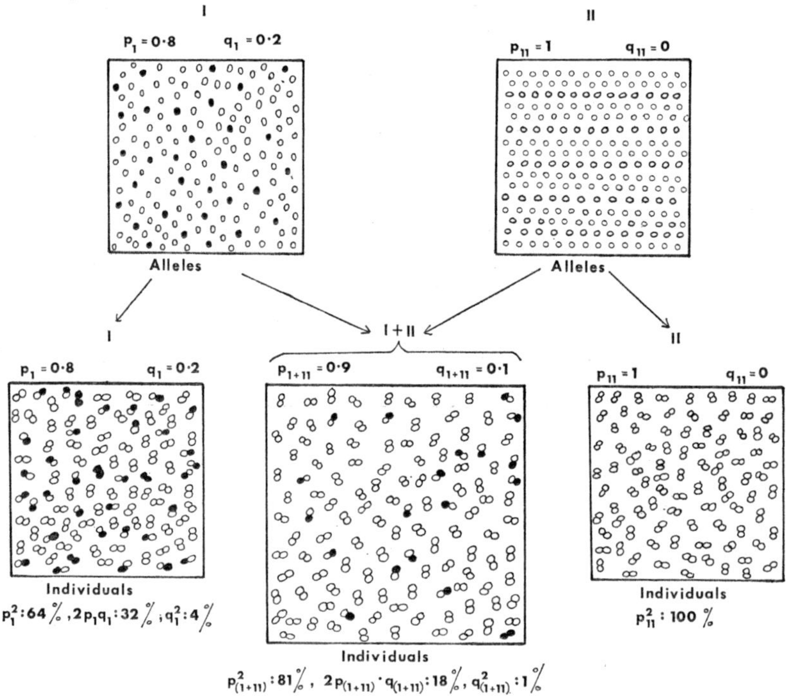

Fig. 5.8.—The genetic consequences of the amalgamation of races or the breakdown of isolates. On the left and right are two races A and B with different allele frequencies $q_I = 0{\cdot}2$ and $q_{II} = 0$ which produce two populations with 4 per cent and 0 per cent homozygous recessive individuals respectively. In the middle a panmictic population derived from I and II produces 1 per cent homozygous recessives. (*From Principles of Human Genetics* (Curt Stern) 2nd. ed., 1960. Copyright, London: W. H. Freeman & Co.)

Earlier in this chapter we considered the Hardy-Weinberg equilibrium and departures from it because of drift and non-random mating in small populations. Each of these processes gave rise to a greater number of homozygotes than expected. Therefore in so far as racial amalgamation is concerned it is, if it is complete, a simple restoration of large population size and random mating. An immediate and very marked reduction of the number of homozygotes therefore occurs, thus reversing, in one generation, the trends towards greater homozygosity that we have previously discussed. As we previously found, so far as any one genetic locus is concerned, one generation of random mating is sufficient to restore Hardy-Weinberg equilibrium. Racial amalgamation is rarely completed in one generation and mating usually remains non-random for considerable periods of time (*see* Chapter 8). The idealized situations analysed in *Table* 5.8 and *Fig.* 5.8 are therefore rarely, if ever, attained. Nevertheless racial amalgamation will always be a move toward greater heterozygosity and therefore less expressed genetic damage. In Chapter 6 we shall find that the races differ in the frequencies of many particular genetic traits and defects, although their overall levels are about the same. Those traits controlled by dominant genes will show an intermediate frequency in racial hybrids; those controlled by recessive genes will show some reduction in hybrids below the interpopulation mean.

One rather special case which can be discussed in a little more detail is the Rhesus incompatibility. This, as we mentioned in Chapter 2, results from the formation of an antibody of Rh^- in women bearing children by Rh^+ fathers. It is therefore the presence of both the alleles controlling these specificities which is necessary for the occurrence of erythroblastosis foetalis. Therefore Europeans, with frequencies of Rh^- varying around 15 per cent, and Negroes, with about 10 per cent Rh^- genes, show characteristic frequencies of erythroblastosis foetalis. On the other hand Mongoloids appear to be entirely Rh^+ and therefore the occurrence of this trait is virtually unknown among them.

If racial amalgamation occurs between Caucasoids and Mongoloids no erythroblastosis foetalis will occur among first generation marriages between Mongoloid women and Caucasoid men, because all Chinese women are Rh^+. Marriages between Mongoloid men and Caucasoid women, however, will show some incidence of the disease among their offspring, since some of the women will be Rh^- and they will be marrying Rh^+ men. This frequency will, in fact, be higher than that found in the unmixed Caucasoid population, although, since it is only confined to half the matings, the net frequency is not higher. Among later generations in this now mixed population, the relative frequency of erythroblastosis foetalis will be lower than in the first generation and lower than in the unmixed Caucasian population. This is primarily, as we have discussed, because of a reduction in the relative frequency of the Rh^- allele.

If, as is usually the case, racial mixing is a gradual process with a steady introduction of genes from one race to another, we can see that this will result in an introduction into the Mongoloid population of a childhood condition which was previously unknown. It is in this respect that one can refer to detrimental effects of racial mixing—in this case detrimental to the Mongoloids. Considered on the larger scale, though, the incidence of the trait will have declined and therefore the outcome for the new mixed population can be considered favourable, relative to the Caucasoids.

In our discussion of the Hardy-Weinberg law and of its use in the analysis of hybridization, we have concentrated on single genes with segregating alleles.

These, we found, will attain equilibrium after one generation of panmixia, a situation illustrated in *Fig.* 5.9. However, the attainment of equilibrium after

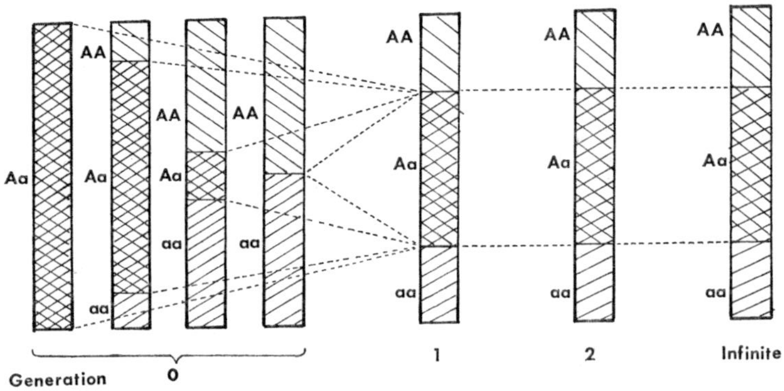

Fig. 5.9.—From whatever starting-point with regard to either gene frequencies or genotypes, equilibrium is attained in one generation for any one gene segregating two alleles. (*From Principles of Human Genetics* (Curt Stern) 2nd. ed., 1960. Copyright, London: W. H. Freeman & Co.)

random mating is not normally accomplished in one generation by populations segregating at more than one locus. And since, of course, races usually differ at many loci we must consider the implications of this more closely.

A population of genetic constitution AABB will produce gametes AB and a population of aabb will produce gametes ab. A population composed of these two in equal numbers and mating at random will produce offspring of type AABB, AaBb, and aabb in the ratio 1:2:1. All four possible types of gametes AB, Ab, aB, and ab will be produced by this first panmictic generation and they will give a second generation in which all nine possible genotypes occur, but not at random frequencies. As illustrated in *Fig.* 5.10 there will be a gradual

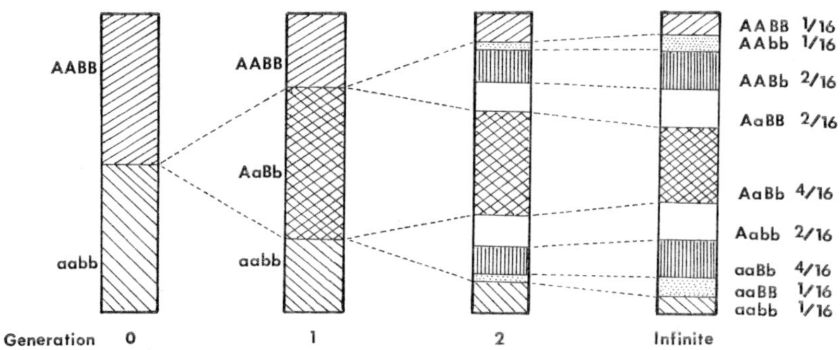

Fig. 5.10.—The approach to equilibrium under random mating of a population consisting in generation 0 of equal numbers of AABB and aabb individuals. (*From Principles of Human Genetics* (Curt Stern) 2nd. ed., 1960. Copyright, London: W. H. Freeman &Co.)

approach to equilibrium. The rate of approach to equilibrium for 2 genes segregating 2 alleles is comparatively rapid, provided the genes concerned are not closely linked. With a larger number of segregating genes, and/or tighter linkage between them, the attainment of equilibrium is considerably delayed.

This phenomenon can often be useful in studying populations for signs of racial intermixture. In other words variations at two or three or more characters are considered together and particularly frequent or infrequent associations are looked for, as indications that equilibrium between the genes concerned has not been reached. Alternatively, when the population is known to be hybrid and some traits which are non-randomly associated in the unmixed groups are randomly associated, it can be assumed that not enough generations have elapsed for equilibrium to be reached. This number of sufficient generations is often amenable to theoretical prediction, in turn providing an estimation of the time of mixing. Rife (1954a, b) studied Sudanese and American Negroes both of whom are Negroid/Caucasoid hybrids. He found among these peoples a non-random association between the degree of pigmentation and specific dermatoglyphic patterns. Since both of these traits showed a random association with the ability to taste PTC, Rife suggested that the genes involved in pigmentation and palm pattern are linked.

5.3.c. *The Variability of Racial Hybrids*

In a randomly breeding population the genotypic frequencies are, as we have seen,

$$p^2 + 2pq + q^2.$$

To find the mean and variance of a population segregating for alleles controlling a continuously varying character, we can set one of the homozygous classes equal to 0 and the other equal to 1. The heterozygotes can then be assigned a value of h from 0 to 0·5 according to the degree of dominance, 0 representing complete dominance, 0·5 exact intermediacy.

For the mean we then have

$$p^2(1) \qquad \text{and} \qquad 2pqh + q^2(0)$$

which gives

$$p^2 + 2pqh.$$

The variance, the mean squared deviation from the mean, is given by

$$p^2(1^2) + 2pqh^2 + q^2(0^2) - (p^2 + 2pqh)^2$$

which reduces to

$$\sigma^2 = p^2 + 2pqh^2 - (p^2 + 2pqh)^2.$$

The great H. J. Muller, who suggested this approach, then used this formula to construct a number of tables showing the means and variances of populations having values of p from 0 to 1 with various values of h (Muller, 1936).

For purposes of illustration an extract of the table so constructed is shown in *Table* 5.9. This table can then be used to construct further tables showing the

Table 5.9.—VARIANCE CAUSED BY PRESENCE IN POPULATION OF RECESSIVE AND PARTLY RECESSIVE GENES OF DIFFERENT FREQUENCY
(*Extracted from* Muller, 1936.)

p	$h = 0.0$	$h = 0.2$	$h = 0.3$	$h = 0.5$
0.0	0.0000	0.0000	0.0000	0.0000
0.2	0.0384	0.0419	0.0503	0.0800
0.4	0.1344	0.1137	0.1108	0.1200
0.5	0.1875	0.1475	0.1350	0.1250
0.6	0.2304	0.1713	0.1492	0.1200
0.8	0.2304	0.1572	0.1271	0.0800
1.0	0.0000	0.0000	0.0000	0.0000

values of the variance of a mixed race as compared with the average variance of the original races, in the case of the different possible values of p and h.

We are concerned with the increase or decrease of variability of a mixed race relative to the original races. Muller suggested that the variance the mixed race would have can be calculated using the straightforward relationship:

$$p_m = \frac{p_1 + p_2}{2}.$$

From this value is then subtracted the value representing the average variance of the original races. The resulting value is then divided by the average variance of the original races to give a value r: *the relative increase of variance of the hybrid* caused by the gene pair in question. The other important information needed is how much variance the original races had, on the average, manifested as a result of their heterozygosity for this particular locus, and the extent to which the mean of one original race had differed from the mean of the other. For the selected values of *Table* 5.9, these calculated relationships are shown in *Tables* 5.10(a), (b), and (c). Within each cell in these tables is shown, at the top, r the relative increase in variance of the mixed race resulting from the crossing of the races having these two respective values of p. The middle value is the average variance of the original races caused by this locus, and the lowest figure is the difference between the means of the original races. Only half of each table is shown, the omitted half being identical except for reciprocal values of $p1$ and $p2$. The diagonal line bisecting these tables is called by Muller 'the line of zero differences' whereas the regions farthest from this line are the regions of maximum differences.

Table 5.10(a) shows the values under the condition of complete dominance ($h = 0$). In about one-third of the different possible combinations there is a

Table 5.10.—THE EFFECTS OF GENES OF VARIOUS FREQUENCIES (p_1 AND p_2) AND DEGREES OF DOMINANCE IN RACE CROSSES

(The upper figure is the increase in variance relative to the average variance of the original races, *r*. The lower figure shows the difference in the means of the original races.) (*Extracted from* Muller, 1936.)

	(a) h = 0·0						(b) h = 0·2						(c) h = 0·5					
p_1 \ p_2	0·1	0·3	0·5	0·7	0·9	1·0	0·1	0·3	0·5	0·7	0·9	1·0	0·1	0·3	0·5	0·7	0·9	1·0
	−0·496	−0·468	−0·375	−0·141	1·097	∞	−0·227	−0·285	−0·206	0·074	1·592	∞	0·059	0·215	0·500	1·166	4·500	∞
0	0·01	0·090	0·250	0·490	0·810	1·0	0·046	0·174	0·350	0·574	0·846	1·00	0·100	0·300	0·500	0·700	0·900	1·0
	0·0	−0·165	−0·169	0·033	1·288	41·2	0·0	−0·083	−0·058	0·180	1·536	20·50	0·0	0·067	0·235	0·600	1·779	4·50
0·1	0·0	0·080	0·240	0·480	0·800	0·990	0·0	0·128	0·304	0·528	0·800	0·954	0·0	0·200	0·400	0·600	0·800	0·900
		0·0	−0·03	0·130	0·955	4·95		0·0	0·012	0·162	0·925	3·62		0·0	0·043	0·190	0·600	1·166
0·3		0·0	0·160	0·400	0·720	0·910		0·0	0·176	0·400	0·672	0·826		0·0	0·200	0·400	0·600	0·700
			0·0	0·053	0·465	1·625			0·0	0·053	0·425	1·318			0·0	0·043	0·235	0·500
0·5			0·0	0·240	0·560	0·750			0·0	0·224	0·496	0·650			0·0	0·200	0·400	0·500
				0·0	0·141	0·605				0·0	0·128	0·516				0·0	0·067	0·216
0·7				0·0	0·320	0·510				0·0	0·272	0·426				0·0	0·200	0·300
					0·0	0·143					0·0	0·129					0·0	0·058
0·9					0·0	0·190					0·0	0·154					0·0	0·100
						0·0						0·0						0·0
1·0						0·0						0·0						0·0

negative value of r. The variability of the hybrid population is actually lower than the average variability of the original populations. The limiting case is $r = -0{\cdot}5$; in other words the variance is halved in the hybrid. This reduction in variability occurs where the frequency of the recessive gene is low in both races and where it is, at the same time, rather lower in one of the races than the others. It must be remembered, though, that when the frequencies of p are low in both of the original races the absolute decrease of variability in the hybrid could not possibly be very great.

When the gene frequencies are similar in the two races, then the variability is about the same in the hybrid population. If the gene frequencies are identical, then so too, of course, is the variability in the hybrid. The maximum increases in variability of the hybrid population occur when the two original populations are homozygous, or nearly so, for opposite alleles of the locus concerned.

Turning to *Table* 5.10(c) which presents data on the opposite extreme, i.e., when there is no dominance at all, we see that there are no negative values of r. In other words there are no conditions in which hybrid variability would decrease. As before, however, the values approach zero as the gene frequencies in the two parental populations approach one another. When the gene frequencies are very different in the two races then variability shows a marked increase in the hybrids. There is one important difference in this case compared with that of complete dominance. In the first case if one or more genes controlling a particular character were at very different frequencies in two populations, then the hybrids would be very variable. However, if other loci also controlling this character were present at low frequencies they would serve to compensate for this, since they tend to cause a reduction in variability in the hybrids. On the other hand, with no dominance there is never any reduction in variability, and therefore no chance for any compensation. Variability will always be increased.

Intermediate values of dominance show intermediate results. Only in the case of $h = 0{\cdot}1$ or $0{\cdot}2$ is there any appreciable degree of decrease of variability. When $h = 0{\cdot}3$, r never attains a negative value greater than $-0{\cdot}12$, which will not effect much compensation for an increase in variability due to the original races being relatively homozygous for opposite alleles at other loci.

In summary, then, we conclude that it is only in cases of marked dominance (and recessiveness) that an actual decrease in variability can be produced, and then only when the recessive genes causing the variation in one race have low frequencies and are even lower in the other.

The implications, then, are that in those cases where racial crosses produce hybrid populations showing little increase in phenotypic variability, most of the intraracial variation of the original races is caused by genes which are to a greater or lesser extent recessive and which individually, at each locus, are rather rare but considered together must be rather numerous. Secondly, it would appear that most of this variation is due to different loci in different races, even though similar phenotypic variation results. This interpretation is fundamentally similar to that advanced in s. 5.*3*.*b*, in the discussion primarily concerned with single gene effects, where it is shown that traits due to recessive genes are likely to be rarer in hybrid populations than in the parental ones.

Part of the intraracial variance of the parental races is suppressed in the hybrid race, its place being taken by the segregation of interracial differences. The fact that increases or decreases of variation are small in the hybrids as compared with the parental races, shows that the inter-individual differences within a race are of similar magnitude to the interracial differences between the original major races. This latter conclusion suggests that recombination and segregation within hybrid populations are no more likely to be disadvantageous than those arising among intraracial populations.

In Chapters 6, 7 of this book we consider in some detail the expression of particular characters in racial crosses. It is worth our while to present briefly some of the general conclusions which J. C. Trevor (1953) arrived at in his painstaking analysis of metrical characters in racial hybrids. *Table* 5.11 shows

Table 5.11.—COMPARISON OF MEAN UNWEIGHTED STANDARD DEVIATIONS FOR PRESUMED NON-EUROPEAN AND EUROPEAN PARENTAL AND HYBRID SERIES* (*From* Trevor, 1953.)

CHARACTER	MALES			FEMALES		
	Non-European	Hybrid	European	Non-European	Hybrid	European
Stature	5·79(7)	6·33(7)	6·38(4)	5·06(3)	5·41(4)	6·07(3)
Head length	6·02(7)	6·72(7)	6·53(4)	5·16(3)	6·11(3)	5·55(3)
Head breadth	4·74(7)	5·08(7)	5·33(6)	4·85(3)	4·89(4)	4·41(3)
Bizygomatic breadth	4·96(6)	5·53(6)	5·23(6)	4·72(3)	4·86(4)	5·05(3)
Nasal breadth	2·69(6)	3·09(6)	2·67(4)	2·76(3)	2·83(3)	2·48(3)
Cephalic index	3·08(8)	3·14(8)	3·31(5)	3·10(4)	2·90(5)	3·08(4)
Facial index	4·75(4)	5·14(5)	5·85(5)	4·90(3)	4·81(5)	5·11(4)
Nasal index	7·84(5)	7·66(6)	6·83(5)	7·94(3)	7·49(5)	7·12(4)

* The number of series is shown in brackets after each mean.

the unweighted standard deviations for the European, non-European, and hybrid series respectively. In both males and females it is the Europeans who are showing the greatest variability for most characters. In exactly half the cases the hybrid population shows a variability intermediate between European and non-European and in 2/16 comparisons the hybrids are lower than both. In the remaining 6/16 comparisons the hybrids are the most variable. Our conclusion that increased variability is not an inevitable accompaniment of racial hybridization, is therefore confirmed by these data.

Finally we might look at the form of distribution of metrical characters in hybrid populations. If it is argued that extreme variability and/or unfavourable gene combinations typify racial hybrids, then the frequency curve for the character concerned would be expected to be significantly non-normal. This possibility was tested by Trevor by fitting normal curves to the distribution of certain characters and testing the likelihood of conformity to such a curve by the χ^2 test. The results are shown in *Table* 5.12. For the six characters selected there was, in many cases, a clear distinction between the means in the corresponding parental groups. All distributions numbered more than 75 individuals and in only 1 case did the *P* value fall below a formal level of statistical acceptability, 0·046 for head breadth in American Negroes. The high *P* values elsewhere do not prove that the distributions of these metrical characters in hybrid populations are normal, but are good evidence against any important departures from normality.

5.4. An Illustrative Example

One of the best examples to illustrate the points made in this chapter is that of the study of the genetic effects of immigration into a rural community of the state of São Paulo, Brazil, published by P. H. Saldanha (1962b). This particular situation, in fact, hardly involves racial integration at all, since it is primarily one of European migration into an already established mainly European population. This in itself is significant, in that it suggests that the genetic phenomena involved are not uniquely associated with race mixing and that race mixing is not a process distinct from all others involved in ongoing human evolution. As we shall see, the original populations in this particular community are racially mixed and the immigrants, in so far as they can be distinguished at all, can be regarded as members of different *local races* and *micro-races*.

The village chosen by Saldanha is that of Capivari founded in 1832 about 165 km. north-west of São Paulo. Catholic marriage records for the area between 1822 and 1826 show that among 272 individuals married during this period, about 50 per cent were born in the community and 40 per cent in neighbouring villages. At the general census of 1950 the Parish of Capivari had 14,052 inhabitants, of whom 89 per cent were White, 2 per cent Mulatto, and 9 per cent Negro. The immigration into this community is shown in *Table* 5.13.

One of the first general points we made regarding the outcome of the breakdown of isolates is that there would be a decrease of random inbreeding rates. This, as we have seen in Chapter 2, can be measured in man as a declining frequency of cousin marriages and in this part of Brazil, where Catholics constitute over 95 per cent of the population, this figure can be accurately determined.

In *Table* 5.14 are shown frequencies of different types of cousin marriages in this community, and the inbreeding coefficient and isolate size derived from

Table 5.12.—VALUES OF *P* FOR HYBRID SERIES* (*From* Trevor, 1953.)

HYBRID SERIES	CHARACTERS					
	Stature	Head Length	Head Breadth	Bizygomatic Breadth	Cephalic Index	Nasal Index
Male hybrid American Negroes	0·758 (225) (12)	0·764 (254) (12)	0·046 (254) (11)	0·502 (254) (9)	0·775 (254) (9)	—
Male Jamaican Browns	—	—	—	—	0·140 (93) (7)	0·831 (80) (8)
Male half-blood Sioux	0·133 (77) (12)	0·664 (77) (12)	0·833 (77) (8)	0·330 (76) (9)	0·297 (77) (9)	—
Male-Ojibwa Whites	—	—	0·146 (80) (8)	0·199 (80) (7)	—	—
Female Ojibwa-Whites	—	—	0·049 (86) (6)	0·459 (86) (8)	—	—
Female Rehoboth Bastaards	0·150 (84) (8)	—	—	—	0·843 (90) (12)	0·247 (88) (6)
Male Kisar Mestizos	0·532 (79) (6)	—	—	—	0·424 (80) (7)	0·145 (78) (6)
Male Anglo-Indians	0·455 (145) (9)	0·075 (145) (10)	0·365 (145) (15)	— —	0·651 (145) (14)	0·237 (145) (14)

* The figures in brackets denote respectively the size of a sample and the number of groups used in the χ^2 test.

these figures on certain assumptions. The frequencies of cousin marriage are extremely heterogeneous over this 120 years, and *Fig.* 5.11 shows the detailed

Table 5.13.—THE FREQUENCY OF NON-BRAZILIANS MARRYING INTO THE COMMUNITY OF CAPIVARI DURING DIFFERENT PERIODS (*From* Saldanha,. 1926b)

PERIOD	*N*	ITALIAN (per cent)	SPANISH (per cent)	PORTUGUESE (per cent)	OTHER (per cent)	TOTAL (per cent)
1830–35	140	0	0	0	0	0
1860–65	306	0	0	1·96	0	1·96
1890–95	1110	22·70	1·08	1·26	2·17	27·21
1920–25	1262	5·70	2·22	0·87	0·47	9·27
1950–55	1618	0·12	0	0·06	0·25	0·43

Table 5.14.—FREQUENCIES OF DIFFERENT TYPES OF COUSIN MARRIAGE, MEAN COEFFICIENT OF INBREEDING AND ISOLATE SIZE ESTIMATED FROM THE FREQUENCIES OF FIRST COUSINS IN DIFFERENT PERIODS (*From* Saldanha, 1962b.)

PERIOD	*N*	TYPE OF CONSANGUINEOUS MARRIAGE (per cent)								
		½C	1C	1½C	2C	2½C	3C	Total	∝ (× 10⁵)	Isolate size
1830–35	70	1·4	4·3	—	—	—	4·3	10·0	463	558
1860–65	164	—	9·8	2·4	2·4	3·0	9·8	27·4	786	245
1890–95	563	0·9	2·3	0·7	1·6	0·9	1·4	7·8	315	1043
1920–25	631	—	0·8	0·5	—	—	—	1·3	64	3000
1950–55	810	—	0·7	0·5	1·4	—	—	2·6	83	3428

½C Uncles and nieces, aunts and nephews
1C First cousins
1½C First cousins once removed
2C Second cousins
2½C Second cousins once removed
3C Third cousins

temporal trend of the frequencies of first-cousin mating. This shows that there was a continuous uniform decline in the frequencies of cousin mating after 1900, just as immigration into the community became heavier.

Saldanha himself subjects these and similar data to a good deal of further critical analysis, but the straightforward conclusion which can be drawn from them, that inbreeding declines as immigration increases, is upheld.

One result which should follow from this, is a decrease in the frequency of homozygotes. We might find this expressed as a decrease in the frequency of rare hereditary abnormalities. Saldanha investigates this possibility by studying the records of a Public Health Post for children established in the community. The records of a total of 7746 different children examined by a paediatrician at this centre were studied, and since both the mother's and the father's name were recorded, the incidence of congenital malformation among the offspring of both 'pure' and 'mixed' marriages could be determined. On a mixture of

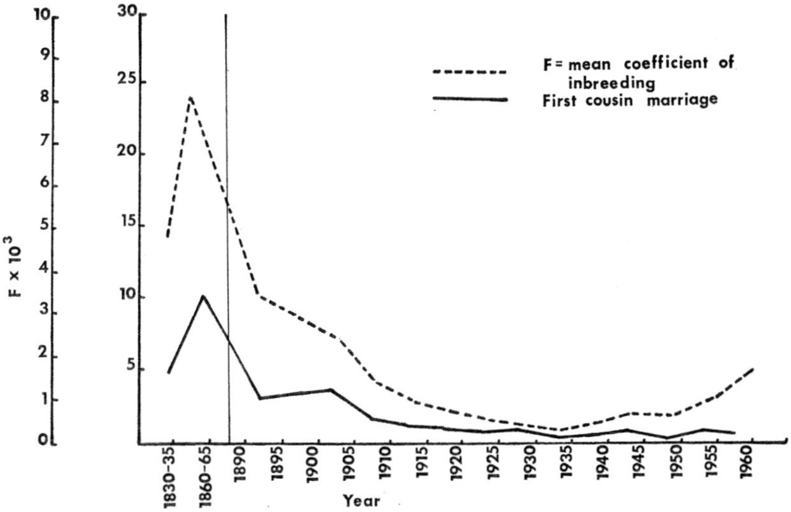

Fig. 5.11.—Changing frequencies of first-cousin mating and of the coefficient of inbreeding between the years 1880 and 1960 in the parish of Capivari, Brazil. (*From* Saldanha, 1962b.)

clinical, genetical, and statistical criteria the following abnormalities were chosen for study:

Hare-lip and cleft palate,
Malformations of limb extremities,
Congenital heart disease,
Mongolism,
Epilepsy,
Inguinal hernia,
Gross mental retardation.

Table 5.15 shows the frequencies of the seven congenital abnormalities among 6742 White individuals classified according to parental origin. A sample of 1024 Negroes was excluded, in this case, because they belong to a distinct population. Saldanha found that the total frequencies of the seven congenital abnormalities among the different groups are significantly heterogeneous. Furthermore, a comparison between the overall frequencies found among the 'pure' groups and that observed in the 'hybrid' group was also significant.

The prediction based on the theory of the breakdown of isolates is again confirmed, in that the frequency of congenital malformations among the hybrids is significantly lower than among the parental populations.

The genetics of human stature is not very well understood and is possibly rather complex. We have already examined in Chapter 4 some possible examples of heterosis manifested in this character, and in Chapter 6 we will examine

Table 5.15.—Incidence (per thousand) of Seven Congenital Abnormalities in the Present Population of the Parish of Capivari, according to the Parental Origin of the Individuals (*From* Saldanha, 1962b.)

Group (according to parental origin)	Total No. of Individuals	Hare-lip and Cleft Palate	Malformations of Extremities	Congenital Heart Disease	Mongolism	Epilepsy	Inguinal Hernia	Mental Retardation	Total
Descendants of Brazilians	2405	1·3	1·7	0·8	0·4	0·4	0·4	0	5·0 (12)*
Descendants of Italians	2103	0·5	0·5	0·5	0	0	1·0	1·0	3·3 (7)
Descendants of Brazilians and Italians	1658	0	0	0	0·6	0	0	0	0·6 (1)
Descendants of other nationalities (pure or mixed)	576	0	0	1·7	0	3·5	1·7	0	6·9 (4)
Total	6742	0·6	0·7	0·6	0·3	0·4	0·6	0·3	3·6 (24)

* Actual number of individuals observed.

some of the racial differences for stature, which are very difficult to disentangle both from environmental influences and the effects of a seemingly almost world-wide secular trend towards an increase in height this century. The results which Saldanha obtained in his detailed study of height among residents, migrants, and hybrids in Capivari are shown in *Table* 5.16. The analysis of var ance for the mean among the three groups is highly significant.

Table 5.16.—MEANS AND VARIANCES FOR THE DISTRIBUTION OF STATURE AMONG MALES IN THE PARISH OF CAPIVARI, ACCORDING TO PARENTAL ORIGIN (*From* Saldanha, 1962b.)

GROUP	*N*	RANGE	MEAN AND STANDARD ERROR	VARIANCE
Brazilian	760	141–186	162·94 ± 0·28	61·81
Brazilian and Italian	501	140–190	166·08 ± 0·31	49·02
Italian	1318	137–185	166·47 ± 0·21	61·14

The mean stature for the hybrid group is quite similar to that of the Italians and is intermediate between them and the Brazilians. Perhaps the most important thing about the stature of these three populations, however, is that the variance of the hybrid group is significantly less than that of the unmixed Brazilians and Italians. These three groups of people have lived in similar environmental and socio-economic conditions and this finding is therefore an important one. These data are consistent with the expected effects caused by the breakdown of isolates, and Saldanha suggests that some of the normally present variance in human stature can be explained, in part at least, by the action of dominant genes.

The data on fertility are unfortunately not very numerous. From a sample of 247 families Saldanha obtained the following figures for number of children per family: 60 Brazilian families had a mean of 3·88 children/family; 72 Italian families had a mean of 2·26 children/family; 59 'Hybrid' families had a mean of 3·52 children/family. Largely because of the smallness of the samples these differences are not significant.

The net result of all this, then, is that this particular population shows in a fairly clear fashion the genetic effects of the breakdown of an isolate. Inbreeding falls, the frequency of homozygotes falls, variance is not increased, and metrical characters are modified by the influx of new genes. Nothing in the way of genetic harm results from this mixing.

Of course it can be argued that this small insignificant population is of no relevance for man's evolution, nor for the more general problem we are considering in this book, but Saldanha quotes details of a number of studies, some of which we have mentioned in Chapter 2 and earlier in this chapter, suggesting that the processes operating in Capivari can be 'frequently found at present in different places in the world'.

Among Saldanha's conclusions are that 'inter-migration and admixture should probably be the most important evolutionary factor which should presently be pushing back the intergroup differences made up by selection during a very long time. . . . Since the selective coefficient of many human genes has approached the neutrality level, the mean effective population size has increased and mutation is a rare event, evolutionary changes should be determined proportionately more by migration and admixture in the future.' We present in Chapter 8 details of recent racial integrations and short-term future trends which amply confirm this suggestion. On the basis of what we have learnt in Chapter 3 it can be argued, though, that Saldanha rather underestimates the importance of racial mixing in the past history of man.

CHAPTER 6

RACIAL DIFFERENCES AND THEIR EXPRESSION IN HYBRIDS

6.*1*. Introduction

The literature and tabulated information on racial differences in man are enormous and cover almost every aspect of biological form and activity. We have already indicated in the first three chapters many of those characters which are of particular significance in racial classification. In this chapter we are more concerned with those characters of general biological importance such as fertility, susceptibility to disease, intelligence, etc. Within this meaning of biological importance we include those traits of principally psychological and sociological importance such as skin colour, facial form, and behavioural differences.

There are three main criteria we shall use for choosing those traits worth close investigation in this chapter. Firstly the incidence in different races of those traits which, while not necessarily harmful in themselves, might be incompatible with or combine imperfectly with traits present in another race. Examples include the incidence of Rhesus-negative genes and the polygenically controlled character complexes such as bodily proportion. Into this category also fall those cases in which it might be feared that the break-up of gene complexes would lead to a range of detrimental types being produced.

The second group of characters are those which are undoubtedly detrimental, at least in some environments and which are present in one race to greater or lesser extent than in others. Examples here would be the incidence of congenital malformations such as hare-lip and cleft palate, and the frequency of abnormal haemoglobins such as that producing sickle-cell anaemia.

Finally there are a number of characters such as intelligence, skin colour, hairiness, and temperament, about which controversies rage. The controversies are concerned variously with the presence or absence of racial differences, the genetic mechanisms underlying the characters, and the importance or otherwise of the characters in biological terms anyway.

The circumstances in which it is possible to investigate the behaviour of these characters during racial amalgamation vary for each of these three categories. By and large, if we do not understand the genetic mechanisms controlling a character we can produce only trivial observations and very little analysis of its likely manifestation in racial hybrids. On the other hand, if we are dealing with characters controlled in a relatively simple way the consequences of racial amalgamation can be predicted fairly accurately, and it merely requires large enough samples to confirm or disprove our predictions. Indeed, once the modes of inheritance of certain traits are well understood they become the markers whereby the age and extent of racial hybridization can be assessed. Such information is necessary for the analysis of the effects of the first group of characters we mentioned, those which might produce incompatibility.

It is these simply controlled characters, then, which are potentially the most informative and the most easily studied. In order to arrive at any meaningful comparisons regarding the genetical and biological effects of racial amalgamation on those characters such as physique which are controlled in a complex way, we must have data that relate not only to the hybrid group itself but also to both its parental populations. There are hardly any cases of race crossing on a large scale where this requirement can be adequately met; either amalgamation began a long time ago, or we do not know the exact geographic origins of one of the parents, or the genetic contributions from the parental races were unequal. In other words, for one reason or another we do not have fully comparable data from each of the contributing populations and it becomes exceedingly difficult to investigate the consequences of racial amalgamation on the expression of complex characters in any meaningful way. J. C. Trevor, in his classic monograph *Race Crossing in Man*, published in 1953, solves this problem by choosing what he feels are 'modern series which may be presumed suitable if the characters used are not subject to secular change', to represent the parental groups. Unfortunately, in many characters we know that secular changes have occurred, particularly as dietary improvements have spread around the world in the last few generations. Secondly, we have seen that in many cases environmental conditions can modify the expression of particular characters. Therefore to compare a hybrid group in, say, the West Indies or on Norfolk Island with a parental population from Britain or U.S.A., as Trevor does, may introduce certain inaccuracies.

Nevertheless if we wish to investigate these complex characters at all we must have large representative samples, and we really have no option but to follow Trevor's lead. As Trevor says, 'the best comparisons which can be made in the circumstances are evidently very imperfect'. Nevertheless he goes on to say that the conclusions 'arrived at are so consistent that (the disadvantages of the method and the defects of the material) are of less consequence than was feared'.

6.2. Race Crossing and Physique

6.2.a. *Height and Weight*

We have already seen that, on average, there is some association between human physique and environment. Roberts (1953) has shown that there is an inverse relationship between mean body-weight and mean environmental temperature, although stature appears not to show any marked environmentally associated variations around the world. There is, furthermore, no marked correlation between these two characters, so that the magnitude of difference in physique between different races is never so great as is suggested by the extremes for each. The lowest body-weights known in man are among the Negroids of southern Africa, which are under 44·5 kg., whereas the highest among some northern Europeans and Polynesians average more than 69·5 kg. This difference is rather marked but, when allowance is made for the marked environmental correlation, we can obtain an estimate of the genetic component

Table 6.1.—Mean Weights of Groups adjusted for Temperature and Stature (in kg.) (*From* Roberts, 1953.)

South African	51·79	Melanesian	59·84
Australian	51·90	American	59·92
African	53·92	European	59·93
Indian	54·52	Central Asian	62·20
East Mongoloid	54·69	Polynesian*	71·82

* Based on 2 samples only.

of this variation. The results of making such an allowance are shown in *Table* 6.1.

Weight, however, is not an anthropological character which is often measured because of the enormous variability within races. Some data from Dunn and Tozzer (1928) on various races and their hybrids in Hawaii may briefly be referred to. Dunn and Tozzer recorded the weights of 62 Hawaiian males, 6 Chinese males, 4 Hawaiian/Chinese hybrids, 13 Hawaiian/European male hybrids, and 4 mature male triple hybrids. Their results are:

	Hawaiians	Hawaiian/ Chinese	Chinese	Hawaiian/ European/ Chinese	Hawaiian/ European	European (average figure)
Weight	168·62 lb.	142·75 lb.	133·33 lb.	144·16 lb.	151·76 lb.	152·90 lb.

We may note the general intermediate level of racial hybrids and refer to *Fig.* 6.3a, b, wherein are plotted weight-gain curves for Caucasoid/Aborigine hybrids. These take up an intermediate position between the parental races which are, from *Table* 6.1, racial groups towards the extremes of human weight

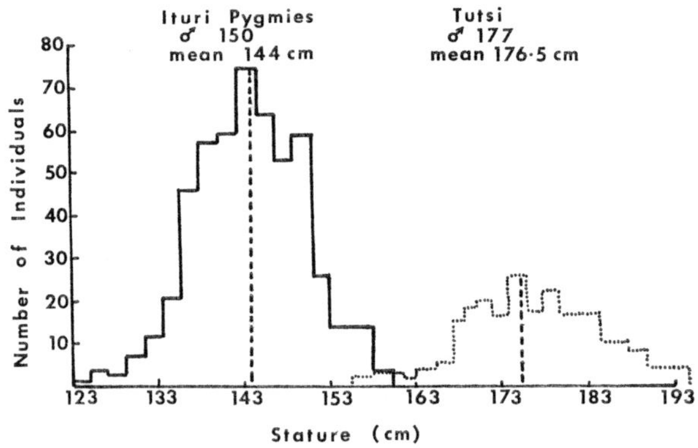

Fig. 6.1.—The distribution of stature in male Ituri Pygmies and male Tutsi. (*From* Harrison and others, 1964.)

distributions. We may suppose then, that differences in body-weight between hybridizing races confer no marked disadvantages upon their offspring.

Stature, on the whole, does not appear to be associated with climate. Probably the shortest people in the world are the Bambuti group of Pygmies from the Ituri forest region in the Congo Republic. Among the tallest are the Dinkas, a pastoral people from the Sudan. The difference between the mean stature of male Pygmies (143 cm. = 4 ft. 7 in.) and the Dinkas (180 cm. = 5 ft. 10 in.) is 37 cm. or 1 ft. 3 in. and yet, as *Fig.* 6.1 shows, the ranges of these two distributions overlap, so that the tallest Pygmies and the shortest Dinkas have similar heights.

There are no data available on crosses between these two races, but Gusinde (1942) has studied crosses between the Congo or Bambuti Pygmies and Forest Negroes with which they are associated. These crosses normally take place in one direction only. The Negro males may take a Pygmy wife, but she becomes a member of the Negro tribe and her children live as Negroes and backcross into the Negro race. The mean stature of the Negroes which Gusinde measured was 158·6 cm. for males and 148·0 cm. for females. The Bambuti males averaged 144·9 cm. and the females 137·2 cm. Unfortunately no standard deviations can be calculated from his data for the parental races. However, the

Table 6.2.—MEAN STATURE OF A NUMBER OF HYBRID POPULATIONS AND THEIR PARENTAL POPULATIONS (MALES)

PEOPLES INVOLVED AND SOURCE	EUROPEAN STATURE	HYBRID STATURE	NON-EUROPEAN STATURE
Negroes and Whites in North America (Trevor, 1953)	174·3 ± 0·22 (727)	171·7 ± 0·56 (87)	170·4 ± 0·56 (102)
Sioux Indians and Whites in North America (Trevor, 1953)	174·3 ± 0·22 (727)	173·5 ± 0·78 (77)	172·4 ± 0·24 (537)
Maya Indians and Spaniards in Mexico (Trevor, 1953)	166·5 ± 0·28 (416)	156·4 ± 0·19 (865)	155·1 ± 0·60 (77)
Hottentots and Dutch in South-West Africa (Trevor, 1953)	173·2 ± 0·87 (70)	168·4 ± 0·83 (69)	162·4 ± 0·76 (73)
Kisarese and Dutch in the East Indies (Trevor, 1953)	173·2 ± 0·87 (70)	164·9 ± 0·66 (79)	162·3 ± 1·03 (39)
Tahitians and English on Norfolk Island (Trevor, 1953)	172·1 ± 0·05 (5426)	174·0 ± 0·79 (64)	171·4 ± 0·61 (85)
Indians (Pods) and English in India (Trevor, 1953)	172·1 ± 0·05 (5426)	166·1 ± 0·54 (145)	162·6 ± 0·40 (150)
Hawaiian and North European in Hawaii (Wissler, 1930)	167·33 (3)	171·21 ± 0·81 (70)	169·51 ± 0·42 (205)
Hawaiian and European in Hawaii (Shapiro, 1930)	172·1 ± 0·05 (5426)	173·48 ± 1·22 (10)	171·31 ± 0·40 (70)
Negroes and Whites in Cuba (Laska-Mierzejewska, 1967)	168·04 ± 1·07 (49)	170·12 ± 1·18 (36)	171·08 ± 1·12 (333)
Hawaiian and South Chinese in Hawaii (Shapiro, 1930)	Chinese 165·2 ± 1·1 (9)	Hybrid 165·6 (4)	Hawaiian 171·31 ± 0·40 (70)
Forest Negro and Bambuti Pygmy in Central Africa (Gusinde, 1942)	Negro 158·6 (?)	Hybrid 149·0 (9)	Pygmy 144·9 (?)

F_1 characters were intermediate, with males ranging from 141·0 cm. to 157·2 cm., averaging 149·0 cm., and with a coefficient of variation of 4·13. The females ranged from 136·5 cm. to 146·0 cm., averaging 14 cm., and with a coefficient of variation of 2·66. The backcross of these hybrids to Negroes showed, as expected, mean stature closer to that of the Negroes and a rather wider range of variation, indicative of genetic segregation. The mean stature of the male backcrosses was 152·3 cm., with a range of 142·9 to 168·3 cm. Females averaged 144·4 cm. and showed a range from 135·4 to 149·2 cm. The coefficients of variation for the two sexes were respectively 6·39 and 3·87. The samples from which these figures are derived are not large, 14 F_1 hybrid and 16 backcross individuals, nevertheless the variability in height is virtually the same as that for other racial groups and their hybrids, despite the fact that we are dealing with the extremes of human stature. We can therefore confidently conclude that disparity in height of contributing races does not in itself cause any problems to their hybrid offspring. *Table* 6.2 shows some more extensive data which are available on stature in racial hybrids, taken mainly from Trevor's (1953) monograph with some additional figures computed from other available data. The table shows only data on male stature, since the data on females are essentially similar but less abundant. We are able to make twelve three-way comparisons and in only three of them do the hybrids not have intermediate values between the two parental populations—Norfolk Islanders, Hawaiian/North European hybrids, and Hawaiian/European hybrids.

In the case of the Norfolk Island hybrids there is no significant difference between the mean stature of the males of the presumed parental populations, Society Islanders and Englishmen, while the hybrids themselves are just significantly taller than either of them. The three female samples from these peoples are statistically homogeneous. The other cases in which the hybrids exceed both of their parental populations are the Hawaiian/North Europeans which are compared with a very small sample of Europeans in this particular survey, and in any case are not significantly different from the Hawaiian. Note that if a more reliable statistic of European stature is substituted here, the hybrids become intermediate. The figures in Shapiro's study for Hawaiians, Europeans, and their hybrids do not differ significantly.

From the previous chapter it is evident that a more important comparison is the degree of phenotypic variability shown by the various hybrids compared with their parental populations. The coefficients of variability for stature in each of these populations are shown in *Table* 6.3. For each population trio Bartlett's test for homogeneity of variance has been carried out, and in no case are there significant differences among these variances. Furthermore in the majority of cases the coefficient of variability of the hybrids falls midway between the values of parents just as do the absolute magnitudes of each character. In our previous discussion of hybridization of Pygmies it was implied that the magnitude of the difference between the parental populations

Table 6.3.—COEFFICIENTS OF VARIABILITY FOR STATURE IN SOME HYBRID POPULATIONS AND THEIR POPULATIONS (MALES)

	EUROPEAN POPULATION	HYBRID POPULATION	NON-EUROPEAN POPULATION
Negroes and Whites in North America	3·33	3·77	3·32
Sioux Indians and Whites in North America	3·33	3·92	3·16
Maya Indians and Spaniards in Mexico	3·42	3·45	3·39
Hottentots and Dutch in South West Africa	4·18	3·91	3·98
Kisarese and Dutch in the East Indies	4·18	3·55	3·96
Tahitians and English on Norfolk Island	3·93	3·65	3·45
Indians (Pods) and English in India	3·93	3·88	3·31
Hawaiian and European in Hawaii*	2·90	4·00	3·60
Hawaiian and South Chinese in Hawaii	2·76	2·58	2·55
Negroes and Whites in Cuba	1·625	1·560	1·481

* Variance of females.

might have an influence on the degree of variability of the hybrids. The correlation coefficient between the degree of variability of these hybrids for stature and the size of parental difference has therefore been calculated. The same calculation has been carried out for a number of other physical characteristics where sufficient data are available. The figures which are arrived at are in four cases negative, although not a great deal of weight can be attached to these results in view of the heterogeneous nature of the data involved and in some cases the smallness of the samples. It is nevertheless rather interesting that the greater the difference between the parental races, the less variable are the hybrids even though they are, in general, still intermediate in expression.

	Correlation coefficient	Number of Comparisons
Stature	−0·274	12
Bizygomatic breadth	−0·784	5
Head breadth	0·182	6
Nasal breadth	−0·810	5
Cephalic index	−0·504	7

We have already mentioned that adult body-weight is much more affected by environmental factors than stature, and it is no doubt partly for this reason that the amount of comparative data that has been gathered for body-weight is very limited. We cannot therefore do the same analysis for weight as for height. There is no reason to suppose, however, that the results would be any different. We might draw attention again at this juncture to Trevor's findings: firstly that of 100 different comparisons involving two parental races and their hybrids, in only 11 was a character appreciably more variable than either of the parental series; and, secondly, in only 27 cases was the mean absolute measurement of the character in the hybrid outside the range defined by the mean of the two parental races.

But of course hidden beneath these simple figures for height and weight there might be important differences in relative growth rates or bodily relationships which could perhaps produce upsets in the offspring of interracial crosses. These may be considered under four headings: possible differences in skeletal development and ossification; possible differences in birth weight and the relationship with maternal pelvis size; possible differences in the shape of the infantile and juvenile growth curves; possible differences in relative bodily proportions leading to ill-adapted combinations either during growth or in the adult. We shall investigate each of these possibilities in turn.

6.2.b. *Skeletal Development*

Comparative data on skeletal development are very fragmentary. For most of the races of the world we have hardly any information and therefore our knowledge of what might happen on hybridization is virtually non-existent. We do have some information concerning the three major racial groups, however, albeit from rather restricted samples, and this allows us to draw some general conclusions.

The stage of skeletal ossification in Negroes seems always to be ahead of Whites at birth. This is probably the result of an inherited difference in hormone secretion during the late foetal period, because Negroes' permanent teeth almost always erupt earlier despite enormous dietary and environmental differences in most cases. The Negro child may maintain his relative advancement for up to 3 years, it being reflected in a greater rate of growth in length and weight and a greater maturity of motor development and behaviour (*see* s. 6.4.*d*, i, ii). Certain post-natal centres of ossification unquestionably appear earlier in American coloured babies (Kelly and Reynolds, 1947; Christie, 1949) and in African Negro children. A rigorous comparative study of foetal Negro and White skeletons (Choi and Trotter, 1970), including data on both length and weight of the bones, showed that there is indeed an overall difference between the races in skeletal parameters. This conclusion was arrived at as the result of a discriminant analysis of the skeletons using the seven ratios between individual bones which contributed most of the calculable differences between the races and the sexes. The statistic which Choi and Trotter calculated is

Mahalanobis' D-square, which is a measure of the squared distance between the populations classified. The D-square statistic enables one to test for differences between two groups, and its square root, D, can be used to calculate the relative distances between groups. These D values and the significance levels of the distances are shown in *Table* 6.4.

Table 6.4.—DISTANCES BETWEEN GROUPS (D) AND THE SIGNIFICANCE LEVEL (*P*) OF EACH BASED ON A STATISTICAL STUDY OF FOETAL SKELETONS (*From* Choi and Trotter, 1970.)

GROUPS	D	*P*
White and Negro	0·71	N.S.
Male and female	0·99	< 0·005
White male and White female	1·40	< 0·005
White male and Negro male	0·79	N.S.
Negro male and Negro female	0·98	N.S.
White female and Negro female	1·15	< 0·05

The difference between the races is found to be not significant, whereas the difference between the sexes is significant, a value of 1·39 when compared to a racial distance of 1·0. The relative distances shown in *Table* 6.4 correspond to the differences due to race and sex found in a study of density of selected bones of the adult skeleton (Trotter, Broman, and Peterson, 1960). So far as longitudinal growth curves are concerned, Trotter and Peterson (1970) present some graphs which are fully in line with these results. Both for combined length of femur + tibia and total skeletal weight the difference between the sexes was as great or greater than that between the races. Furthermore the amount of variation among individuals of a given age appears to increase with age, and overall it seems that skeletal changes with age appear to resemble the pattern shown by the increases in stature of living children. The similarities of skeletal growth curves between the races and this relationship with stature clearly suggest that, like stature, the pattern of growth of the skeleton in hybrids will be intermediate between that of the parental populations, and that there will be no problems regarding incompatibilities or differential times of development. Further evidence bearing on this important point will be presented in the section on growth rates later in this chapter (s. 6.*2.d*).

Differences between Hong-Kong (Chinese) children and White Americans are in the opposite direction to that shown by Negroes and Whites. Between the ages of 6 and 15 years in boys, and 6 and 12 years in girls, skeletal maturity of Hong-Kong children is retarded relative to the Greulich-Pyle standards. At adolescence, however, changes in growth rate occur and the Chinese develop more rapidly so that they become more advanced at comparable ages than Americans. The general level of skeletal development in young children appears, therefore, to be Negro > White > Mongoloids. It is therefore of some importance to study skeletal ossification and its effects in interracial hybrids, since this could be an important source of disharmony. The only direct data available are those of Gedda, Serio, and Mercuri (1961), who studied a number of war babies in Italy who had Italian mothers and American Negro fathers. It does not appear, in fact, that it is likely to be a major source of incompatibility. There is also a certain amount of subsidiary evidence available from dental studies and from consideration of orthodox growth curves. We shall examine this evidence separately.

6.2.*c*. *Birth-weight, Pelvis Size, and Obstetric Risk*

Some of the different mean birth-weights for different racial groups are shown in *Table* 6.5*c*. By no means all of this variation shown is due to genetic factors. It is known, however, from studies on correlations between relatives,

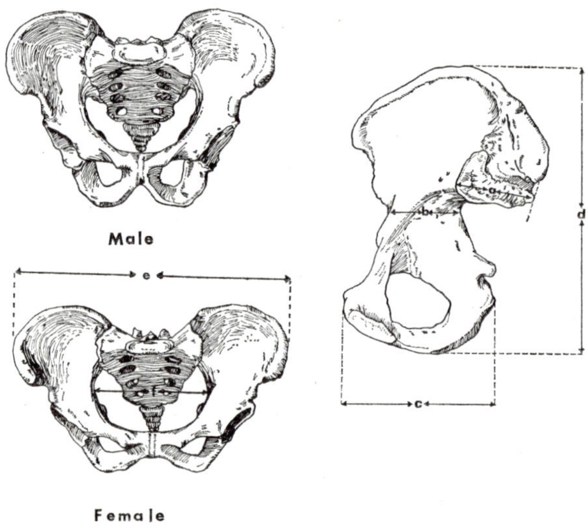

Fig. 6.2.—Male and female pelvis and some of the measurements which are commonly made of the pelvis. **a**, Maximum length of auricular surface; **b**, Minimum ilium breadth; **c**, Maximum ischio-pubic diameter; **d**, Maximum height of os coxal; **e**, Bi-iliocristal diameter; **f**, width of pelvic inlet.

that there there is some genetic component determining birth-weight, although it is small. Parsons (1965) has suggested that the rather small pelvic girdle and narrow hips of Australian Aborigines (Davivongs, 1963) may be relevant in explaining the low birth-weight of Aboriginal babies. If this is correct it suggests an important possible hazard for interracial matings, particularly when a 'large-baby' race marries a 'narrow-pelvis' race. This hazard has indeed been postulated by some people, and we must therefore examine the possibility with some care.

One of the most appropriate parameters for comparisons of pelvic girdles between different races is the ischium-pubic index, which compares the pubic length with the ischial length (*see Fig.* 6.2). *Table* 6.5*a*, from Davivongs (1963) shows this index for a number of races and *Table* 6.5*b* shows a related anatomical

Table 6.5*a*.—THE ISCHIUM-PUBIC INDEX FOR FEMALES IN VARIOUS RACES. (*From* Davivongs, 1963.)

RACE	NO.	MEAN	RANGE
American White	100	99·5	91–115
Eskimo	95	98·8	91–109
Bantu	70	98·1	87–107
American Negro	50	95·0	84–106
Australian Aborigine	72	92·7	81–107

Table 6.5*b*.—MEAN HIP WIDTH FOR FEMALES OF VARIOUS RACES (*From* Meredith, 197.)

RACE	NO.	WIDTH (Bi-iliocristal Diameter)
American White	1978	28·4 cm.
Eskimo	140	28·1 cm.
American Negro	117	26·8 cm.
Bantu	99	25·8 cm.
Australian Aborigine	20	24·6 cm.
Bambuti Pygmy	116	22·6 cm.

Table 6.5*c*.—THE MALE BIRTH-WEIGHT IN VARIOUS RACES.

RACE	BIRTH-WEIGHT	REFERENCE	COMMENTS
American White	3457 g.	Comas, 1960	Mean of 2 samples
American Negro	3226 g.	Comas, 1960	Mean of 2 samples
Bantu	2965 g.	Comas, 1960	Mean of 2 samples
	3067 g.	Hollingsworth, 1965	In Nigeria
Australian Aborigine	2949 g.	Propert and others, 1968	7 tribes
Bambuti Pygmy	2640 g.	Meredith, 1970	Mean of male and female births

measurement, hip width, with which, not surprisingly, there is a high correlation. Selecting those data for birth-weight which correspond with this list allows us to decide whether there is an association between pelvis dimension and birth-weight. A Spearmans Rank Correlation Coefficient calculated from the data of *Table* 6.5 suggests that there is such an association, although probably a rather low one, because of the major environmental contribution to birth-weight. This therefore suggests that any hazard from this mismatching is not likely to be an important one.

The birth-weights for West Australian Aborigine/White hybrids are significantly higher than West Australian full bloods (*Table* 6.6). It is not clear whether

Table 6.6.—Birth-weight for West Australian Aboriginal Full Bloods and Mixed Bloods compared with an English Sample. (*From* Propert and others, 1968.)

Sample	Number	Males	Number	Females
Full Blood	92	105·6 ± 22·7 oz.	81	104·6 ± 18·5 oz.
Full Blood	209	106·7 ± 22·0 oz.	175	102·3 ± 18·5 oz.
Mixed Blood	117	116·9 ± 19·3 oz.	155	111·3 ± 21·4 oz.
Mixed Blood	155	113·2 ± 22·0 oz.	196	109·3 ± 20·6 oz.
English	6697	117·6 ± 18·7 oz.	6419	114·1 ± 17·6 oz.

this is due to genuine heterotic effects or merely maternal effects. In fact Morton and others (1967) show that paternal race is as important as maternal race in influencing birth-weight in a variety of interracial matings in Hawaii. They also find that there is no significant effect of hybridity either on increasing or decreasing the birth-weight. Hybridity, furthermore, had no effect in their study on increasing the frequency of deviant births, either those extremely light or those extremely heavy. This latter finding is clearly an important one in suggesting that risks to hybrid offspring at birth are no greater than among non-hybrid babies.

We are left then with the possibility that abnormal labour or birth injury might be increased in women who are married to men of different races. There are three investigations on this point: Bowles (1957) and Morton and others (1967) in Hawaii, where many races are involved, and Sargent, Westfall, and Adams (1958) in Japan, who concern themselves entirely with Japanese women married to Caucasoid men.

One of the problems of determining frequencies of complications of labour is that of imprecision of definition and incompleteness of reporting. Morton, Chung, and Mi included in their definition of abnormal labour: cephalopelvic disproportion, malposition, and malpresentation; abnormality of forces of labour; operative delivery; maternal injury during labour; and some abnormal placental conditions and behaviour. In Hawaii they found, overall, a frequency of 7·6 per cent abnormal labour. They found that there was no increase in the incidence of abnormal labour with hybridity of the mother, father, or child. This conclusion was upheld in those racial crosses in which the effect might have been expected to have been most pronounced. They found that the recorded frequency of abnormal labour for Japanese women with Caucasian husbands, for instance, was 7·3 per cent, compared with 7·1 per cent for those with Japanese husbands. For Filipino women with Caucasian husbands the frequency was 8·7 per cent, compared with 8·5 per cent for those with Filipino husbands. Neither of these differences approaches significance. Similarly negative results were found when the effect of hybridity on birth injury to the foetus was examined. With an overall frequency of 2·3 per thousand deliveries, birth injury did show an increased frequency in rural as opposed to urban deliveries and also the expected decrease with birth order, but no change with hybridity of the mother, father, or child. Clearly, therefore, maternal/foetal disproportion is not increased on interracial crosses.

The same conclusion was arrived at by Bowles (1957) and Sargent and others (1958) who state: 'The bony pelvis of the Japanese women was found to be remarkably consistent and surprisingly adequate when considering the small pelvic girth'.

6.2.d. Growth Rates in Racial Hybrids

The genetics of human growth is one of those fields most seriously handicapped by lack of data. As Garn (1962) points out, 'not one normal human growth attribute is as yet analysed as to mode of inheritance'. He goes on to indicate most of the obvious difficulties in the field which explain the paucity of data. Although he mentions that two longitudinal studies in the U.S.A. are in their second generation of data collection, they can contribute very little to our requirements at the moment. There are, however, a number of straightforward comparative studies which at least allow us to decide if there are any major disturbances in the growth patterns of racial hybrids.

Kirchner (1952) made a study of 44 war children (21 male and 23 female) aged up to 6 years from German mothers and American Negro fathers. His results showed that the male hybrid children were larger up to the age of 4 years and then smaller than White children in Berlin. The females were smaller up to 2 years old and the same size as the Whites after the third year. Sieg (1955) made an extensive series of measurements on another group of children who were the offspring of German mothers and American Negro fathers, in his case 48 boys and 52 girls. The hybrid children, especially the girls, exceeded in total stature, sternal height, and shoulder height a comparative group of 100 German children of similar ages. In both this study and Kirchner's the hybrid girls began to grow faster from birth than the boys, and this itself is a situation slightly different from that usual in Europe where the increased growth rate of females normally occurs rather later. But the most valuable studies from the present point of view are those of Wissler (1930) in Hawaii, Laska-Mierzejewska (1967), in Cuba, Da Costa Martins (1971) in Moçambique, Ashcroft, Heneage, and Lovell (1966) in Jamaica, and Propert, Edmunds, and Parsons (1968) in Australia, who each provide reasonably comprehensive data on fairly large samples showing comparative growth rates for parental populations and their hybrids. Other useful studies are those of Hiernaux and Heintz (1967) on Franco-Vietnamese hybrids, Benoist (1963) on hybrids from Martinique, and Little (1943) on Anglo-Negro hybrids, although in each of these cases we have only graphical comparisons of growth-rate curves and we cannot make any comparisons of the relative variability, for instance, of the hybrids compared with either or both of their parental populations.

In *Fig.* 6.3A, B the growth rates for West Australian full-blood Aborigines, hybrid or mixed-blood Aborigines, and White Australian infants are compared. The weight curves up to 150 days for each of these groups are very similar, but then the curves for the full bloods begin to fall below those of the Melbourne sample. The curves for the mixed-bloods remain similar to the Whites until

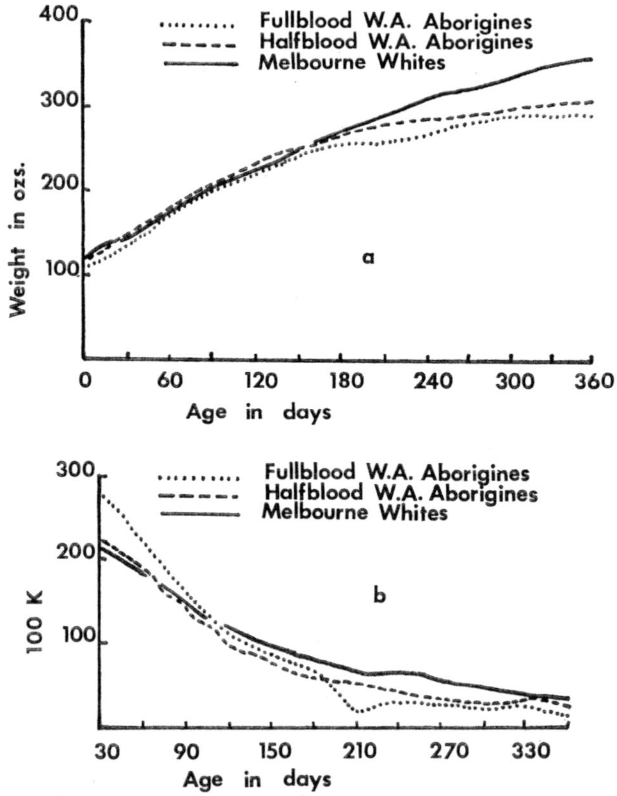

Fig. 6.3.—a, Weights of male infants up to 360 days in comparative samples of Australian Whites, Aborigines, and White/Aborigine hybrids. b, Percentage growth rates corresponding to the growth curves of a. (*From* Propert and others, 1968.)

200 days, when they begin to follow the same pattern as those for the full blood Aborigines but are intermediate in position between the full-bloods and the White Australian (Melbourne) sample. It is obviously difficult to decide what are the genetic and what the environmental contributions to the differences we here observe. Some work by Elliot, Maxwell, and Vawser (1967) on the prevalence of lactose maldigestion among Aboriginal children suggests that genetic factors may play a part, but the obvious important contribution of

environmental deficiencies is highlighted by Jose and Welch (1970). This latter paper has further evidence of the intermediate nature of mixed-blood growth patterns in Aboriginal/European hybrids, their data on this point being shown in *Table* 6.7.

Table 6.7.—HEIGHT DISTRIBUTION ACCORDING TO RACE OF CHILDREN AGED 5 TO 12 YEARS AT KURANDA STATE SCHOOL, AUSTRALIA (*From* Jose and Welch, 1970.)

RACE	NUMBER	MEAN HT INDEX	PERCENTAGE WITH HEIGHT INDEX BELOW 10TH PERCENTILE
Full-blood Aborigine	72	74	67
Mixed-blood Aborigine	54	92	20
Caucasian	59	112	5

Ht Index = $\frac{\text{Ht Age (per cent)}}{\text{Chronological Age}}$ where Ht Age = age at which the measured height or weight corresponds to the fiftieth percentile curve of standard Caucasian growth charts.

A series of curves showing increases of stature and weight with age, which has been constructed from Wissler's data on various races from Hawaii and their hybrids, shown in *Fig.* 6.4A–D. Clearly we are justified in suggesting that racial hybridization has no major effect on the average rates of growth of the offspring. Indeed, so far as the Hawaiians and Asians are concerned, where the differences between the races are greater, the hybrids are intermediate in both height and weight throughout development. Most of Wissler's data, at least over the middle age ranges between about 7½ and 16½ years, are based on samples of about 50 children. *Fig.* 6.5A, B show curves published by Dr. K. Little in 1943 for growth of English and Anglo-Negro children, based on surveys of English dockland communities. His sample sizes vary between 5 and 20 and he is unable to provide comparisons with both parental races and so his data are intrinsically less useful. Nevertheless his results provide confirmatory evidence of the identity of the shape of the growth curves for Caucasoid and Caucasoid-Negroid hybrids.

Other comparative data on the development of Caucasoid/Negroid hybrids come from the large-scale studies undertaken in Jamaica by M. T. Ashcroft and co-workers. In particular, Ashcroft and others (1966) recorded the height and weights of 5,000 children aged between 11 and 17 years who attended eight fee-paying secondary schools in Kingston and were therefore, in general, from the middle and upper socio-economic classes.

Ashcroft and others classified their children into those who were three parts or more African, those who were three parts or more European, and those who appeared to have equal contributions of each race. In the schools they visited, these groups comprised 47 per cent, 18 per cent, and 24 per cent respectively of

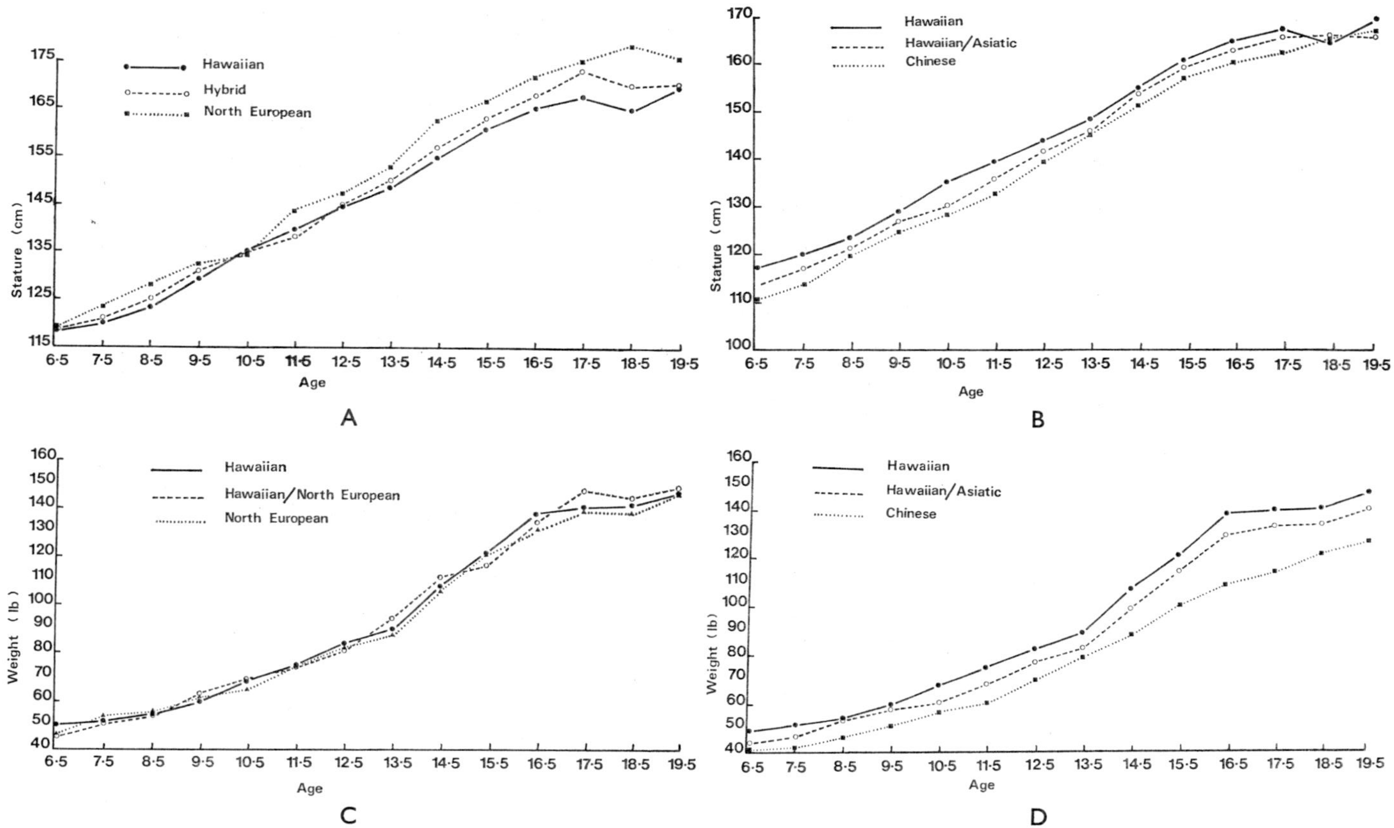

Fig. 6.4.—A, Increase in height of Hawaiians, North Europeans, and their hybrids between the ages of 6½ and 19½ years. B, Increase in height of Hawaiian, Chinese, and their hybrids between the ages of 6½ and 19½ years. C, Increase in weight of Hawaiian, North Europeans, and their hybrids between the ages of 6½ and 19½ years. D, Increase in weight of Hawaiians, Chinese, and their hybrids between the ages of 6½ and 19½ years. (*From data of* Wissler, 1930.)

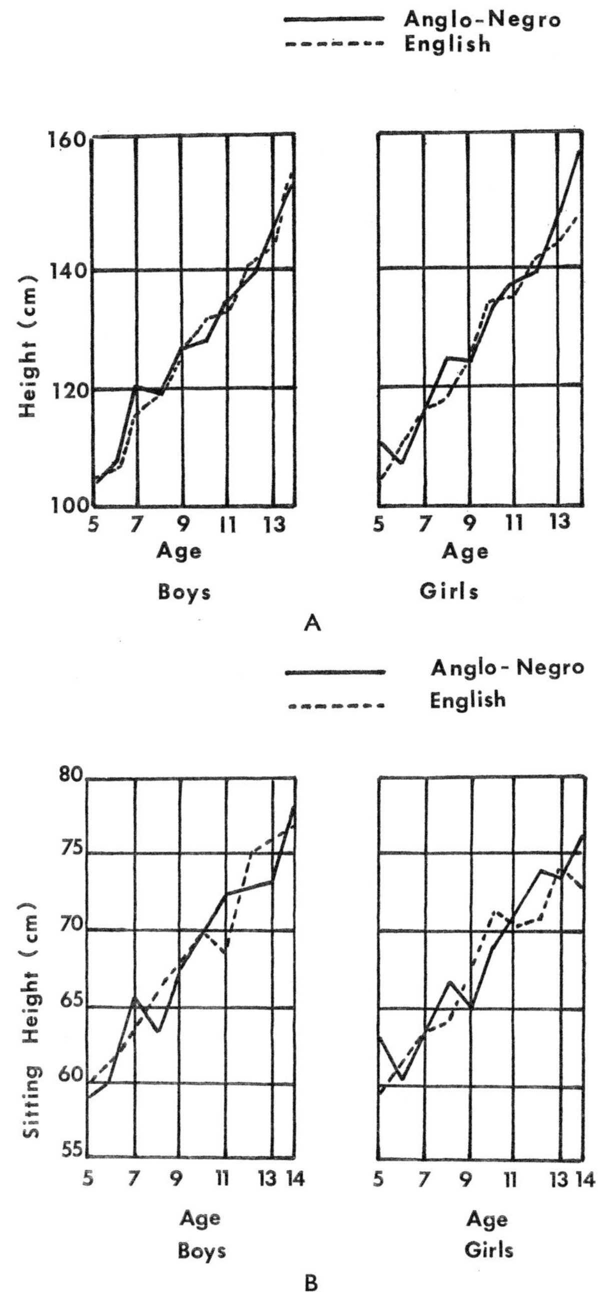

Fig. 6.5.—A, The average height at each year of age for English and Anglo-Negro children. B, The average sitting height at each year of age for English and Anglo-Negro children. (*From* Little, 1943.)

the school population, with the rest being Chinese. The data they obtained on the growth rates of the boys of these samples are shown in *Fig.* 6.6A, B. The results they obtained with the girls are not materially different. Ashcroft concluded that: 'With the exception of African boys, who were a little less heavy at most age groups, the average heights and weights of African, Afro-European, and European boys and girls did not differ consistently'. Their sample sizes varied between 21 and 317, averaging close to 100 at each age

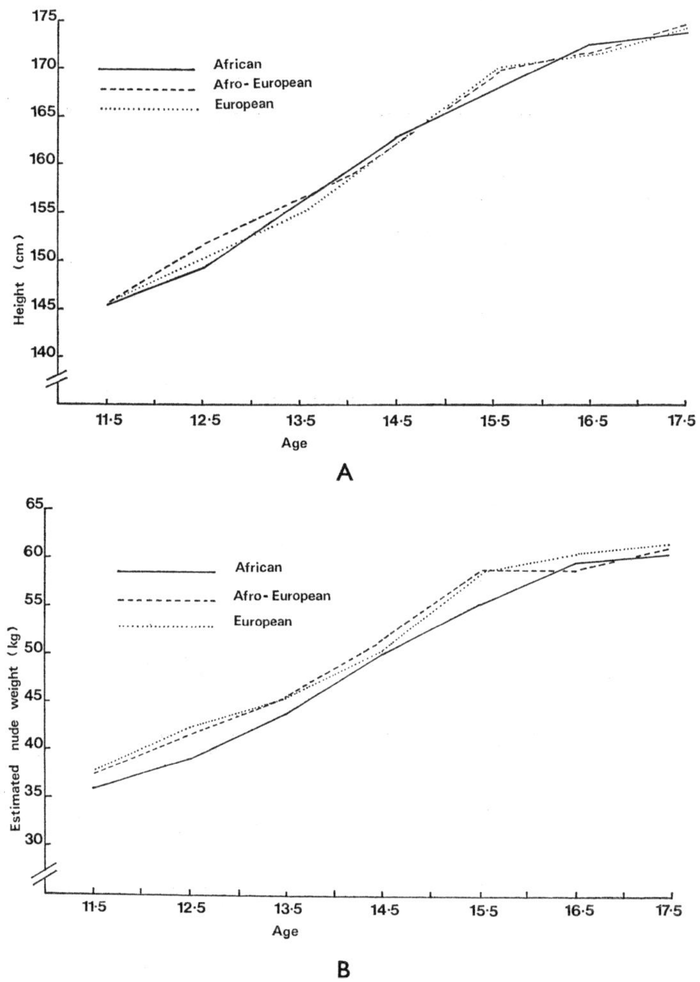

Fig. 6.6.—**A**, Mean heights of developing European, African, and Afro-European male children in Jamaica. **B**, Mean estimated nude weights of developing European, African, and Afro-European male children in Jamaica. (*From* Ashcroft and others, 1966.)

group, so this is a fairly large survey with correspondingly dependable conclusions.

There is some suggestion from the data that the Afro-Europeans may be slightly more variable than either the Africans or the Europeans considered separately. In 7 comparisons of boys heights at four different ages the 'hybrids' are more variable, in 1 case significantly so. The overall tendency throughout, though, is for the hybrids to be intermediate. Among 14 age comparisons for height among both boys and girls, there are 4 where the hybrids are taller, 3

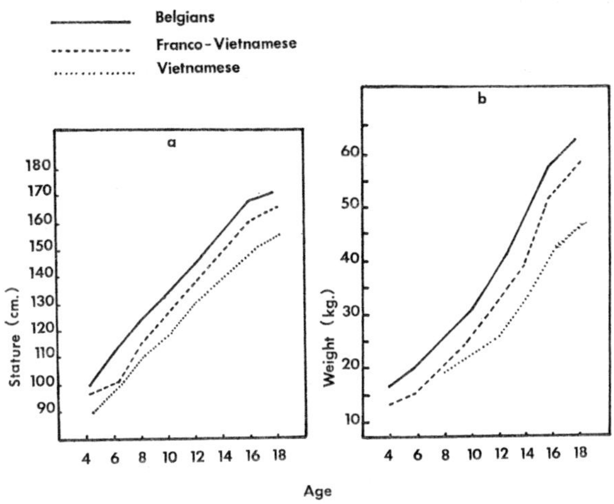

Fig. 6.7.—Age changes in height (A) and weight (B) for Franco-Vietnamese hybrids compared with those in Belgians and Vietnamese, representative of parental populations. (*From* Hiernaux and Heintz, 1967.)

where they are shorter (none of these differences being significant), and 7 where they are intermediate. For 14 weight comparisons, there are 3 cases of hybrid superiority, 6 cases of inferiority (again none significant), and 5 where the hybrids are intermediate.

Similar conclusions emerge from the work of Hiernaux and Heintz (1967) on Caucasoid-Mongoloid hybrids. *Figs.* 6.7A, B show that age changes in height and weight for Franco-Vietnamese hybrids are intermediate between Vietnamese and Belgians at every age between 4 and 18 years. The sample sizes at some ages are very small, and in only a few cases are there more than 20 individuals at any one age, so that the significance of these results should not be overemphasized. However, the Franco-Vietnamese are also either intermediate between, or identical with, the parental population when the height–weight relationship and other body and facial indices are plotted. *Fig.* 6.8 shows several of the facial indices in these hybrids which are discussed further in

s. 6.*3.a*. It can be presumed, therefore, that incompatibilities or incongruities of bodily dimension in these hybrids do not occur.

Benoist (1963) presents diagrams showing gains in weight, height, and chest circumference for White and Coloured Martiniquais of similar socio-economic standing. The growth curves which are illustrated in *Fig.* 6.9A, B are very

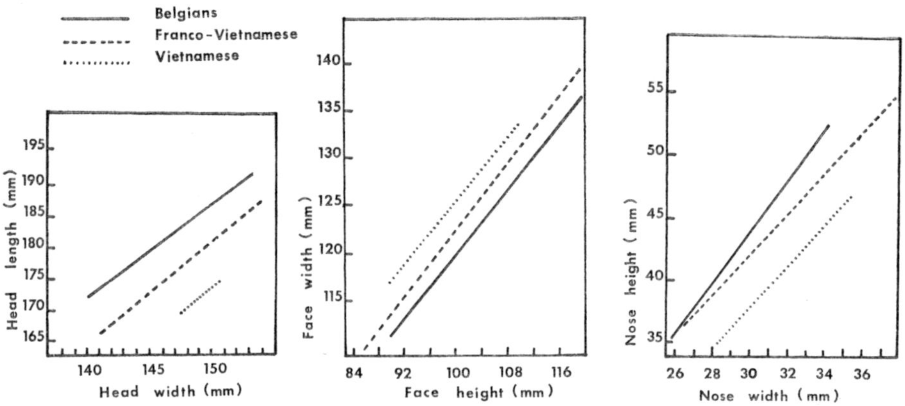

Fig. 6.8.—Some anthropometric indices of Franco-Vietnamese compared with the same indices among Belgians and Vietnamese. (*From* Hiernaux and Heintz, 1967.)

similar for both height and weight, and therefore the conclusion we can draw would seem to be the same as for all other studies, viz., there are no important upsets in growth rates or growth patterns in human racial hybrids. At the ages where the growth curves do diverge slightly it is probably due to a slight difference in time of the growth spurt (Tanner, 1964) in the races.

More positive conclusions can be drawn from the work of Da Costa Martins (1971) in Moçambique, who studied a total of 14,640 White, Negro, Indian, or mixed children in Lourenço Marques. Height, weight, and chest circumferences were measured in children between the ages of 5½ and 18½ years. The findings, which are illustrated in *Fig.* 6.10 (males only), speak for themselves. In all cases at most ages the White children show superiority in that they are heavier, taller, etc. It is suggested that most of this advantage is environmentally mediated, although the fact that the Indians usually show the lowest values argues for some genetic component. The mixed children are intermediate at most ages. Note that the mixed boys still show a marked growth spurt which occurs at the same age as in White and Indian children and a little after that in Negroes. This argues that the growth processes are essentially similar in these different racial groups.

Similar positive conclusions can be drawn from the work of Laska-Mierzejewska (1967). Her measurements are on samples of Whites, Negroes, and

Mulattoes from Havana aged from $6\frac{1}{2}$ to $20\frac{1}{2}$ years. Her average sample sizes are 56, 34, and 39 respectively, which makes this one of the more substantial studies in this field. Graphical representations of her data are given in *Fig.* 6.11A, B, which shows that they are fully in agreement with those from Hawaii, Australia, Vietnam, Jamaica, Moçambique, and England, in that the hybrid growth curve is throughout compatible with, and mostly completely intermediate between, those of the parental populations. What is perhaps more important in all these studies, however, is whether they give any indication of

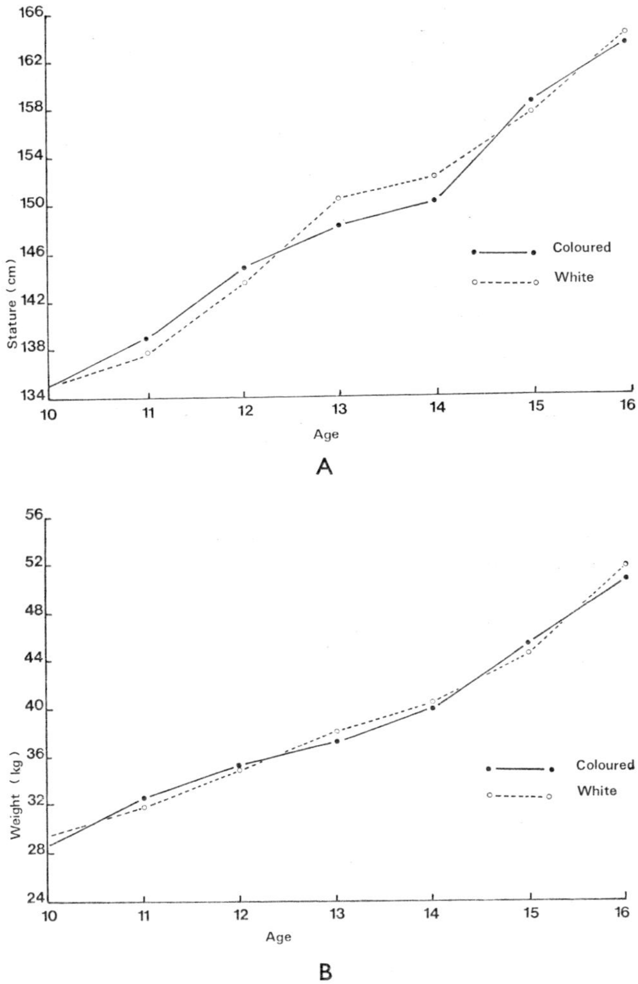

Fig. 6.9.—Change of stature (A) and weight (B) in White and Coloured male Martiniquais of similar socio-economic standing at different ages. (*From* Benoist, 1963.)

significantly increased variability among the hybrids during any stage of growth. Among approximately 200 comparisons using Bartlett's test for homogeneity of variance on samples from Hawaiians, Europeans, Chinese, Aborigines, Negroes, and various of their hybrids, only 17 turn out to be significantly different, a figure hardly greater than that expected on the basis of chance alone. In other words, over the whole of the growth span from 5-year old children to 20-year old adults involving the three major racial groupings of mankind and various

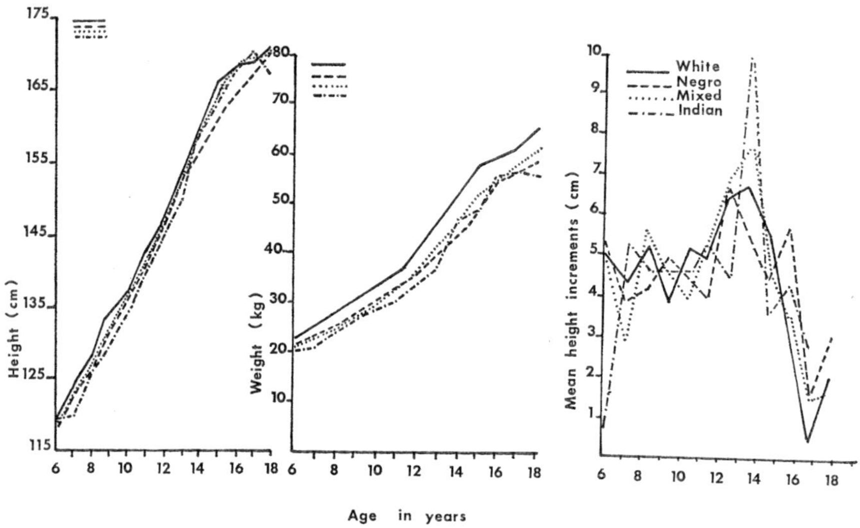

Fig. 6.10.—Heights, weights, and height increments of male school children of Lourenço Marques, Moçambique. (*From* Da Costa Martins, 1971.)

of their hybrids, the increase in variability among these hybrids is negligible and, moreover, shows no tendency to be concentrated during one particular part of the growth cycle.

We cannot restrict our consideration of growth patterns of hybrids simply to height and weight. Problems of relative growth are among the most interesting in all biology and with the possibility that different relative growth is occurring, there exists the possibility that physical disharmonies may arise among growing offspring. In other words, arms or legs might become disproportionately long or short during some stages of growth, or increases in height and weight might occur at different times. Probably the most comprehensive data on this point are those of Wissler (1930), although the findings of many other workers fully confirm his results. As an example of the sort of information we have on this point, *Table* 6.8 shows changes in relative span during growth and the coefficients of variance of each sample at each age for Hawaiian, Chinese, North European, and South European males and each of their respective male hybrids.

In not one single case do the variances at any age differ significantly from one another among these groups.

So far as Hawaiian/North European hybrids are concerned we find that the relative span is intermediate in 10/15 comparisons (including adults); is greater than both parental races in 3/15 times, and less than both parental races in 2/15 times. The coefficient of variation is intermediate in 6/10 comparisons, and

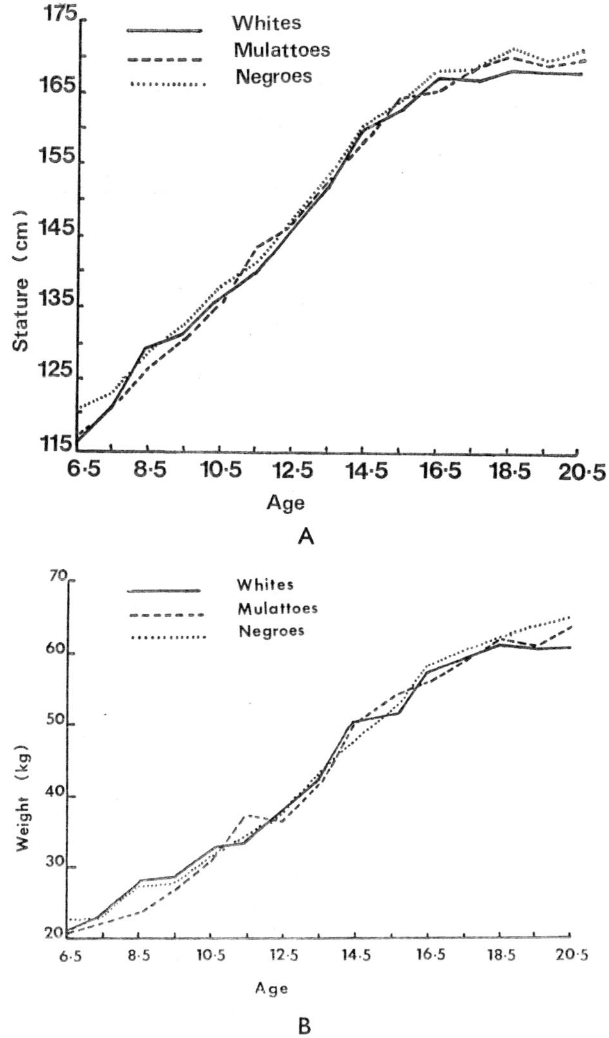

Fig. 6.11.—Increase in height (A) and weight (B) of Cuban males: Negroes, Whites, and their hybrids between the ages of 6½ and 20½ years. (*From data of* Laska-Mierzejewska, 1967.)

Table 6.8.—Age Changes in Relative Span among Various Racial Groups and Their Hybrids in Hawaii

Age	Hawaiian			Chinese			N. European			Hawaiian/ N. European			Hawaiian/ Asian			Hawaiian/ Asian/ European		
	No.	M	V	No.	M	V	No.	M	V	No.	M	V	No.	M	V	No.	M	V
6·5	9	99·55	0·026	49	98·53	0·030	21	98·24	0·022	25	98·80	0·020	23	99·56	0·024	3	100·43	
7·5	19	99·68	0·019	81	98·28	0·022	34	99·65	0·020	33	99·84	0·021	33	99·30	0·019	13	99·08	0·022
8·5	28	100·21	0·022	70	98·67	0·026	32	99·09	0·021	48	99·56	0·019	36	100·00	0·020	14	99·92	0·025
9·5	29	100·34	0·020	76	99·26	0·022	49	99·35	0·023	35	100·08	0·022	32	100·50	0·026	11	100·00	0·017
10·5	37	100·62	0·015	73	99·08	0·022	50	100·76	0·017	45	100·33	0·017	37	100·94	0·022	11	100·64	0·022
11·5	34	100·85	0·022	79	99·87	0·023	40	101·20	0·017	56	100·60	0·021	26	101·04	0·022	11	101·82	0·019
12·5	26	101·85	0·021	56	100·54	0·023	33	100·54	0·024	44	101·45	0·024	24	101·54	0·019	11	102·10	0·019
13·5	22	102·14	0·018	66	100·79	0·020	37	101·40	0·022	40	102·22	0·019	22	102·00	0·028	10	100·30	0·021
14·5	31	102·77	0·014	66	101·42	0·022	37	101·60	0·022	36	102·66	0·023	26	102·00	0·014	12	102·42	0·018
15·5	17	102·77	0·017	50	102·12	0·025	38	102·26	0·023	31	102·87	0·020	15	102·93	0·028	3	101·88	
16·5	10	104·3		45	101·22	0·018	23	102·56	0·017	29	103·20	0·018	8	104·00		9	102·76	
17·5	8	103·5		36	101·67	0·017	16	102·00	0·027	18	103·66	0·024	5	102·60		4	103·75	
18·5	5	105·4		30	102·00	0·022	4	100·42		18	103·16	0·022	8	102·78		1		
19·5	4	104·25		10	102·60	0·021	6	102·76		13	103·08	0·020	8	103·16		1		
Adult	145	104·34	0·023				3	103·56		51	104·20	0·019	26	103·35	0·019	1		

exceeds or is less than the variation of both parental races 2/10 times in each case. The mean relative span of the Hawaiian/Asian hybrid is intermediate in 9/14 possible three-way comparisons and exceeds both of the parental races in 5/14 cases. The coefficient of variation is intermediate in 4/10 cases and both exceeds and is less than both parents 3/10 times. In the case of the triple hybrids, where we might expect any variability or disparities to be magnified, the mean falls within the range defined by the parental races 6/12 times, exceeds all three parental races 4/12 times, and is less than the three parental races 2/12 times. The coefficient of variation is intermediate in 6/8 comparisons and is actually less than all three parental races in 2/8 comparisons.

There is clearly no evidence in this situation that individuals are likely to be at any disadvantage owing to incompatibilities of arm length or stature. If anything the reverse is true. Relative span is slightly greater in hybrids more frequently than it is less, and variability is reduced more often than it is increased.

It is clear that the evidence from all these examples is overwhelmingly against the idea that problems of physical incompatibility and so on might be important during development. We must now investigate the possibility, using a wider range of characters, that they might be important in fully grown adults.

6.2.b. *Bodily Proportions in Racial Hybrids*

We have already examined something of this in our examination of growth rates and physiques and concluded that any disharmonious combinations are unlikely. In this section we merely compare a large number of statistics gathered in appropriate anthropological surveys. The parameters chosen are span and relative span, which are obvious and significant possibilities for disharmonious combinations. Indeed it has been claimed in the past (Davenport and Steggerda, 1929) that human hybrids are often at a disadvantage just because of the inheritance in some of them of long bodies and short limbs, or long legs and short arms, or other unlikely combinations. When these characters and their variabilities in various Hawaiian hybrids, and similar parameters in Cuban hybrids between Negroes and Whites, are compared with their respective parental populations, or when sitting height or relative sitting height, which are other characters for which data exist in both groups, are compared, there is no evidence whatsoever of disharmony. Since concrete information exists on this point concerning the three major racial groups of man and many hybrid groups throughout development, we can completely dismiss it as a possible phenomenon.

Of course from the evidence presented in Chapter 4 it would appear most unlikely that any such disharmonious combinations could possibly occur. To repeat W. E. Castle's (1930) conclusions from his experiments in racial crosses with rabbits, he found correlation coefficients for various limb-bone lengths of between 0·9 and 0·93; in other words almost complete correlations between long or short limbs among his hybrids. In fact the independent inheritance of unit characters of this sort is an entirely fallacious idea anyway. Virtually all the anthropological evidence and the experimental evidence from other organisms shows that a common influence, termed the fundamental rate of growth, is the master controller of all ontogenetic processes from early cleavage stages on. Exceptions to this are mainly due to single genic pathologic conditions such as achondroplasic dwarfism. We discuss the incidence of such rare genes in a later section of this chapter (s. 6.5).

Nevertheless claims that badly proportioned offspring do occur with some frequency in racial hybrids have been made in the past by reputable scientists, presumably with the idea of discouraging miscegenation by providing evidence of its harmful effects. Thus in the study by Davenport and Steggerda (1929), *Race Crossing in Jamaica*, it is stated that the Whites have relatively short legs and long bodies while the Negroes have relatively long legs and short bodies. 'Some of the Mulattoes have an unexpected combination of long legs and long body and others of short leg and short body'. It is further stated that arms and legs are both long in Negroes and shorter in Whites and that some of the crosses have 'the long legs of the Negro and the short arms of the White which would put them at a disadvantage in picking things up from the ground'. *Table* 6.9, taken from Castle (1930), provides detailed statistics on this point,

Table 6.9.—SOME STATISTICS ON THE RACES OF JAMAICA
(*From* Castle, 1930.)
(Adult male Blacks, Browns, and Whites)

		BLACKS	BROWNS	WHITES
Leg length	Mean	92·5	92·3	92·0
	Variance	4·8	4·8	4·2
	Coefficient of Variation (per cent)	5·2	5·2	4·6
Stature	Mean	170·6	170·2	172·7
	Variance	6·8	6·7	6·9
	Coefficient of Variation (per cent)	3·96	3·97	39·7
Arm length	Mean	57·3	57·9	56·8
	Variance	3·1	3·1	2·9
	Coefficient of Variation (per cent)	5·5	5·4	5·2

which put into perspective the generalizations distilled from Davenport and Steggerda's data. In each case there were approximately 50 Blacks and Whites measured and about 90 Browns.

We see that, on average, the legs of the Blacks are just 0·5 cm. longer than the Whites, with the Browns intermediate between the two. Neither are the Browns more variable, since the standard deviation and coefficient of variability are exactly the same for Blacks as for Browns. Variability is no greater for stature among the Browns and therefore the problem seems unlikely to be one of relative leg length. The difference in mean arm length between the Blacks and the Whites is just the same as that for leg length, 0·5 cm. There is presumably, as Castle points out, a close correlation between arm length and leg length in man. The statement regarding inheritance of the 'long legs of the Negro and the short arms of the White' which, it is claimed, would put the hybrids at a disadvantage is seen to involve, on average, just 1 cm.—hardly an important consideration.

Benoist (1963) investigated the inhabitants of Martinique, who are long-established Negro-White hybrids. When comparing the coefficient of variability

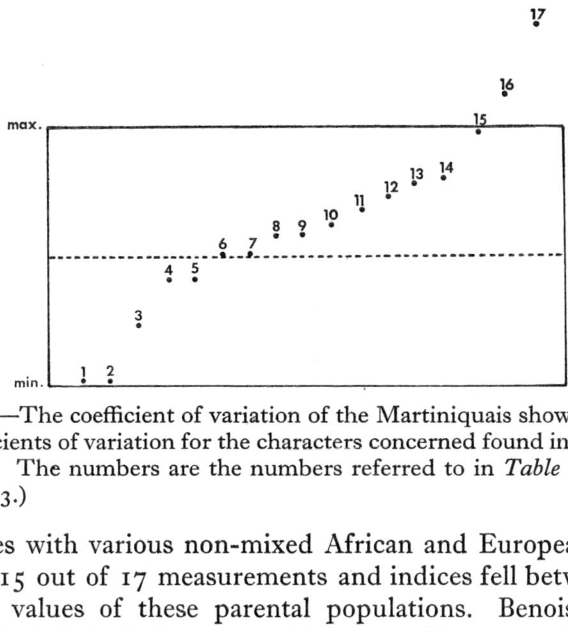

Fig. 6.12.—The coefficient of variation of the Martiniquais shown in relation to the coefficients of variation for the characters concerned found in the parental populations. The numbers are the numbers referred to in *Table* 6.10. (*From* Benoist, 1963.)

of these peoples with various non-mixed African and European populations, he found that 15 out of 17 measurements and indices fell between maximum and minimum values of these parental populations. Benoist also studied a number of physiological characters which normally are much more variable than anthropometric ones. In 4 cases out of 9 the Martiniquais were of intermediate relative variability, in 4 cases they were below the lowest value of the non-mixed race in relative variability. And in only 1 case, the amount of total blood cholesterol, were they more variable than either of the non-mixed races.

The relative variabilities of the anthropometric characters in Martiniquais with respect to the unmixed races, as calculated by Benoist, are shown in *Fig.* 6.12. His detailed results for both anthropometric and physiological characters are shown in *Table* 6.10.

Table 6.10.—COMPARISON OF THE COEFFICIENTS OF RELATIVE VARIABILITY OF MARTINIQUAIS AND VARIOUS AFRICAN AND EUROPEAN UNMIXED POPULATIONS (*From* Benoist, 1963.)

CHARACTER	MARTINIQUAIS	UNMIXED AFRICANS AND EUROPEANS	
		Minimum	Maximum
Measures and Indices:			
1 Head width	3·4	3·9	6·3
2 Index	2·6	2·6	3·1
3 Ear index	7·0	6·1	9·8
4 Height	3·0	2·4	5·0
5 Weight	9·8	5·6	17·4
6 Bizygomatic width	3·6	2·4	4·8
7 Facial index	6·3	4·9	7·7
8 Head length	3·3	2·2	4·2
9 Ear length	6·2	5·0	7·1
10 Ear width	5·8	4·4	6·7
11 Cephalic index	3·9	2·7	4·0
12 Bipalpebral width	9·4	7·2	10·3
13 Mouth width	7·2	5·1	7·9
14 Facial height	5·8	3·5	6·3
15 Nasal width	7·6	4·2	7·7
16 Nasal index	10·6	6·8	9·8
17 Nasal height	6·6	4·2	5·6
Physiological Characters:			
1 Respiratory rate	14·9	14·2	24·6
2 Heart rate	12·8	10·0	14·9
3 Systolic blood-pressure	12·0	11·1	13·4
4 Diastolic blood-pressure	12·3	11·9	16·3
5 Haematocrit	4·4	6·8	8·3
6 Blood sugar	9·2	10·8	12·1
7 Total cholesterol	17·9	10·6	17·2
8 Urea	20·7	21·4	26·8
9 Blood calcium	3·5	4·3	10·8

There is in fact only one area in which there is positive evidence for possible disharmonious combinations appearing in racial hybrids, and that is in the rather special case of the jaws and the teeth. We shall examine this evidence in a separate section (s. 6.5.*a*).

6.3. RACE CROSSING AND SUPERFICIAL CHARACTERISTICS

In this section we shall investigate a number of facial characteristics such as facial width, nasal width, hair form, etc., and we shall also consider that most emotive of all characters, skin colour. The unifying feature about these characters is that while they are relatively unimportant biologically, they assume enormous psychological and sociological importance. In this section we are, in effect, answering the question what will man actually look like in the future if racial hybridization continues at its present rate, or even increases.

6.3.a. *Facial Characteristics*

One of the most extensive and instructive studies of facial characteristics among racial hybrids is that of Herskovitz (1930), who measured more than 500 Negro male hybrids, principally from Howard University, Washington, D.C. Herskovitz recognized eight subgroups with various degrees of White and Indian ancestry. Following Trevor (1953) we shall consider only those with White and Negro contributions. The White group which Herskovitz used for

Table 6.11.—TRENDS OF MEAN VALUES OF SOME FACIAL CHARACTERISTICS IN NEGRO-WHITE HYBRIDS (*From* Trevor, 1953.)

	N	NNW	NW	NWW	W
Head Length					
m.	198·0 ± 0·63	196·9 ± 0·62	196·1 ± 0·66	195·9 ± 1·27	197·1 ± 0·22
s.d.	6·54 ± 0·65	7·06 ± 0·44	6·40 ± 0·46	6·98 ± 0·90	6·04 ± 0·16
Head Breadth					
m.	151·9 ± 0·5	151·7 ± 0·57	151·4 ± 0·49	153·2 ± 1·23	153·9 ± 0·19
s.d.	5·23 ± 0·35	6·44 ± 0·44	4·78 ± 0·35	6·76 ± 0·87	5·20 ± 0·14
Facial Height					
m.	123·9 ± 0·54	122·7 ± 0·55	122·6 ± 0·62	120·1 ± 1·08	119·3 ± 0·43
s.d.	5·59 ± 0·38	6·20 ± 0·59	6·03 ± 0·44	5·93 ± 0·77	6·72 ± 0·30
Nasal Height					
m.	53·1 ± 0·30	53·3 ± 0·33	53·7 ± 0·33	53·7 ± 0·63	53·5 ± 0·21
s.d.	3·16 ± 0·21	3·71 ± 0·23	3·24 ± 0·24	3·45 ± 0·45	3·36 ± 0·15
Nasal Breadth					
m.	43·4 ± 0·27	41·3 ± 0·30	40·0 ± 0·31	37·5 ± 0·70	36·1 ± 0·16
s.d.	2·81 ± 0·19	3·44 ± 0·21	3·01 ± 0·22	3·86 ± 0·50	2·51 ± 0·11
Mouth Width					
m.	55·0 ± 0·36	52·9 ± 0·38	52·4 ± 0·33	52·1 ± 0·63	53·7 ± 0·24
s.d.	3·77 ± 0·26	4·23 ± 0·27	3·21 ± 0·23	3·46 ± 0·45	3·78 ± 0·17

comparison is that of the 'Old Americans' studied by Hrdlicka (1925) and comprises subjects whose grandfathers at least were all born in the United States.

Table 6.11 shows some of the data available from this valuable survey. Both the means and standard deviation are given with their respective standard errors. It is apparent that there is a general tendency, illustrated by each of these characters, for there to be a gradual transition from the unmixed Negroes at one end to the unmixed Whites at the other in the expression of particular phenotypic traits. This table shows that some of these characters differ significantly between the races and that others differ hardly at all. Nasal width and facial height show a regular gradation in values, a situation which is illustrated in *Fig.* 6.13. Nasal height and head length, on the other hand, show neither

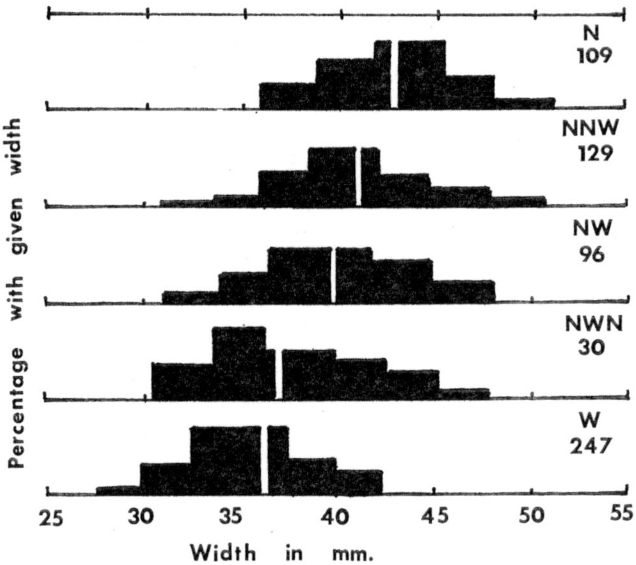

Fig. 6.13.—Changes in the mean value of the nasal breadth in three series of hybrid male American Negroes. The mean value is indicated by the white line in each case. (*From* Trevor, 1953.)

significant nor systematic differences. A character such as mouth width shows a gradation in value for the groups N, NNW, NW, and NWW, although the W group departs from this progression. However, the four Negro groups were measured by one anthropologist and the Whites were part of an independent investigation by another. The comparison as a whole strikingly suggests that there is a tendency for the average measurements of a hybrid population to approach the values of the stock which has made the greater contribution to its ancestry. Trevor's monograph, which brings together results from a number of comparative surveys, and one or two other small investigations confirm the fact that some facial characters vary between the races while others do not, and that those characters which do differ between the races take up predictable intermediate values in the hybrids. Bearing this in mind, then, we can confidently predict that hybrids are in general likely to be intermediate between their parents so far as facial characters go, and that the possibility of disharmonious combinations can be virtually eliminated. Within the collected data available on this point there is really only one possible exception to this generalization, and this concerns the head width of the Yucatecans, hybrids between Spaniards and Maya Indians. The head width of the hybrids appears to be significantly greater than that of Maya Indians, which in turn is significantly greater than the Spaniards. The difference amounts to 1·7 mm., but the Maya and Yucatecan series were measured independently by different investigators and part at least of the recorded difference in head width may be differences in technique used by the different workers. It seems unlikely, therefore, that this is a genuine exception to the previously stated generalization. Hiernaux and Heintz (1967) studied a number of superficial characteristics and *Fig.* 6.8 shows their results for cephalic, facial, and nasal indices. There is no doubt whatsoever that for these characters the hybrids are intermediate in all situations.

Our overall conclusion, therefore, is that these continuously distributed facial characteristics are controlled in a similar way to those of physique and, in a similar manner, will produce predictably intermediate offspring without any marked increase in variability and without any risk of disharmonious or incompatible combinations. For those facial characters controlled by just a few genes the situation may be rather different. We may consider the situation with regard to hair form first. Fischer's (1939) model for the genetics of hair form has not as yet been formally superceded, although it is now generally recognized to be probably too simple an interpretation. Fischer suggested that three genes control hair form: *Ll*, with *L* being dominant for straight (Mongolian) hair; *Cc* which generally controls wavy (Australian) type hair; and *Sc* with *S* being dominant for spiral twisting. Smooth European hair would then be *llccss*, woolly Negro hair *llCCSS* and Chinese *LLccss*. It is assumed that *L* is generally epistatic to *C* and *S* and that *S* cannot work in the absence of *C*. A fourth gene *P* is necessary for peppercorn hair which would be the homozygous dominant. Hottentot hair would then probably be *PpllCcSs* (Gates, 1957).

Given this relatively straightforward system it is fairly easy to predict the hair form of hybrids. It may be, of course, that more than two alleles at each of these loci are contributing to hair form, in which case the distinctions between the various hair types, especially in the hybrids, will become blurred. More recent work with a wider variety of races and racial hybrids has suggested that overall there are a greater number of genes for hair form than Fischer suggested. Thus Gates (1962) suggests that between the peppercorn hair of the South-African Bushmen and the woolly hair of the Bantu Negro there are two intermediate hair forms known as tufted and matted. This series of four types, peppercorn-tufted-matted-woolly, can now be accounted for on the basis of probably two or three genes showing additive effects and without dominance.

Without pretending that we know all the details we can still say that the overall situation is relatively simple and straightforward and should allow fairly good predictions concerning the hair form of racial hybrids. The published investigations suggest that this in fact is the case. Thus, for example, Fleming (1939) found that among 81 Negro/White hybrids 72·7 per cent showed the characteristic

Negro hair form and 46/49 offspring resulting from backcrosses to Negroes showed the Negro hair form, while 14/15 White backcrosses showed essentially the White hair form. Data such as these come, in general, from the more frequent and therefore the more well-documented racial crosses. Some of the rarer crosses, which are known from only a few families with a very limited number of offspring, suggest that further additions to the number of genes controlling hair form will have to be made. For instance Gates (1958a) has examined a few offspring from Zulus and East Indians in Natal. The hair form in the F_1 hybrids and in later generations is quite different from that in Negro/White offspring. It is much more abundant and with finer crinkles, indicating that East Indian hair is different genetically from European hair.

The general situation with regard to hair form and racial hybridization would therefore seem to be that a number of different genes are known to effect hair form, which may or may not interact with one another, but which definitely have major effects as opposed to biometrical or polygenic effects. Racial mixing will therefore bring about the genetic segregation of these different types which may remain rather distinct from one another. Such mixing will certainly introduce rather greater phenotypic variability into many races than exists at present, and the frequency of these new hair forms will obviously be directly proportional to the number of new genotypes introduced by such racial mixing.

The comparative genetics of mammalian hair colour have been extensively reviewed by A. G. Searle (1968), who discusses the five principal loci controlling hair colour, Agouti, Brown, Albino, Dilution, and Extension, together with a number of less important loci including both structural genes and modifiers. There is little doubt that the Albino and Brown loci are represented in man. The Brown locus probably has many alleles and albinism is probably caused by a number of different loci.

The albino phenotype is an interesting one in man. Witkop has recently divided it into a number of different syndromes which have been conclusively shown to be due to different genes (Witkop, 1971). As already mentioned (Chapter 2, s. 2.2.*d*), some of these show curious racial differences in frequency, others show marked local variations in frequency, presumably mainly due to chance effects in the isolates in which they were sampled. In view of the marked selective disadvantage of most albinos owing to visual defects, deafness, and other traits, particularly photophobia, it is surprising that there is some evidence that the frequency of albinism is higher in Negroes than in Caucasoids (*see Table* 6.12). For instance, in a survey of the North Carolina population Witkop (1971) found that tyrosinase-positive albinism was significantly higher in Negroes than Whites. A number of surveys reported by Waardenberg (1961) in which the type of albinism was not determined, give figures of about 1:20,000 in many countries of W. Europe, compared with 1:3,000 or 1:4,000 in parts of Africa.

Some of the known information on types of albinism and their frequencies are shown in *Table* 6.12. Albinism has been reported from all the continents and all the major races including, recently, the Australian Aborigines (Walker, 1969). So far as race crossing is concerned, Witkop (1971) reports albino offspring from an albino Negro mother and a normally pigmented European father, of albino children from a Chinese father and a Filipino mother, and an albino son

Table 6.12.—SOME DIFFERENT TYPES OF ALBINISM AND THEIR RACIAL FREQUENCIES (*After Witkop*, 1971.)

TYPE OF ALBINISM	PHENOTYPIC FREQUENCIES		
	Negroid	Caucasoid	Mongoloid
Oculocutaneous Albinism			
Tyrosinase test negative			
1. Tyrosinase-negative oculocutaneous albinism	1:28,000*	1:39,000* 1:15,000†	
2. Hermansky-Pudlak syndrome			
Tyrosinase test positive			
1. Tyrosinase-positive oculocutaneous albinism	1:15,000* 1:2,858§	1:40,000*	1:143 —1:6,500‡
2. Chediak-Higashi syndrome			
3. Cross syndrome			
Tyrosinase test variable			
1. Yellow mutant			
Ocular Albinism			
Cutaneous Albinism			
1. Without deafness e.g., Menkes syndrome			
2. With deafness e.g., Waardenburg syndrome			

* North Carolina (Witkop, 1971).
† Ireland (Froggatt, 1960).
‡ Various N. Amerindian tribes (Witkop, 1971, Woolf and Dukepoo, 1969).
§ Negroes of S. W. Nigeria (Weiner, 1951).

from a Caucasian French father and a Malgache mother from Madagascar. This was long taken as evidence of the world-wide distribution of a single allele for albinism. More recently Witkop has reported families in which both parents were albinos but one or more offspring completely normal, clearly indicating non-allelic genes.

This information in *Table* 6.12 is given, not because it is itself very relevant to studies of racial crossing, but because it shows how complex this seemingly simple trait of albinism has become with recent detailed study. The implication is that many of the other traits discussed in this and other chapters are of similar complexity, and that to make sweeping generalizations regarding the genetic constitution of this or that race and the implications for racial hybrids completely misrepresents the real facts of racial biology.

As a very brief indication of the reality of this, Witkop quotes some data on phenotypic effects of known albino heterozygotes. A common finding in the past has been that albino heterozygotes very frequently showed marked iris translucency; for instance, Froggatt (1960) found that 68 per cent of the heterozygote parents of Irish albinos had abnormal iris translucency. However in the Brandywine triracial isolate in the United States there were no abnormal effects in heterozygotes and it became clear, after comparing heterozygotes from the Brandywine triracial isolate with Caucasian Negro, and Amerindian heterozygotes, that this phenomenon is associated only with the tyrosinase-negative forms of albinism, and in fact is only reliably present in albino heterozygotes from lightly pigmented populations.

Red hair is probably distinct from the Brown series and is usually inherited as a recessive. It occurs only infrequently in races other than Caucasoids. The other loci have not been recognized yet in man. In most cases of racial crosses darker hair colours are much more prominently represented among the offspring although there is usually a noticeable blending effect, as is shown for instance in *Table* 6.13.

Table 6.13.—THE PERCENTAGE OF DIFFERENT HAIR COLOURS AMONG ENGLISH AND ANGLO-NEGRO CHILDREN (*From* Little, 1943.)

		Medium Red	Light Red	Fair Red	Fair	Light Brown	Medium Brown	Dark and Very Dark Brown
Anglo-Negro	Boys	0·9	—	—	0·9	0.9	10·0	87·3
	Girls	—	—	—	2·0	1·0	11·6	87·5
English	Boys	4·5	1·8	—	39·1	40·9	8·2	5·5
	Girls	1·7	1·7	7·7	26·5	28·1	19·7	14·5

There are a number of other characters inherited in a similar relatively simple manner to hair form. We may briefly mention the epicanthic fold of the eye, eye colour, differences in the form of the ear lobe, and the presence or absence of the chin. The epicanthic eyefold, that fold in the upper eyelid characteristic of Mongoloid races, varies considerably between these local races. The Mongolian proper, Amerindian, and African eyefolds are all different. Hooton (1947) recognized the complete Mongolian fold, the internal and external epicanthic folds, and the median fold occurring among different groups. The usual condition in European eyes is no fold at all. Young White children, however, frequently possess an epicanthic fold, which is stated to be present in the foetus of all races. The nature of the eyelid and epicanthic fold in different races is shown in *Fig*. 6.14.

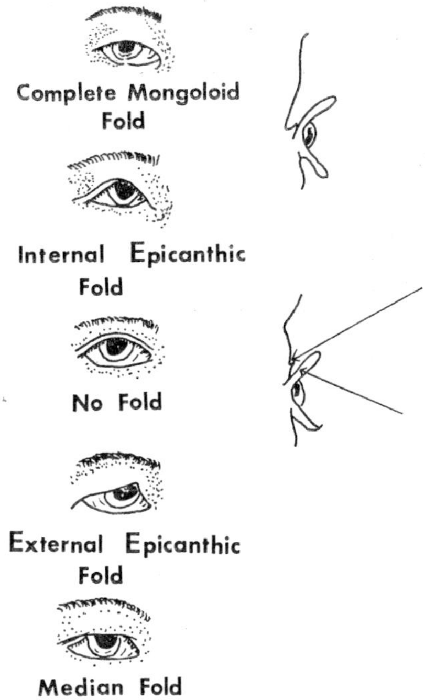

Fig. 6.14.—Varieties of skin fold over the upper right eyelid. (*From* Montague, 1960.)

So far as the inheritance of the trait is concerned, it has been found to be present in offspring of crosses between Mongoloid and all other races so far examined, although usually not fully developed. Thus in hybrids between White Americans and Japanese, many of the offspring showed the inner fold half or completely present. Negro fathers and Japanese mothers had children with the inner fold more often fully present, although the top fold was usually only half present. *Table* 6.14 from Wissler (1930) shows the situation with regard to the epicanthic fold in crosses involving Hawaiians, which again shows the general tendency of the eyefold to be inherited in a dominant manner, although almost always with some modifications in interracial crosses such that it is expressed to a lesser degree.

Eye colour in man is controlled in a rather complex way, but a great deal of the observations on the inheritance of eye colour can be explained on the basis of a single gene with three alleles E^{bl}, E^{gr}, and E^{br}. The homozygous $E^{bl}E^{bl}$

is blue eyed, the homozygous $E^{gr}E^{gr}$ is grey eyed, and the homozygous $E^{br}E^{br}$ is dark brown eyed. Heterozygotes $E^{bl}E^{gr}$ will be grey eyed and heterozygotes $E^{bl}E^{gr}$ and $E^{gr}E^{br}$ will be various shades of green, hazel, and light brown.

By far the commonest eye colour in man is dark brown, and this is also the universal eye colour of the great apes. A heavily pigmented dark brown iris will only allow visible and ultra-violet light to enter the eye through the pupil. This gives a clear image and would appear to be at a marked advantage in very

Table 6.14.—THE PERCENTAGE OCCURRENCE OF THE EPICANTHIC EYEFOLD IN VARIOUS RACES AND THEIR HYBRIDS

RACE	ABSENT		TRACE		MEDIUM		MARKED	
	Male	Female	Male	Female	Male	Female	Male	Female
Hawaiian	50	46	34	42	12	11	4	1
Chinese	1	1	4	4	72	72	23	23
Japanese	1	1	2	3	67	65	30	31
N. European	96	98	2	1	1	1	1	0
S. European	95	96	5	3	0	1	0	0
Haw-Asiatic	9	12	23	26	46	47	22	15
Haw-N. European	75	77	18	17	6	5	1	1
Haw-S. European	72	77	22	14	5	8	1	1
Haw-Asiatic Eur.	32	30	37	39	20	23	11	8

bright areas. Thus dark-coloured eyes are common or universal in the tropic zones, whereas blue and grey eyes are restricted to the cloudy temperate zones.

Reports on eye colour in racial hybrids are scattered through the literature, but the general impression of dominance of the alleles for brown and dark brown eyes is apparent from the reports of Fleming (1939) and Little (1943), who each found in excess of 95 per cent of Anglo-Negro hybrids with brown or very brown eyes. The comparative English sample (Little, 1943) shows 2 per cent (boys) and 6 per cent (girls) with blue eyes, 7 per cent (boys) and 14 per cent (girls) with blue-grey eyes, and about 24 per cent of each sex with grey eyes.

Among Hawaiian hybrids a similar picture emerges as is shown in *Table* 6.15.

One of the characteristics of many African races is the absence of an earlobe and the general smallness of the ears. In a survey of Japanese war children of both White and Negro fathers, Gates (1958b) found the distribution of earlobe types which is shown in *Table* 6.16.

Gates concludes that 'It is clear that a small lobe or none is more frequent in the children of a Negroid father, whereas a medium-sized or large lobe is much more common in the children of a White father. Probably not more than two or at most three cumulative genes are involved'.

Finally we might mention that fairly important facial feature—the chin. This, in its well-developed form, is most common among Europeans and *Table*

Table 6.15.—THE PERCENTAGE OF DIFFERENT EYE COLOURS AMONG EUROPEAN AND HAWAIIAN CHILDREN AND THEIR HYBRIDS (*Compiled from the data of* Wissler, 1930.)

	BLACK/ DARK BROWN	LIGHT BROWN	BLUE	GREY	BLUE/BROWN OR GREY BROWN
North European	11·5	18·5	38·5	5	26·5
Hawaiian	92·5	7·5	—	—	—
North European/ Hawaiian hybrids	56·5	30	4	1·5	8

Expressed as percentages.

Table 6.16.—THE DISTRIBUTION OF EARLOBE TYPES AMONG HYBRID JAPANESE CHILDREN (*From data of* Gates, 1958b.)

	NONE	SMALL	MEDIUM	BIG
Children of White fathers	11 (12·2 per cent)	18 (20 per cent)	49 (54·4 per cent)	12 (13·3 per cent)
Children of Negro fathers	8 (25 per cent)	8 (25 per cent)	13 (40·6 per cent)	3 (9·4 per cent)

Table 6.17.—THE FORM OF THE CHIN IN DIFFERENT RACES AND THEIR HYBRIDS (*After* Wissler, 1930.)

	PROMINENT (per cent)	MEDIUM (per cent)	RECEDING (per cent)
Hawaiian	3·5	48·5	48·0
Chinese	1·0	42·0	57·0
N. European	50·0	41·5	8·5
S. European	18·5	62·0	19·5
Hawaiian/Asiatic	3·0	46·5	50·5
Hawaiian/N. European	17·5	58·0	24·5
Hawaiian/S. European	8·0	51·0	41·0
Hawaiian/Asiatic/European	6·0	57·0	37·0

6.17 shows that on crossing them with Hawaiians and Asians, in whom the chin is not prominent, it appears to behave as a biometrically controlled character. In other words, hybrids are intermediate between the parental races.

6.3.b. *Skin Pigmentation*

One of the problems in analysing human skin coloration in the past has been that of finding an objective method of measurement. Prior to 1950 the work on skin colour in man involved the use of a colour chart which had to be carefully painted by an artist. The skin of the subjects under scrutiny had then to be compared with this colour chart and an assignment to a particular colour category made in this way. However, Weiner (1951) and others showed that portable reflectance spectrophotometers could be used in the field for obtaining objective and repeatable measurements of pigmentary variation on a continuous scale of reflectance. The principle employed is essentially very simple. Light is emitted from a standard source, an electric bulb, and passes through interchangeable filters, with characterized transmission curves on the surface being measured. The amount of light reflected from this surface, as compared with

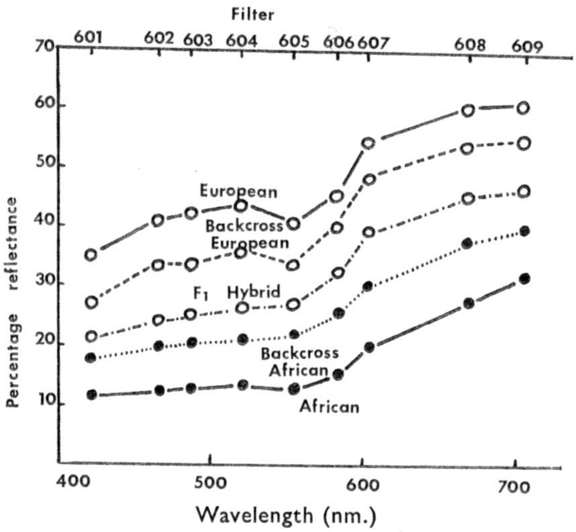

Fig. 6.15.—Mean reflectance curves of Europeans, Africans, and various hybrids from England. (*From* Harrison and Owen, 1964.)

the amount reflected from a pure white standard, is measured by a photocell and recorded in a galvanometer unit. Variations of this method are now universally used by anthropologists working in this field. Although a great deal of work was done on the inheritance of skin coloration prior to 1950 (*see*, for example, Davenport, 1913b; Gates, 1962) and the introduction of reflectance spectrophotometry, we will restrict our discussion almost entirely to the later work using this new method. Even then comparisons between one report and another are still difficult because the position of measurement on the body has to be accurately known, differences in environment (particularly the amounts of ultra-violet radiation received by the measured part of the body) must be allowed for, and the full value of the method can only be realized if reflectance curves over the whole of the visible spectrum are obtained.

Illustration of this last point is provided by Harrison and Owen (1964), who studied the inheritance of skin colour in a group of Africans, Europeans, and their first and second generation hybrids in the port of Liverpool in England. Some of their results are presented in *Fig.* 6.15. These curves show the characteristic absorption of melanin at shorter wave-lengths and the absorption band of haemoglobin at 545 nm. which becomes increasingly prominent in those individuals with a lower amount of melanin in their skin. The F_1 hybrid curve is approximately intermediate between the two parental groups, although it is clear that the exact relationship changes with wave-length. The curve is nearer that of the African parent than the European at 424 nm., while at 685 nm. the relationship is reversed. The European backcrosses are roughly intermediate between the curves for the F_1 hybrid and the European, but in the case of the backcross African the curve lies nearer the F_1 hybrid than the African parent.

Now Harrison and Owen showed in 1956 that *in vitro* melanin concentration is linearly proportional to the reciprocal of the reflectance values, and therefore they transformed some of their *in vivo* data to reciprocals. The results of this transformation are shown in *Table* 6.18, which clearly indicates that the F_1

Table 6.18.—RECIPROCAL TRANSFORMATION OF SKIN REFLECTANCE (R) AT DIFFERENT WAVELENGTHS IN EUROPEANS, AFRICANS, AND THEIR HYBRIDS (*From* Harrison and Owen, 1964.)

WAVELENGTH	EUROPEAN	F_1 HYBRID	F_2 HYBRID	AFRICAN	B_E HYBRID	B_A HYBRID
$\frac{1}{R} \times 1000$						
425 nm.	28·2	50·6	47·4	85·4	39·4	58·5
545 nm.	24·7	36·5	34·0	74·0	29·9	44·1
685 nm.	16·1	19·5	18·9	29·5	17·5	21·3

hybrid is more similar to the European than the African parent in terms of melanin concentration. The numbers involved in Harrison and Owen's study are small (105 Europeans, 166 West Africans, 94 F_1 hybrids, 30 backcross Europeans, 26 backcross Africans, and 14 F_2 hybrids) and they experienced considerable difficulty in interpreting their results in a genetic way. They were, however, able to conclude that there were three or four effective factors contributing to skin colour differences between these two races, a figure which broadly agrees with the conclusion of Stern (1970) that only 3 or 4 gene pairs are responsible for pigmentation variability in the American Negro. Earlier work by Gates (1953, 1958a) had also suggested that three or four genes were responsible for controlling human skin coloration. Gates, however, suggested that three of these, which he called R, S, and T, had respective melanin values of 6, 2, and 1 and the fourth, Q, was largely responsible for the mahogany skin of the Pygmies in Africa, a colour unique to that race. This, of course, is in direct contrast to the assumption of equal and additive gene effects which both Stern and Harrison and Owen make. Nevertheless this latter assumption does allow a model to be constructed which conforms quite closely with reality.

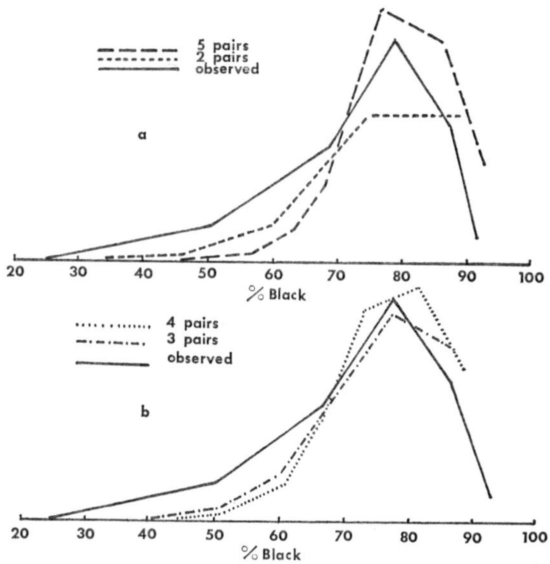

Fig. 6.16.—Frequency distributions for skin pigmentation observed and expected based upon (a) models of 2 and 5 gene pairs, (b) models of 3 and 4 gene pairs. The frequency distributions are generated from the binomial $(0{\cdot}8N + 0{\cdot}2C)^{2n}$ where N are African alleles, C are Caucasian alleles, and n the number of gene pairs.

Fig. 6.16 shows the terms of the binomial $(0{\cdot}8N + 0{\cdot}2C)^{2n}$ where N are African alleles and C are Caucasian alleles and *n* the number of gene pairs. These particular terms for the distribution were chosen by Stern, since Reed (1969a, b) had shown that the Negro American population is likely to contain about 20 per cent Caucasian alleles.

The problem remains of deciding whether this small number of effective genetic factors can account for world-wide variation in skin pigmentation. Certainly a greater degree of variety in skin coloration than would be suggested by the existence of just 4 genes with 2 alleles, has been found in different parts of the world, for example by Barnicot (1958) among neighbouring tribes such as the Yoruba and the Ibo in Nigeria, among various Indian populations (Kalla, 1969), and in the different coloured peoples of New Guinea (Walsh,

1971). However, the complex environmental effects make it very difficult in these cases to decide how many genes might be involved. The results of comparative studies on Indians, Europeans, and Indo-European hybrids by Tiwari (1963) are similar to those of Barnicot (1958) and Harrison and Owen (1964) in studies of Afro-European crosses, as shown in *Fig.* 6.17. This similarity extends down to the fact that the mean curves of the hybrid children lie

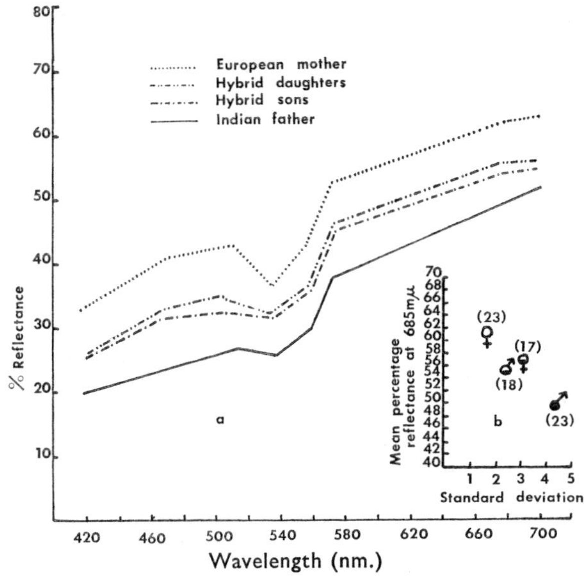

Fig. 6.17.—a, Reflectance curves of forearm in European women, Indian men, and their hybrid sons and daughters. b, The relationship between variability and percentage reflection at 685 nm. of the parents and their hybrid children. (*From* Tiwari, 1963.)

in an intermediate position over most of the reflectance range, although they are distinctly nearer the darker, Indian, parents at the short-wave end of the range. It may therefore be assumed that the same genes are involved in Indian as in African pigmentation.

A small number of mixed families in Papua were studied by Walsh and Price (1963). The crosses were mainly half-European/Papuan, half European/quarter European, with some other mixtures. Among the offspring they found some evidence of a bimodal distribution of skin coloration and while on average it could be said that the offspring were between the parental values, the two modal groups were often as dark or as light as their parents, sometimes even exceeding them for reflectance recordings on some parts of the body. Here, then, is evidence for a small number of genes segregating in complex crosses.

In point of fact, of course, from what we know of the biochemical synthesis of melanin and the structures involved in its production, we would expect the genetic control of pigmentation to be broadly the same in all races. The major enzyme involved is tyrosinase and one at least of the loci determining skin pigmentation probably involves variations in this enzyme. Since heterozygosity for most enzyme variations results in enzyme activity closer to the more active variant, it would be expected that this locus would have some dominance for darker skin colour associated with it (Livingstone, 1969d). We have seen that this is indeed the case in some circumstances.

In addition to this, potent tyrosinase inhibitors are known to be present in Caucasian skin (Kugelman and van Scott, 1961), and Livingstone points out that if these are genetically comparable to repressors then one might expect some dominance at these loci for light skin colour. Again this is a situation which seems to occur in some circumstances. It seems, therefore, as if we may be dealing with a situation of balanced dominance for skin colour loci. This would mean that hybrid populations would perhaps be likely to show a greater degree of variety for this phenotypic characteristic than they do for many others which are also continuously distributed, and in fact Walsh and Price (1963) suggested a bimodal distribution among segregating offspring of various race mixtures in New Guinea. It is possible to decide if greater than usual phenotypic variability is in fact manifested by skin coloration by examining a fairly long-established hybrid population with a relatively high degree of admixture. A population such as that of southern Brazil for example. This population is compounded essentially of African Negroes and European Whites, with perhaps a small amount of Amerindian admixture which might be expected to increase observed variability even farther. Harrison and others (1967) studied a number of children aged from 7 to 16 years in Porto Alegre, the capital of the state of Rio Grande do Sul. In all cases measurements were taken with an EEL reflectance spectrophotometer at 425 nm., 545 nm., and 685 nm. on the medial aspect of the upper arm, the medial aspect of the forearm, and the forehead. Now Tondo and Salzano (1962) and Salzano and Hirschfield (1965) had already examined this hybrid population to determine the amount of Negro-White intermixture, using a number of single gene systems. If the three estimates which these workers arrived at, based on R^0, Hp^1, and Gc^1 (Rhesus, Haptoglobin, and Group Specific Component), which are the most reliable indicators, are given equal weight, the arithmetic mean for the degree of White admixture is 46 per cent. The estimate of the degree of White admixture based on skin colour was obtained by expressing the differences between the Brazilian Negro and the Brazilian White as a percentage of the differences between the African Negro and the Brazilian White. There was some variation in the estimates of admixture obtained, partly dependent on the wavelength at which reflectance was measured, partly upon the particular African sample chosen for comparison and partly upon sex. Harrison and others decided that their most reliable

African comparison was with the Kuangali Okavanga Bantu tribe (Weiner, Harrison, Singer, Harris, and Jopp, 1964). This comparison suggested 47 per cent White admixture from the male sample and 52 per cent admixture from the female sample. In either case, then, there is close agreement with the other estimate of genetic admixture based on single gene frequencies.

In addition to this is Tiwari's (1963) finding that Anglo-Indian F_1 hybrids are of intermediate variability between the highly variable Indian fathers and

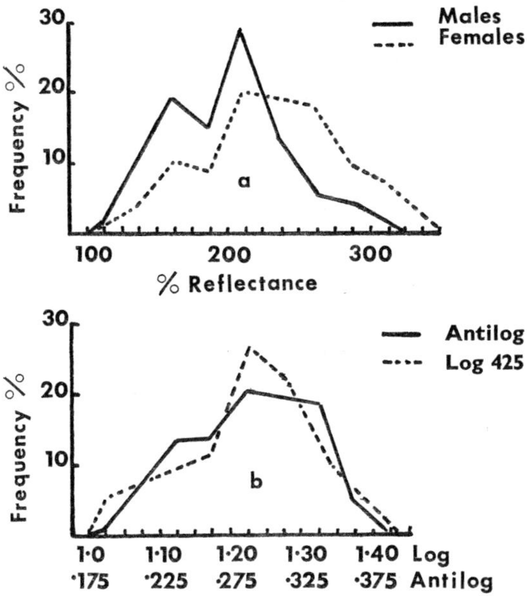

Fig. 6.18.—a, The observed frequency distributions of skin colour in a hybrid Brazilian population as measured by reflectance at 545 nm. b. Observed frequency distributions of female skin colour in a hybrid Brazilian population as measured on transformed scales. (*From* Harrison and others, 1967.)

moderately English mothers (*Fig.* 6.17b). It may be supposed, therefore, that skin colour behaves in the same additive way as virtually all the other continuously varying characters we have examined.

To further examine this possibility that skin colour in hybrid populations might yet be more variable overall than other continuously varying characters, we can examine the frequency distribution of skin colour in this S. Brazilian population which Harrison and others determined (*Fig.* 6.18a, b). These curves are based on 216 White males, 268 White females, 68 hybrid males, and 117 hybrid females. The distributions have been standardized for area and are exactly comparable. There is an indication of bimodality in the distribution of reflectance values at 545 nm., and on each of the transformed scales there is some evidence of negative skewing, which is consistent with bimodalism.

Inequality of gene effects, such as we have already mentioned, could account for these phenomena, although the other possible explanation is that mating is not at random within the hybrid population. At any rate this study confirms earlier findings that only a few genes are responsible for skin colour differences between the races. More importantly it shows that variability for continuously varying characters among a long-established hybrid population is no greater than among the contributing parental populations, and that skin coloration can be included in this category. We may compare this with some of the theoretical predictions to the same effect that we have already made in Chapter 5.

Since the genetics of skin coloration is reasonably well understood and the proportion of Negroes and Whites in the U.S.A. is known, it is possible to

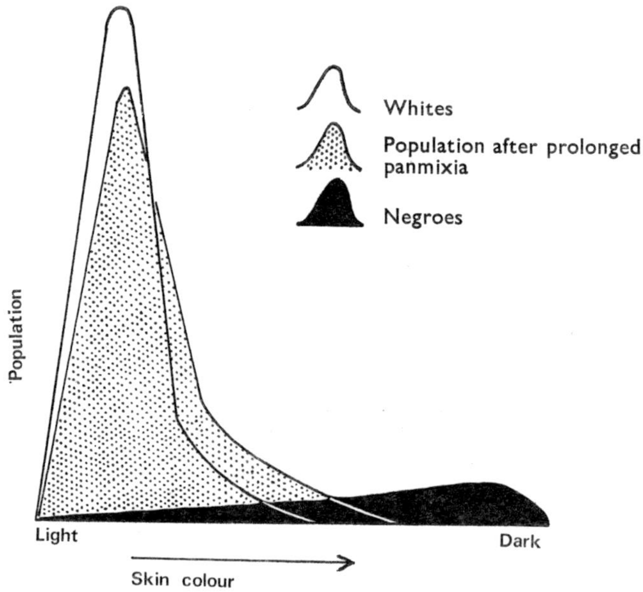

Fig. 6.19.—Frequency distributions of skin colours in the United States, now and after prolonged panmixia.

predict what the phenotype of the United States population would be after long continued panmixia. If the frequency of African alleles among Negroes is of the order of 80 per cent and among Whites 0 per cent, and if the frequency of Negroes among the American populations as a whole is 10 per cent, then the frequency of African alleles overall is $q_n = 0{\cdot}08$ and European alleles is $p_c = 0{\cdot}92$. At equilibrium the nine colour types with from zero to eight African alleles are obtained from the binomial $(0{\cdot}92C + 0{\cdot}08N)^8$, on the assumption of four loci each with two alleles of equal effect. The terms of this binomial show that at equilibrium slightly over one-half of the population will have only

White genes, and over a third will have only one African gene. The above argument is that of Stern (1960) and *Fig.* 6.19 shows the change which would come about in the skin colour of the United States population if mating was completely at random with regard to skin colour: the Negro or black skin coloration would completely disappear and there would be a small and barely perceptible darkening of the average White man's skin coloration. It is possible to carry out similar calculations in other parts of the world, both with regard to

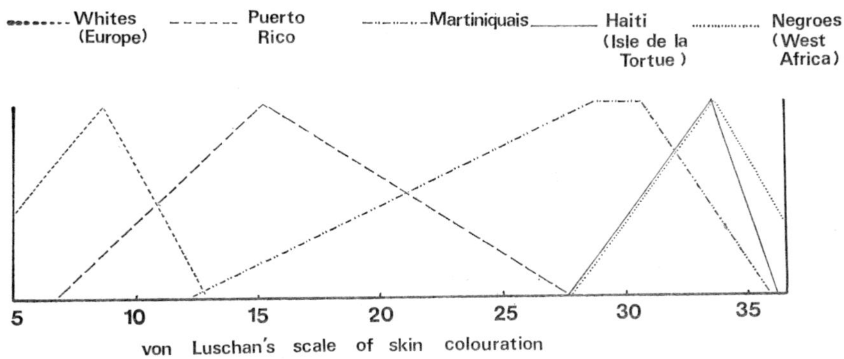

Fig. 6.20.—Comparison of skin pigmentation of some Negroes and Whites and their hybrids. (*From* Benoist, 1963.)

skin colour and other traits, provided the number of genes concerned and their relative frequencies in the amalgamating populations are known.

In a number of cases the distribution of pigmentation in long-mixed populations has been determined using the colour scale of von Luschan, and it is found that mixed populations are considerably more variable than European or West African populations. This is shown in *Fig.* 6.20 taken from the lengthy paper by Benoist (1963) on the people of Martinique. The populations concerned, those of Puerto-Rico and Martinique, have certainly not approached panmixia, though, and the difference in the means and shape of distribution of these two populations is a reflection of their relative racial contributions.

6.3.c. *Dermatoglyphics*

To conclude this section on superficial characteristics, we may briefly consider the dermatoglyphic patterns of racial hybrids. We have already discussed some of the racial variation characteristic of these traits, and mentioned that their significance from the present point of view is that they are almost entirely under genetic control. Furthermore they are very sensitive markers of the overall genetic stability of the individuals concerned (Cummins and Midlo, 1943; Holt, 1968). They are particularly diagnostic of abnormalities such as schizophrenia and chromosome imbalance, and they are also recognizably different from normal in individuals suffering from neurofibromatosis, a heritable disease characterized by the formation of multiple tumours of nerves, and possibly other partly genetically controlled traits. The stability or otherwise of dermatoglyphics in racial hybrids might therefore be considered a fairly sensitive indicator of biological stability in these hybrids.

Davenport and Steggerda (1929) included a study of dermatoglyphics in their work on the races in Jamaica. For finger-prints they studied 124 Blacks, 213 Browns, and 47 Whites, and they found that for both whorls and arches and for overall pattern intensity there was a graded series as shown in *Table* 6.19.

Table 6.19.—DERMATOGLYPHIC PATTERNS AMONG JAMAICANS
(*From* Davenport and Steggerda, 1929.)

	WHORLS	ARCHES	PATTERN	INTENSITY
Blacks	30 per cent	11 per cent	12·01	11·76
Browns	25 per cent	10 per cent	11·63	11·39
Whites	22 per cent	7 per cent	11·69	11·19

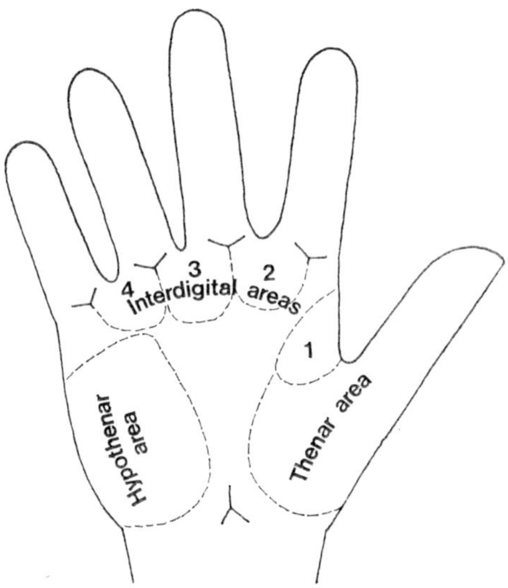

Fig. 6.21.—Map of the six chief dermatoglyphic areas of the palm.

Intermediate values in pattern frequencies of the two palmar areas, the hypothenar and fourth interdigital (*see Fig.* 6.21), were also found, but the three other configurational areas showed a different relationship. Too much cannot be read into these results because of the small numbers involved and also because of the possibility of admixture even among the Blacks.

Cummings and Midlo (1943) report on some analyses of Mexican Indians, Mexican Spanish, and their hybrids. Again with regard to finger-print traits the hybrids were found to be intermediate. With regard to frequencies of palmar, hypothenar, and thenar/first interdigital patterns, the hybrids are intermediate between the Spanish and Indians. The third interdigital area, repeating the trend noted in connexion with Negro-White hybrids, showed a higher pattern frequency than either of the two original stocks.

Tiwari (1963) studied the dermatoglyphics of Indians and Europeans and their hybrids. His data confirm the fact that there are no important departures from normality among hybrids, although he too found that in one or two cases the prints of palm or finger lay outside the values of the parental races. For instance, although the European mothers show almost double the number of loops in the hypothenar area as do their Indian fathers, the incidence in both male and female hybrid offspring is greater than in either of the parents, and in general it seems that loops are more frequent in the hybrids than in their parents except in the second interdigital area.

In the case of finger-prints the frequency of whorls in male hybrids is even higher than in the Indian fathers, who in turn show higher values than the European mothers. The female hybrids show intermediate values. In analysing ridge counts certain sex factor corrections have to be made (Holt, 1952). Tiwari first calculated the total ridge counts of twenty-three families of Indian-European marriages with one or more children. His findings were:

	Mean Ridge Count	Standard Deviation
23 Indian fathers	158·91	55·06
23 European mothers	132·74	52·85
18 Male hybrids	159·94	37·14
17 Female hybrids	147·71	45·50

A more accurate approach is to take the ridge count separately for each child and to calculate a weighted average, correcting the resulting daughters' mean and mothers' mean for sex. Tiwari's results are shown in *Table* 6.20. The observed value for the hybrids lies outside the range defined by the parents, but it differs from the expected intermediate value by only 7·8, which is a value lower than the error for either parent.

Similar studies (DeSá, Benevides, and Salzano, 1969) in Brazil could not come to any very clear conclusion with regard to the manifestation of finger-prints in Caucasoid/Negroid hybrids. For many specific features the Brazilian Negroid samples fell outside the range given by the Portuguese and African frequencies, although the overall differences between the races were not large.

In summary, then, so far as what we have termed 'superficial characteristics' are concerned, we can draw a twofold distinction. There are firstly a number of racial characteristics which are continuously varying and are controlled by a fairly large number of genes. These include such physical features as nose width, mouth width, width of lips, facial height, cranial index, form of chin, and other less commonly recorded parameters. These all have intermediate values in racial hybrids and usually accurately reflect the relative racial contributions to the hybrids. Within this category we can include skin colour, although

Table 6.20.—WEIGHTED CORRECTED AVERAGES FOR FINGER RIDGE COUNTS IN INDIAN MALES, EUROPEAN FEMALES, AND THEIR HYBRIDS. (*From* Tiwari, 1963.)

No.	INDIAN FATHER (Father's Mean)	HYBRID		EUROPEAN MOTHER (Mother's Mean)*
		Son's Mean	Daughter's Mean*	
18	161·05	160·05	—	149·27
17	146·94	—	159·80	151·92
Total 35	154·20	159·92		150·54
		Expected intermediate value 152·37		
		Difference 7·8		
	S.D. 55·1			52·8
	S.E. 9·5			9·1

* Corrected for sex.

this shows a slight but definite tendency to be closer to the European, or light-coloured parental race, than the coloured race in racial hybrids.

The second group are those controlled by one or a few genes—eye colour, hair colour, and hair form—which on the whole show a greater or lesser degree of dominance for the non-European form.

This fairly simple division serves to explain and predict the behaviour of most superficial characters on racial mixing. Within the overall limits set down by the homeostatic growth mechanism, to which we have already referred, we may get a mixture of racial characters. But there is no evidence of disharmonious arrays of characters or that any disadvantageous features result.

6.4. RACE CROSSING AND INTELLIGENCE

This topic is one of the most important, and at the same time one of the most controversial, in the whole field of racial comparisons. The genetical control of intelligence is incompletely understood, partly because intelligence itself cannot as yet be tested in isolation from cultural and environmental influences. Although it would be generally agreed among geneticists and psychologists that heredity contributes between 60 per cent and over 80 per cent of the determination of man's intelligence, there is still considerable latitude in these estimates and the different consequences which result if either of these two extremes are correct are quite profound (Erlenmeyer-Kimling and Jarvik, 1963). Many respected workers in this field claim that with our current knowledge and the methodology available to us at the present time, it is impossible to add precision to these estimates. For the same reasons these authorities suggest that it is impossible to decide with any certainty whether there are or are not any genetically based racial differences in intelligence. Other workers state that there are definitely differences between the races in intelligence, even after allowing for the obvious cultural and attainment inequalities. One of the best reviews of both sides of this debate can be found in the *Harvard Educational Review* for 1969, which contains a long paper by Arthur Jensen presenting the case that the average observed difference in I.Q. between Whites and Negroes in the U.S.A. of about 15 points is almost entirely due to genetic differences between the races. The same volume then publishes a number of commissioned and uncommissioned replies to, and rebuttals of, all or part of Jensen's paper from geneticists and psychologists. Other more general contributions to this debate are by Lewontin (1970a, b), Jensen (1970; being a reply to Lewontin's severe criticisms), Eysenck (1971), and Bodmer and Cavalli-Sforza (1970). The present situation seems to be that only a few experts in this field would agree with all of Jensen's arguments, although conversely, only a few would deny that in many areas he makes a strong case. After briefly looking at some of the evidence in this debate relevant to our particular purpose, we shall conclude that many of the concepts so freely quoted are but poorly understood, and that the evidence so laboriously gathered within the framework of these concepts is often heavily biased in a manner which it is difficult to avoid, using the methodology open to us at present. We must, in other words, suspend final judgement and say we just don't know whether there are any genetically determined differences in intelligence between the races. Hopefully we may be able to say that, if indeed there are differences, then the magnitude of such differences must lie between certain limits and have the more modest aims, at the present time, of defining such limits. This would allow us to say, if racial mixing is going to take place, that it might or might not have some predictably small overall effects on the general levels of intelligence in the hybrid population.

If indeed there are important differences between the races in what we might term 'fundamental intelligence potential', then we might expect the hybrids to be intermediate between the two parental races, as we have found to be the case for all other quantitative characters we have so far examined. And if there are significant genetically determined differences in intelligence between the races then clearly it could be argued that it would be detrimental in the long run for a race with superior intelligence to hybridize with one possessing inferior intelligence. Such an argument would, in fact, rest too heavily on typological and genetic uniformity ideas of racial classifications to be very important, but the point does have to be answered. Our prime purpose in this section, therefore, will be to examine the evidence for possible differences in intelligence between the races. To this end we shall consider, firstly, possible differences in brain size and structure, since these are characters which are, or should be, capable of objective accurate examination. We will leave aside for the moment the fact that the relationship between these features and intelligence as such may be by no means a direct one. Secondly, we will examine specific aspects of brain activity such as perceptual and spatial ability, colour blindness, musical aptitude, and so on, which, while not part of intelligence *sensu stricto*, are nevertheless possible contributors to what we might term 'intelligent behaviour' and are certainly important therefore from a social point of view. Thirdly, we will investigate possible differences in the rates of development and overall levels of sensorimotor and perceptual motor abilities for similar reasons. Finally, we will consider I.Q. itself, the best measure that psychologists have given us to date of the efficiency of our own higher sensory systems. Only when we have considered each of these will it be meaningful to consider intelligence in racial hybrids.

6.4.a. *Racial Differences in Brain Size*

We may note initially that all comparisons of cranial capacity as indicators of brain size are totally inadequate. Within the endocranial area, in addition to brain tissue, there are the roots and intracranial trunks of twenty-four cranial nerves, the thick outer brain covering, dura mater, the two thinner inner coverings, the arachnoid and pia mater, the subarachnoid space and its enlargements containing fluid, blood-vessels, and cerebrospinal fluid. Estimates of the cranial capacity that is taken up with brain tissue vary from 10 per cent (Brandes, 1927) to 33 per cent (Mettler, 1955). This proportion is not constant anyway, since the brain shrinks with age and during certain illnesses. Our indicator of brain size should therefore be the weight of the brain itself. Some figures for average brain weights among different races are given in *Table* 6.21.

Table 6.21.—AVERAGE BRAIN WEIGHTS (*From* Tobias, 1970.)

RACE	NO.	g.
Kenya Africans	389	1276
Africans		1293
American Negroes	139	1355
American Whites	2752	1301
11 White Series		1301–1455
3 Mongoloid Series		1360–1375

Figures such as these are often taken to prove that the average brain weights of European Whites are between 8 per cent and 10 per cent greater than African Negroes. However Tobias (1970) summarizes evidence that brain size varies with:—

1. Sex;
2. Body size;
3. Age at death;
4. Nutritional state in early life;
5. Source of sample (i.e., from pauper hospitals, etc.);
6. Occupational group;
7. Cause of death;
8. Lapse of time after death;
9. Temperature after death;
10. Anatomical level of brain severance;
11. Presence or absence of cerebrospinal fluid;
12. Presence or absence of meninges;
13. Presence or absence of blood-vessels.

Tobias writes, 'It must be concluded that among the widely varying estimates for Negro brain size, none reflects the normal brain size of a healthy Negro group' and 'These considerations largely invalidate most interracial comparisons so far published'.

It is furthermore well known that many eminent men of considerable academic achievement have shown enormous variations both above and below mean brain weights; conversely, men of modest achievement have also possessed very large and very small brains.

6.4.b. Racial Differences in Brain Structure

The only important difference in brain structure which has been suggested exists between the races is in the thickness of the grey matter of the cerebral cortex. Vint (1934) claimed that the thickness of the cerebral cortex of Kenyan Africans was, on average, reduced by 14·8 per cent compared with the European brain. There are a number of other considerations, however, which suggest that this claim and the accompanying one that the component supragranular layer of the cortex is reduced by 10 per cent cannot be upheld. Firstly Vint himself measured only Negro brains and compared his results with those of von Economo (1929), who had measured only European brains in Europe. This indirect comparison introduces a number of sources of error; first with regard to exactly where the sections of the cortex should be cut; secondly, just where the measurements should be made; thirdly, how the measurements should be taken; and fourthly, how the brain should be prepared for study. Each of these variables can bias comparisons between different workers and laboratories to a very great extent. They apply, moreover, to measurements made of the thickness of the whole cortex ranging from between 1·886 mm. and 3·006 mm. Vint himself was concerned with a component part of this layer, the supragranular layer, which he found to vary in average thickness from

0·338 mm. to 0·922 mm. in six different parts of the cortex. It seems clear that there is no basis in this 30-year-old piece of work, an isolated non-comparative study, for concluding that there is any difference between the races examined so far as brain structure is concerned, and certainly we cannot extrapolate to any other races. In the words of Tobias (1970), 'there is a need for more carefully controlled studies'.

Brain functioning, in so far as it is measured by electroencephalograms, seems to be similar in United States Negroes and Whites. Merrill and Cook (1957) and Johnson and Corah (1963) found no difference in electroencephalograms in these two peoples.

6.4.c. *Racial Differences in Special Aptitudes*

Although the investigation of simple sensory differences between the races would seem to be a promising approach to understanding more complex psychological differences between the races, there seems to have been relatively little work done on special aptitudes during the last few years. Much of the work we shall therefore discuss was carried out a long time ago, and no doubt the data could be considerably improved if extensive surveys of this kind were carried out again today. Dreger and Miller (1960, 1968) in two reviews cover much of the recent work in this field, but the bulk of their discussion is concerned with intellectual functioning, educational attainment, and personality. More recent work on sensorimotor and perceptual-motor development and abilities we will consider in the next section.

One of the earliest comparisons of races with regard to behavioural traits was carried out by an American clinical psychologist (Bache, 1895). A number of Caucasians, Amerindians, and a group of mixed African-Caucasian descent were compared with respect to reaction time to visual, auditory, and tactile stimuli. The Amerindians showed the quickest reaction times, followed by the African-Caucasian hybrids, with the Caucasians being the slowest to react. Bache's data are shown in *Table* 6.22, and in the second part of this table are shown some later results gathered in Hawaii by Livesey and Louttit (1930), who tested 253 university students of different racial and national origins. The differences inreaction time illustrated in this latter study are small and insignificant, and do not consistently favour any one group over any other. We may note that in most cases the part-Hawaiian individuals are intermediate between the Caucasians and non-Caucasians.

Other early work included that of Rivers and others on Torres Strait and Murray Islanders, and Rivers on the Todas of southern India (Rivers, 1901, 1905; Myers, 1901). The general findings of this work seem to be that the Murray Islanders are possibly superior to Europeans in visual acuity, and generally inferior in auditory acuity and tone discrimination; 'their sense of touch is twice as delicate as that of Englishmen while their susceptibility to pain is hardly half as great'. With respect to reaction times European subjects seemed to be faster in responding to visual stimuli, although there was no group difference in reaction time to auditory stimuli. The Todas were superior in visual acuity and tactile discrimination compared to Caucasians, but inferior

Table 6.22.—SOME DIFFERENCES IN REACTION TIMES OF DIFFERENT RACIAL GROUPS AND THEIR HYBRIDS.
(*Note that the different experiments are not necessarily comparable.*)

GROUP	NO.	AUDITORY R.T. (millisec.)		VISUAL R.T. (millisec.)
a. Bache (1895)				
White	12	146·9		164·8
Amerindian	11	116·3		135·7
Negro/Whites	11	130·0		152·9
b. Livesey and Louttit (1930)				
Caucasian	14	181		250
Chinese	29	205		295
Japanese	69	205		288
Part Hawaiian	15	182		272
Caucasian	45	217		297
Chinese	42	222		300
Japanese	41	244		345
Part Hawaiian	31	231		300
c. Myers (1901)				
Murray Islanders	17	135·7	11	243·8
English	24	141·6	21	222·3
Sarawakians	18	120·7	9	208·0

in auditory discrimination. The third part of *Table* 6.22 shows some of Myer's findings, which are comparable with those of other workers.

The other important early study is that on participants at the St. Louis World Fair in 1904. Bruner (1908) reported detailed findings regarding auditory acuity, and he concluded that Whites have a clear superiority, both in keenness and in the range of the hearing sense, over other races. Woodworth (1910) found that Caucasians were clearly superior to other racial groups in colour discrimination, although Amerindians and Malaysians were superior to Caucasians in visual acuity. A number of recent workers have studied racial and cultural differences in the incidence of various perceptual illusions (including the Muller-Lyer illusion, the horizontal-vertical illusion, the black-white irradiation illusion, and the Sander parallelogram). Campbell (1964) confirms a number of other reports that European subjects are more susceptible to the Muller-Lyer illusion than many other racial groups. The Muller-Lyer illusion is that in which two equal lines are made to appear of different length by the addition of arrowhead lines at their extremities (*Fig.* 6.22). It is argued by some that this is merely a reflection of the very different experiences to which peoples of various cultural levels are exposed. However, the fact that young children are more susceptible to the illusion than older subjects in Western European culture, argues for some genetic component of this trait.

Pettigrew and Nuttall (1963) confirmed the finding that Negroes and Whites differ in perception of the black-white irradiation illusion, but do not differ in response to variously coloured irradiation illusion cards. More Whites report

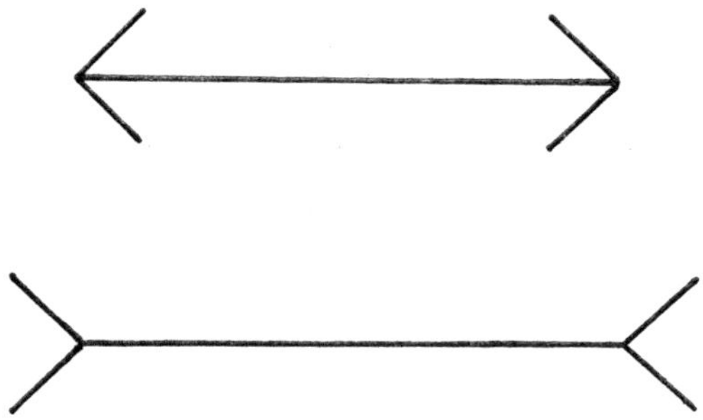

Fig. 6.22.—The Muller-Lyer illusion.

the black-white illusion than Negroes. They believe that the differential illusion response may be due to genetic factors evidenced by the generally greater visual acuity among Negroes.

The evidence for this latter assertion comes from such studies as that of Karpinos (1960), who studied the vision records of more than 150,000 military registrants in the United States, and discovered that 69 per cent of the Whites, as compared with 82 per cent of the Negroes, had 20/20 vision. On the other hand the proportion of correctability was found to be greater for Whites than for Negroes.

Myopia or near-sightedness is a trait which definitely shows variation in racial frequencies and Post (1971), as we mentioned in Chapter 2, has suggested that the relaxation of selection in more advanced cultures may be responsible for at least part of these differences. According to Mann (1966) six groups of populations in which myopia is rare or absent are: Bushmen, African Negroes, Amerindians, Australian Aborigines, some Indians (except Jews and Parsis), and Melanesians and Polynesians. Myopia is reportedly common in eight areas: China, Japan, Korea, Egypt and Morocco, Iran, Thailand, and all the countries of Europe, particularly Malta and Holland. Post cites this distribution of Myopia in support of his suggestion of relaxed selection.

A few recent studies point to possible auditory differences between races. Post (1964) summarized five large investigations on the difference between Negroes and Whites in this field of sensory response as follows:

1. Mild hearing losses of Negroes are more frequent than those of Whites at three lower pitches (tonal frequencies of 440,880, and 1760 cycles per second)

for four age groups; to a very slight degree in the younger (10–19) and maximally in the older (40–49).

2. At the highest frequencies the reverse situation is found, with mild hearing losses of Whites more frequent than those of Negroes (7040 cps). Results are mixed at the intermediate pitch (3520).

3. Among African Negroes tested in their own culture, where noise exposure has been minimal, the mild hearing losses at all tonal values were very rare.

Neither the evolutionary significance nor the genetic control of these differences have been determined, although Stevenson, Johnston, Golding, and

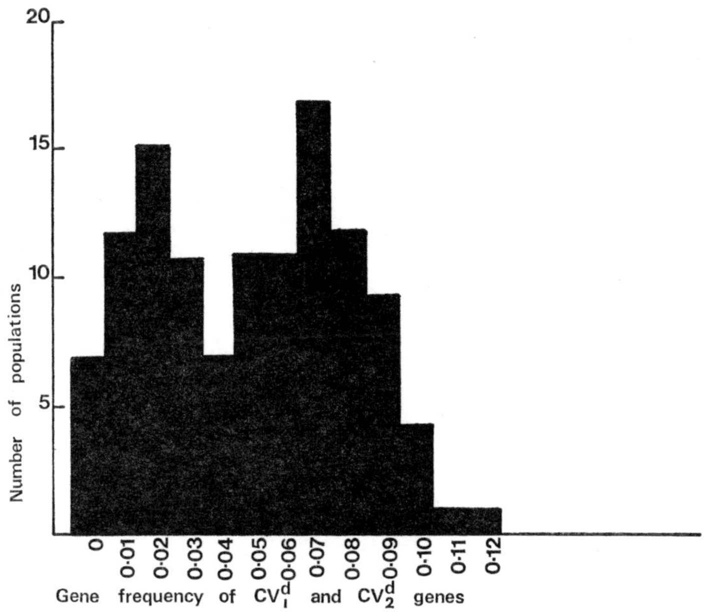

Fig. 6.23.—Distribution of protan and deutan colour-blind genes in samples from 114 populations. (*From* Spuhler and Lindzey, 1967.)

Stewart (1966) have shown that there are a large number of genes probably segregating which may cause hearing defects.

Virtually the only other area in which racial differences with regard to behavioural traits are known to exist, is in the field of defective colour vision. We have already examined this in Chapter 2 and the possible reasons for it. *Fig.* 6.23 shows the distribution of colour vision defect gene frequencies among 114 populations, a distribution which is clearly bimodal. We have already suggested that the selective disadvantage of red-green colour blindness in terms of differential survival and fertility varies today from about 1 in 1,000 to 1 in 10,000. There is little doubt, though, that these figures are only meaningful for the different societies in which these different values occur, and the disadvantage of colour-blind individuals is probably decreasing everywhere. It is not known for certain whether just one genetic locus with 5 different alleles or two closely linked loci each with 3 different alleles are involved in X-linked colour blindness. The consequences of either of these alternatives on racial amalgamation are very similar, in that the hybrid population will have intermediate gene frequencies which will generally mean, in the case of European hybrid populations, a lowered frequency for either or both colour-blind alleles.

This is about the sum total of our knowledge of special abilities and possible racial differences in them. Claims for racial superiority in the field of mathematical ability, literary ability, or musical ability have just not been investigated on a sufficiently quantitative basis to warrant our further attention. We might note that one recent small study did investigate the oft-repeated claim of the Welsh to be more musical than most other peoples and certainly more so than the English. McLeish and Thoma (1970) were able to show fairly conclusively that such claims were ill founded. Similarly the supposed special endowments of the Negro in the field of music, especially rhythmic ability, have not generally been confirmed by reasearch. Conflicting results have been obtained after use of the Seashore Test of musical ability on Whites and Negroes and in one comparison of Negroes with American Indians, who are not usually regarded as being particularly gifted in rhythm, it was found that the Indians were superior (Klineberg, 1935). Woods and Martin (1943) tested Negro and White children using the Kwalwasser-Dykema Music Ability Tests and while the Negroes showed some superiority the results were not conclusive.

Vernon (1966) has admitted that while 'there has been an increasing tendency to break down a single global intelligence or 'g' into a large number of partly distinct ability factors attempts to develop differential tests for abilities along different lines have not in fact met with much success'.

As our overall conclusion we may still agree with Klineberg (1935) concerning the simpler psychological functions that: 'With reference to sense perception in general it may be concluded . . . that there is no difference in sense acuity as such' (between the races), and again with regard to other non-intellectual psychological traits he says after reviewing the evidence that 'it is doubtful whether even a more exhaustive survey would reveal any significant racial differences'. Dreger and Miller (1960), writing twenty-five years later, conclude their review by stating that:—

> '(*a*) There are still wide differences between Negro and White in many areas of psychological functioning and (*b*) a number of differences attributed in times past to heredity have been shown to be the result of social class determination. It is not clear whether some differences adumbrated here, specifically in the intelligence and temperament realms, are genetically based or not.'

It seems unlikely, therefore, in the face of this evidence, even though much of it is rather old, that there are any very significant genetic differences between the races regarding characters such as these. Those differences which do exist appear to involve a wide range of characters but not, so far as we can see at the moment, in any very systematic way. So far as most of these characters are concerned the fact of racial hybridization is irrelevant to their expression and therefore we need consider them no further.

6.4.d. *Sensorimotor Development and Ability Differences*

We have already mentioned in earlier sections of this chapter that there are a number of differences between the races with regard to motor development and growth rates. These differences are not, in general, in dispute but the significance of them most certainly is. Many of these characters, such as the ability to sit unaided, the age of walking, early levels of ossification, etc., are stated (e.g., Eysenck, 1971) to be inversely correlated with I.Q. in Whites and to be expressed before most of the environmental deficiencies which might affect I.Q. become operative. This correlation is said to be the result of the action in man of a general law in biology which states that the longer the processes of development are prolonged in any species, the greater are the intellectual abilities of that species (Eysenck, 1971). The extension of this law to intra-species comparisons cannot really be rigorously tested at our present stage of understanding of this sort of phenomenon, and this idea should therefore be treated with caution. Furthermore, as we discuss in Chapter 9, both the factual evidence on which the comparison was made in the first place and the logic of the argument itself, can be criticized on a number of grounds (*see also* s. 6.4.*d*.i, *below*).

In many of the baby tests of intelligence carried out during the first two years of life, American Negroes have significantly scored higher than White infants during their first year of life. 'For example about 30 per cent of White infants as compared with about 60 per cent of Negro infants between 9 and 12 months were able to pass such tests as 'pat-a-cake' muscular co-ordination, ability to walk with help, to stand alone and to walk alone' (Jensen, 1969). Furthermore since children from lower socio-economic groupings are more alike in I.Q. than are children of these two races belonging to higher socio-economic levels, one would expect, on the basis of the 'extended-infancy' hypothesis, that there would be greater development differences in favour of Negro infants in high socio-economic Negro and White samples, than in samples from low socio-economic groups, and this is indeed just what is found. These early advantages in sensorimotor co-ordination and physical development last for the first few years of life, after which there is some evidence that Negroes are inferior to Whites in this field (*see* s. 6.4.*d*.ii).

There are, in fact, two separate threads running through this argument which we must disentangle. Firstly, what are the differences between races, what abilities do they encompass, and what is the importance of any differences in themselves? Secondly, what is the significance of any difference regarding the idea of extended infancy and the correlations of these early sensorimotor developments with later I.Q.? We shall first examine these early developments and their correlation with later I.Q.

6.*4*.*d*.i. Children's Precocity and I.Q.

Methods of investigating the ideas we have been discussing involve large-scale, longitudinal, multidisciplinary studies encompassing anatomy, genetics, psychology, and sociology. Since many of the problems have only really been identified in the last few years, the number of satisfactory studies completed in the field is rather small. It would be foolish therefore to be too dogmatic in our conclusions. But, as we will see, too simple an adherence to the theory of extended infancy being operative in man is almost certainly wrong.

One of the pioneers in this field is Nancy Bayley, who has worked on the development of young children for almost forty years. During this time she developed a test for use with young children which uses such items as:

says two words,
spontaneous scribble,
builds towers of various numbers of blocks,
places pegs in holes.

She was able to determine infant test score performances during the first few years of life. Later on, at ages 6 and 7 years, subjects were given the Stanford-Binet test. Correlations between scores on the infant tests and on later tests were insignificant (Bayley, 1943). These Bayley scales can be divided into motor and non-motor sections which give respectively, a Developmental Motor Quotient (D.M.Q), and a Developmental I.Q. A number of studies suggested that there was a negative correlation between the Developmental Quotient overall and later I.Q. (Bayley 1965a). Further study of this relationship suggested that it primarily involved D.M.Q. and that it was more pronounced in boys than in girls. The whole situation is very complex, however, in that the correlations vary markedly in sign and magnitude depending on the subject's personality type, parental attitudes, and the interactions between these variables (Bayley, 1966). Since the work was also carried out entirely with White subjects it does not seem that very firm conclusions can be drawn from it. In another report published in 1965 Bayley (1965b) looked more closely at the specific items in the motor scale which showed a clear difference between Negroes and Whites. Of the 60 items she studied, 11 showed a median age placement at least 0·7 month younger for the Negroes. These included:

Holds head steady,
Sits alone steady,
Walks with help and walks alone.

These 11 items are the only ones which showed any important racial differences. Bayley says that 'It would appear that the behaviours which are developing during the first 15 months of life, whether they are motor skills or the early perceptual and adaptive forms of mental abilities, are for the most part unrelated to sex, race, birth order, geographical location, or parental ability'.

Later work by Goffeney, Henderson, and Butler (1971) modifies some of Bayley's conclusions. They tested a total of 626 infants for mental, fine motor, and gross motor scores on the Bayley Infants Scales at the time the infants were 8 months old. These children were then retested at 7 years of age using seven WISC subtests and the Bender-Gestalt test. Examining first their results from the point of view of racial differences, they found that there was little difference between the races at 8 months but highly significant differences between the races at 7 years. Their results are shown in *Table* 6.23.

Turning to the correlations between the 8-month and the 7-year tests, it was found that, although all the 8-month measures for the whole population

Table 6.23.—THE MEAN AND STANDARD DEVIATION ON 8-MONTH AND 7-YEAR EXAMINATION OF SENSORIMOTOR AND INTELLIGENCE LEVELS OF NEGRO AND WHITE CHILDREN
(*From* Goffeney and others, 1971.)

	8 MONTHS		7 YEARS		
	Negro (N = 229)	White (N = 397)		Negro (N = 229)	White (N = 397)
Mental			(FSIQ)		
M	78·1	78·5		90·8	97·9
S.D.	4·24	3·62		11·4	13·0
Fine Motor			(VSIQ)		
M	2·28	2·28		89·3	96·6
S.D.	1·23	1·15		11·3	13·1
Gross Motor			(PSIQ)		
M	6·38	6·11		94·2	99·8
S.D.	2·94	3·26		12·9	13·9

FSIQ = WISC Full scale,
VSIQ = WISC Verbal scale,
PSIQ = WISC Performance scale.

correlated significantly with all the 7-year measures, there was a great difference in correlation between the races, in particular between Negro and White males. No 8-month score for the Negro male group correlated significantly with any later I.Q. or Bender score, whereas there were 9/12 significant correlations among the White males. There were 7/12 significant comparisons among the Negro females and 8/12 significant comparisons among the White females. A summary of the results are shown in *Table* 6.24.

Table 6.24.—CORRELATIONS BETWEEN 8-MONTH AND 7-YEAR TEST RESULTS USING BAYLEY INFANT SCALES AND WISC (*From* Goffeney and others, 1971.)

	NEGRO (229)	WHITE (397)
Mental v FSIQ	0·13*	0·23†
Fine v FSIQ	0·13*	0·24†
Gross v FSIQ	0·11*	0·17†
Mental v VSIQ	0·14*	0·23†
Fine v VSIQ	0·16*	0·24†
Gross v VSIQ	0·09	0·14†
Mental v PSIQ	0·07	0·17†
Fine v PSIQ	0·05	0·17†
Gross v PSIQ	0·09	0·15†

FSIQ = WISC full I.Q.;
PSIQ = WISC performance scale I.Q.;
VISQ = WISC verbal scale I.Q.
* Significant at the 5 per cent level.
† Significant at the 1 per cent level.

If there is, in general, a positive correlation between infants' intelligence and later intelligence in Whites it is hard to see why there should not also be one for Negroes, and indeed there is, just; particularly among females. The fact that there are much lower (but still positive) correlations between infant and childhood I.Q. in Negroes is at least suggestive that environmental factors have intervened between the ages of 8 months and 7 years to a much greater extent in Negroes than in Whites.

Of course there is still a good deal of instability in I.Q. scores between the ages of 8 years and adulthood. But evidence from a longitudinal study (Honzik, Macfarlane, and Allen, 1948) in California shows significant positive correlations in I.Q. between the ages of 2 and 18 years, although this and other work shows that the correlations between the I.Q.s at any two ages becomes smaller as the interval of time between the two tests gets larger.

We conclude, then, that the statement that, because Negro children are more advanced in certain physical and sensorimotor traits than Whites this indicates that they will be intellectually less able as adults, is open to serious question. The question of the extent of environmental interaction is still too difficult to be resolved.

6.*4*.*d*.ii. SENSORIMOTOR ABILITY DIFFERENCES

We must first ask what exactly is meant by the term sensorimotor? It is used by psychologists virtually interchangeably with the term 'psychomotor'. And Noble (1968) suggests that psychomotor skills are used in 'those situations that require the identification and combination of stimulus-organism-response elements into coordinated spatiotemporal patterns of receptor-effector activity as a joint function of practice, repetitions, and reinforcing feedback so as to optimize probability, amplitude, and time scores in acquisition retention and transfer'. Without attempting to interpret fully this technical definition we can see that it encompasses a very broad range of human motor actions and their mechanisms.

Nevertheless comparative studies on possible differences between the races are very few, and Dreger and Miller in their two reviews of Negro-White differences (1960, 1968) do not make any mention of any work in this field at all, other than studies of early development. Noble (1969) discusses four earlier pieces of work which utilized a wide variety of psychomotor skills of different complexities, and which tended to show that Whites were superior to Negroes, especially in more complex tasks. However, it is on the basis of more recent work by Professor Noble himself that we can attach some quantitative values to possible racial differences. Noble (1968) has also made some comparisons of the performance of racial hybrids, and it is this aspect of the problem which is of course the main reason why we are reviewing these experiments in the first place.

The first group of experiments which Noble carried out involved the use of the rotary pursuit apparatus. This comprises a metal disk on a rotating gramophone turntable, and the task is to follow this metal disk with a metal stylus. The score recorded is the length of time the stylus is on target in a 20-second trial. Noble selected 152 right-handed 9–12-year-old rural schoolchildren from Georgia, U.S.A., and divided them into four equal-sized groups matched for age, sex, and conditions of practice (left versus right hand). They were each given a total of 24 20-second trials, with 10 seconds of rest between each trial. The results are shown in *Fig*. 6.24. The differences between right and left hand were, as expected, highly significant, as was the difference between early and late trials. Of most importance is the difference which is apparent between the races, and it is particularly important to note that the Negroes do less well with their right hand than Whites with their left. On a separate comparison of 81 Whites, Mulattoes, and Negroes similarly matched, the mean response time on target for the three groups were 4·6 per cent, 2·6 per cent, and 2·1 per cent respectively.

As Noble himself says, these are 'preliminary empirical findings without theoretical interpretation' and 'these data must be considered quite tentative'. Writing in his 1968 review he states that 'normative molar generalizations are at this time of writing (*a*) scientifically premature to be asserted as inductive laws and (*b*) unsuited anyway to nailing down the relative contributions of genetic and culture factors to the variance in response time on target measures'. He further points out that McNemar (1933) got a correlation coefficient of only 0·17 between tracking and mental age. He quotes McNemar to the effect that 'correlations of the motor tests with intelligence are very low'.

Other work of Noble's (Noble, 1969; Eysenck, 1971) also suggests differences between Whites and Negroes in sensorimotor co-ordination in that Whites learn rather more rapidly on the Discrimination Reaction Time apparatus. There appears to be little initial difference in the speeds at which Whites and Negroes respond to different stimuli, although over intermediate blocks of trials the Whites increase their response speed more rapidly than do Negroes.

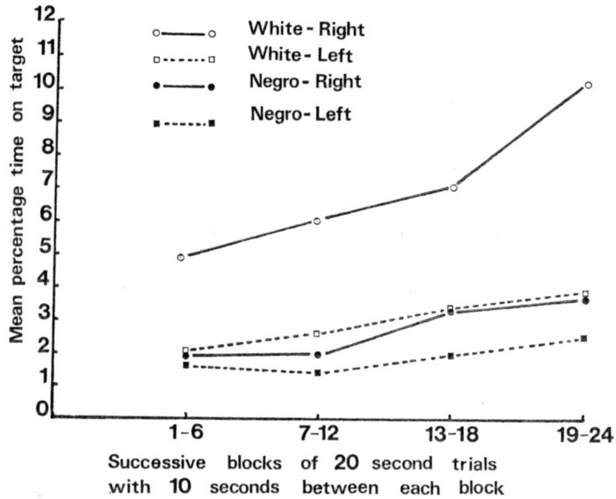

Fig. 6.24.—Differences in pursuit rotor performance with right and left hand respectively of White and Negro subjects. (*From* Noble, 1969.)

On balance it seems that in some areas White exceed Negroes in some sensorimotor or psychomotor skills; the extent or significance of these differences cannot yet be determined and must await considerably more research. Their genetic bases are almost entirely unknown and their manifestation in racial hybrids is therefore almost completely unknown. Neither long-term nor short-term predictions regarding the possible outcomes of racial mixing can therefore be made at the present time. There is very little data available on other races. Noble (1972) quotes some data which were collected in 1904 involving form-board skill (the fitting of nine geometric forms into correct holes) in which 7 African pygmies were, on average, superior to 12 Filippinos, who in turn were superior to 55 Amerindians and Eskimos and they to 74 U.S. Whites. Average error scores fell in the same order and hybrid groups were ranked in between. However, no attempts were made to equate for differences in cultural values found in the different societies from which these very small samples came, and the rankings were found to change for other psychomotor tasks. A more recent study, quoted by Noble, found that a sample of Mongoloids from Saipan and Indians from the U.S.A. exceeded the normal values of Whites on a test of maze-tracing ability. In this case, too, racial hybrids achieved, on average, intermediate scores. Even with our very fragmentary knowledge at present we can come to the fairly firm conclusions that there is no race which is uniformly superior in all psychomotor aptitudes and capacities.

6.4.e. *Race and Intelligence*

The characters we have discussed in the previous sections are controlled in the main by one or only a few genes. Intelligence is controlled by an unknown, probably rather large number. It is also, as we have mentioned, a character rather difficult to define, since it is really the name for the overall efficiency and level of complexity of an individual's cognitive and intellectual processes. Vernon (1966, 1969) summarizes the underlying relationship between the genetic situation and our measurements of intelligence in the following way. The genetic potentiality, which at the present time is largely non-observable and non-measurable, he calls intelligence A. He labels as intelligence B that which we observe in a person's behaviour and thinking, and it is the product of the interaction between intelligence A and environment. So far this is merely the normal distinction between genotype and phenotype, but Vernon suggests that we might usefully distinguish intelligence C which is simply a sample of B which can be scored in I.Q. or other units. 'The tests that we apply may or may not yield a fairly adequate sampling of all those concepts and skills that go to make up B' (Vernon, 1966).

Morton (1963) has estimated that genes at 71 loci are concerned with low-grade mental defect. Presumably, therefore, there are at least this number of genes controlling normal intelligence. Dewey, Barrai, Morton, and Mi (1965) increase this estimate to 114 as the number of loci contributing to severe mental defect. In a recent analysis Jinks and Fulker (1970) conclude that at least 22 loci would seem to be controlling I.Q. and possibly as many as 100. This slightly lower estimate may be merely a result of concentrating on the narrower concept of I.Q.

Normal intelligence, we have already mentioned, is determined primarily, i.e., between 60 per cent and 80 per cent, by the genetic constitution and to a much lesser extent by the environment. What must be remembered, however, is that this particular relationship holds only for the particular environment in which the testing is carried out. The realized I.Q. of different genotypes in different environments cannot be predicted in any simple way. Therefore even supposing that a completely culture-free test of intelligence could be devised (which most psychologists would agree is impossible anyway), we would have to determine mean intelligence levels, racial variation in intelligence and heritability for each racial group within identical environments. By the very nature of the historical and current situation in the U.S.A., where most racial testing has been carried on, such an investigation is impossible of execution. Now there are very many studies showing differences between mean I.Q. levels in different races, particularly between Whites and Negroes in the U.S.A. (*see* Shuey, 1966, and Jensen, 1969, for reviews of the many different surveys reported), but there are few studies of heritability of I.Q. in different races. Since heritability is the amount of the observed variation in a character which is under genetic control this is the crucial statistic to have.

As previously emphasized any estimate of heritability is only relevant to the particular population from which it is taken, and strictly speaking even if two populations differed markedly in a character which showed high heritability in both of them this is no proof that the observed difference is genetic (Lewontin, 1970a, b). *Table* 6.25 shows some of the principal estimates of heritability for intelligence reported in the literature. None come from non-Caucasoid populations. The estimates for the twins, which are mainly from monozygotic twins

Table 6.25.—A COMPARISON OF HERITABILITY ESTIMATES FOR INTELLIGENCE DERIVED FROM VARIOUS STUDIES. (*From* Huntley (1966) and others.)

STUDY	SOURCE OF MATERIAL	TESTS USED	HERITABILITY	
			From Twins	From Sibs
Newman, Freeman, and Holzinger, 1937	USA Whites	Stanford-Binet	67	—
Roberts, 1940	Bath, England	Advanced Otis	—	≦ 82
Maxwell and Pilliner, 1960	Scotland	Terman-Merril—Form L.	—	≦ 76
Shields, 1962	UK	Vocabulary and Dominoes	78	—
Dunsdon and Huntley, 1965	Greater London	Four Vocabulary	60	≦ 90
Juel-Nielsen, 1965	Denmark	Wechsler	68	—
Burt, 1966	Greater London	Group tests and Stanford-Binet	88	—
Erlenmeyer-Kimling and Jarvik, 1963*	Various	Various	80†	—

* This paper summarizes all the twin studies reported in the literature prior to 1963 and should therefore give the most reliable estimate of heritability.

† This is the figure calculated by Jensen (1967) from their data. It is based on comparison of the correlation between MZ and DZ twins and allows for the fact that there is a measure of assortative mating for intelligence.

reared apart, are particularly reliable, though, since monozygotic twins are identical genetically and therefore I.Q. differences between them must accurately reflect the influence of the environment. Jensen (1967) reviews all the major studies of both monozygotic and dizygotic twins and finds values of heritability ranging from 0.42 (Swedish Military Induction Test) to 0.93 (Otis I.Q. test). He concludes that the most reliable estimate would be 0.80.

Against this background of knowledge of heritability of I.Q., *Fig.* 6.25 shows the comparative distribution of I.Q. in Whites and Negroes obtained in one fairly typical survey (Kennedy, van de Riet, and White, 1963) and *Table* 6.27 shows a number of racial comparisons for I.Q. taken from Shuey (1966). The facts of the matter as we know them at present are clear enough; there is a consistent difference in I.Q. between U.S. Negroes and Whites of between 10 and 20 I.Q. points, and estimates of heritability are consistently in excess of 60 per

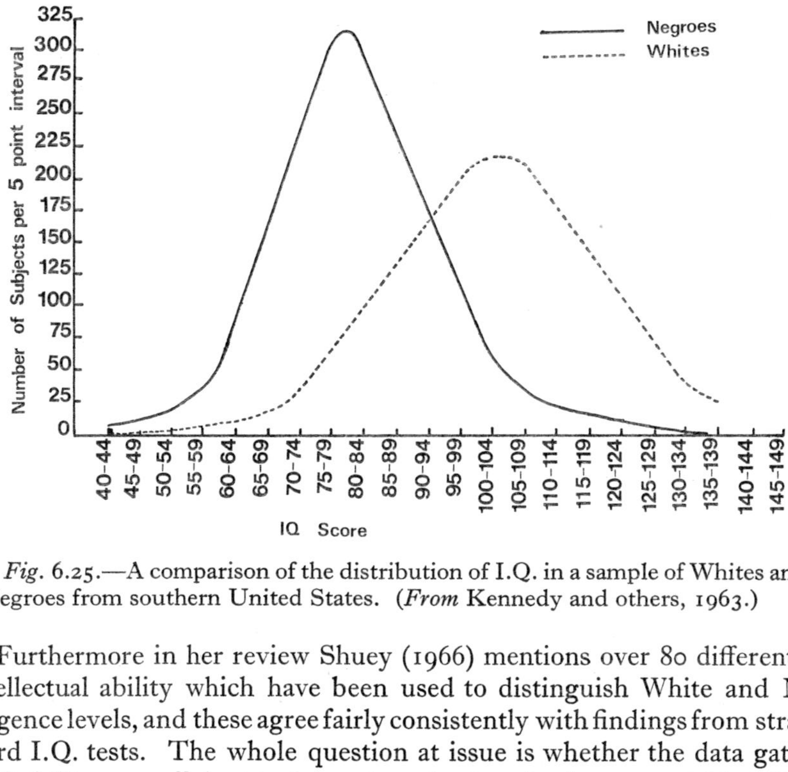

Fig. 6.25.—A comparison of the distribution of I.Q. in a sample of Whites and Negroes from southern United States. (*From* Kennedy and others, 1963.)

cent. Furthermore in her review Shuey (1966) mentions over 80 different tests of intellectual ability which have been used to distinguish White and Negro intelligence levels, and these agree fairly consistently with findings from straightforward I.Q. tests. The whole question at issue is whether the data gathered on heritability are sufficient to determine whether the data on racial intelligence differences are indicative of real genetic differences, or are merely a reflection of the very different environmental conditions and educational opportunities open to different races.

One of the most important arguments in favour of a substantial genetic component in the I.Q. differences is the fact that it persists even when comparisons are made between U.S. Negroes and Whites of the same socio-economic status. In other words when the length and quality of schooling, current occupation, and incomes, i.e., a large part of the relevant environmental variation, are as similar as we can match them between the two groups, then at least a part of the intelligence differences remain. Another strong argument which

Jensen uses in his review is that American Indians, in spite of similar deficiencies in their schooling and environmental conditions, do not show the same lowered intelligence levels as do the Negroes from low socio-economic groups.

Finally there is some important evidence presented by Duncan, Featherman, and Duncan (1968), who show that there is a much greater regression to the mean shown by children of Negroes of high-status occupation than by children of Whites of similar levels. In other words, it is suggested that the genetic baseline of the Negro population must be much lower than that of the Whites.

Needless to say, some compelling arguments have been advanced in favour of there being no genetic differences between the races with regard to overall intelligence levels. Arthur Jensen has lent probably the most weighty support to the idea of inherited differences being all-important, but it has been pointed out that he himself describes how, by setting children at their ease during the testing situation (by playing with them for 2 or 4 half-hour sessions), he obtained an increase in poor children's I.Q.s of from 5 to 8 points. Then he analyses a set of studies which shows that by changing the environment of children for six or nine months one can change their I.Q.s by 5 to 10 points, and finally he admits that compensatory education can add small increments to I.Q. and that if Negro children are tested by Negro testers they are also likely to score slightly higher. As Deutsch (1969) says, if we make allowance for each one of these factors 'we find that all statistically significant differences have been obliterated. Jensen thereby leaves himself no argument'. Stinchcombe (1969) puts it a slightly different way: 'Equalize the amount of fear between races, put Negroes in all-White schools, or schools with the same conditions, improve the standard of Negro families by a standard deviation, then (if the effects are additive) the I.Q.s would be equalized'.

We might also mention that some of the data which Jensen himself quotes are capable of a very different interpretation from that which he puts upon them. He says, for instance, that the hypothesis of genetic difference between the races would predict a greater Negro-White discrepancy at the upper socio-economic levels, since the Negro offspring would be regressing to a lower population than their White counterparts in the same socio-economic level, and consequently a larger proportion of the lower part of the distribution would fall below that which results in phenotypic I.Q.s below 75. Jensen on the basis of the data presented in *Table* 6.26 asserts that this is so. In fact at the highest socio-economic level there is a difference of 2·6 per cent increasing progressively to 35·1 per cent at the lowest socio-economic level. This progression is fully in accord with an environmental explanation for the differential incidence of technical mental retardation.

Dr. Audrey Shuey in her review published in 1966 discusses more than 200 tests of Negro intelligence and comparisons between Negro and White intelligence involving more than 80 different intelligence tests, most of which, as we have mentioned, agreed in suggesting a lower average intelligence level in

Table 6.26.—ESTIMATED PREVALENCE OF CHILDREN WITH I.Q.S BELOW 75 BY SOCIO-ECONOMIC STATUS (SES) AND RACE (GIVEN AS PERCENTAGE) (*From* Jensen, 1969.)

	SES	WHITE	NEGRO
High	1	0·5	3·1
	2	0·8	14·5
	3	2·1	22·8
Low	4	3·1	37·8
	5	7·8	42·9

Negroes. Clearly we cannot evaluate all of this data, nor can we consider morl than a few of the arguments which have been put forward in support or denia of this evidence. Any discussion of this topic in the present context must therefore be highly selective and therefore probably very subjective and inaccurate. To illustrate the sort of results which Shuey presents, *Table* 6.27

Table 6.27.—SOME RACIAL COMPARISONS IN I.Q. (*Compiled from* Shuey, 1966.)

AUTHOR	LOCATION	NO.	TEST	I.Q.	
				White	Negro
W. G. Katzenmeyer, 1962	Jackson, Michigan	W1061 C193	Young children Large-Thorndike Level 1 (a non-verbal test of intelligence)	102·4	83·06
I. J. Semler and I. Iscoe, 1963	Austin, Texas	W115 C108	School children		
			Age 6	104·5	95·0
			Wechsler 7	97·4	90·7
			Intelligence 8	105·5	97·6
			Scale for Children 9	96·3	87·5
			Total	101·4	93·7
R. T. Osborne, 1965	South-East U.S.A.	W5502 C2775	High School students California Test of Mental Maturity 1963	100·3	80·2
			1964	100·3	79·8
			1965	100·3	85·6
E. Stainbrook and P. S. Siegel, 1944	Large Southern University Southern and Negro College	W45 C45	College students	120	99
B. M. Levinson, 1964	New York	W50 C32	Adults Verbal	101·3	85·9
			Perform.	91·6	82·3
			Full Score	96·8	83·8

has been compiled. This comprises six different research projects; the criteria for inclusion simply being that they are the largest post-war projects in each age category which give rise to an orthodox comparison of I.Q. This selection only underlines everything that we have said before about the difference in I.Q. between Whites and Negroes being consistent and significant. They are also open to the same objections that we cannot assess to what extent environmental and educational differences are contributing to the result. There are, nevertheless, some important studies in which these two important variables are largely standardized between Whites and Negroes. One of these studies is that by M. W. Curti of intelligence tests of White and Coloured children on the island of Grand Cayman in the West Indies. As Curti (1960) says, 'Grand Cayman seems to be a natural laboratory for racial testing for several reasons; English is the language spoken; the island is small enough (20 miles long) so that all children of a given age in school could be tested; a sizeable part of the population has remained "pure white" while there are large numbers of brown and black people; and the island at the time of study was quite isolated and all the inhabitants were living in an unusually simple environment. In Grand Cayman there were many white families which have been exposed for some generations to the same simple type of environment as coloured and black families. People of all colour, then, were restricted to a rather narrow range of occupational opportunity'.

Curti employed a number of different tests on junior and senior pupils, including draw-a-man, memory squares, verbal classifications, number completions, analogies, and reasoning. Of the nine tests administered by Curti, two show a significant difference at the 1 per cent level associated with colour and one shows a significant difference at the 5 per cent level. These tests are respectively verbal classification, Opposites, and Analogies. Curti points out that the verbal classification test and the opposites test included such words as 'garage, laboratory, . . . decadence-progress' and she says that despite the simple nature of Cayman society, there was unquestionably a generally higher level of culture among the White than among the Coloured and Black groups. This shows up in the fact that the dark seniors are fully the equal of the Whites in classifying geometrical figures, but rather inferior in classifying words. In the other verbal test, that of reasoning, the words used were not so academic, and in any case were used in sentences where their meaning could be deduced; under these circumstances the significant differences associated with colour entirely disappeared.

With an overall result that three out of five performance tests for juniors showed no significant differences between the races and senior tests showed only three significant differences in eight different tests, which were largely explicable in view of the islands social structure, Curti concluded that 'The present study then does not lend support to the conclusion that coloured inferiority in intelligence tests has a racial basis'.

Another important test running counter to most of the evidence quoted by Shuey and Jensen is that of Burnes (1970). She gave the Wechsler Intelligence

Scale for Children (WISC) to a number of upper middle class and lower class Negro and White boys 'drawn primarily from a large parochial school district which encompasses a metropolitan area in the midwest'. The boys at the time of testing had a mean age of 8 years 5 months. Socio-economic status was determined by the occupation of the head of the household. In the words of the author, 'No important differences were found between the two racial groups, but verbal I.Q., performance I.Q., and full-scale intelligence quotient differences were significant for the subjects grouped according to socio-economic

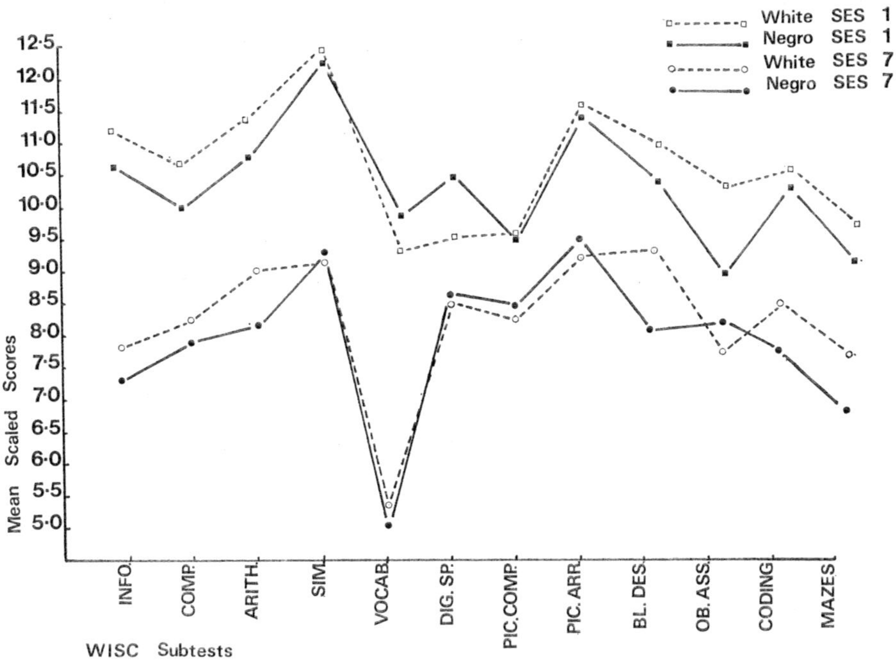

Fig. 6.26.—Group patterns of WISC subtest scores of U.S. White and Negroes from widely varying socio-economic classes. (*From* Burnes, 1970.)

status'. To emphasize the similarities of the two races and the differences of socio-economic status, Burnes's chart of the subtest scores in the WISC test is here presented as *Fig.* 6.26. This shows the profound effect of socio-economic status on overall performance for all parts of this test and the negligible effects of race.

Finally we might mention the study by Golden, Burns, Bridger, and Moss (1971) which shows that the distribution of I.Q. among Negro children of various socio-economic classes is almost identical with that found among Whites of similar socio-economic classes. In a longitudinal study of 89 Black children from different social classes Golden and others found that while there were no significant SES differences on the Cattel Infant Intelligence Scale at 18 and 24 months of age, there was a highly significant 23 point mean I.Q. difference on the Stanford-Binet test at 3 years of age between children from welfare and middle-class Black families. The range in mean I.Q.s of the Negro children in the extreme socio-economic classes (93–116) was almost identical to that obtained by Terman and Merril (1937) in their standardization sample of 831 White children between 2·5 and 5 years of age. The results obtained by Golden and others are shown in *Table* 6.28.

Table 6.28.—I.Q. AND SOCIAL CLASS IN NEGRO AND WHITE CHILDREN (*From* Golden and others, 1971.)

SOCIAL CLASS	*N*	I.Q.
Black children in longitudinal study classified by ABC system		
C (> High school)	37	112
B (≤ High school)	31	102
A (Welfare)	21	93
Black children in longitudinal study classified by modified Hollingstead system		
1. Middle class	16	116
2. Working class	38	107
3. Lower class/non-welfare	13	100
4. Lower class welfare	21	93
Terman and Merril's White children classified by father's occupation ($N = 831$)		
i. Professional		116
ii. Semi-professional/managerial		112
iii. Clerical/skilled trades/retail		108
iv. Rural owners		99
v. Semi-skilled/minor clerical/small business		104
vi. Semi-skilled labourer		95
vii. Unskilled labourer		94

Shuey's book was received with considerable reservation by many geneticists and psychologists when it appeared, and some contrary evidence, such as that presented by Burnes, Curti, and others certainly exists. One of the objections is that it is a rather uncritical compilation and virtually all investigations large and small, old and new, almost regardless of their reliability are given equal weight. Recent improvements in research methodology are not stressed, nor is the fact that many surveys were carried out before the important differences between culture-fair and culture-bound tests were fully recognized. So the virtual unanimity of all the reports mentioned in Shuey's review is less telling than it first appears. Our conclusion at this stage can only be that some workers believe that observed differences in I.Q. between Negroes and Whites are real ones, indicative of important genetic differences, while many believe that such a hypothesis is not yet proven, or rather that the hypothesis of no genetic difference between the races is not yet disproven. There are few comparative investigations of the intelligence of other races, but some of those who do believe in genetic differences between the races tentatively conclude, partly on the basis of sensorimotor development and accomplishments, that the genetic order of intelligence levels might be Mongoloid > Caucasoid > Negroid.

Since we ourselves have reached no definite conclusion on the basis of racial difference in intelligence we can make no predictions on the outcome of racial crosses. There are, however, a number of surveys which show us what actually occurs, although we must remember that socio-economic variables may still be influencing the results in these cases also. Shuey herself reviews 18 different studies. She finds that 'in 12 of the 18 comparisons, the lighter in colour, or those identified as mixed, scored higher than the darker . . . or the unmixed'. She goes on to say, however, that 'the differences are not typically large and . . . make no important contribution to the problem of race differences in intelligence'. We may note that there are only two post-war surveys of intelligence in racial hybrids that Shuey quotes. One of these tests, that of Codwell (1949), seems fairly conclusive in showing a relationship between skin colour and intelligence. Codwell tested more than 500 adolescent boys from one school in Houston, Texas. His subjects were classified as Dominantly Negroid, Intermediate, and Strong Evidence of White. The Dominantly Negroid group had an average I.Q. of 87, the Intermediate 90·9, and the Strong Evidence of White 91·9, these differences being small but statistically significant.

It should be mentioned, however, that Codwell's main interest was in motor function, not intelligence, and his main finding was that 'there is no significant change in the amount of motor function as a composite as the amount of Negroidness increases or decreases in the Negro-White hybrid'.

The other post-war test is that of Grinder, Spotts, and Curti (1964), who studied the relation between skin colour and intelligence as measured by the Goodenough Draw-a-Man Test. The 941 children tested were all from 50 Jamaican schools, and were assigned to one of three colour classes—light, mixed light-dark, and dark. Each child, furthermore, was classified according to a five-point socio-economic scale: lower class, three middle classes, and upper class. The authors noted that very few children rated as dark were in the upper class, and conversely very few children rated as light were in the lower class. They therefore attempted to standardize the effects of social background on I.Q. by comparing skin colour and I.Q. within each of the three middle-class groupings. The results they obtained suggested that not one of these three comparisons between similar socio-economic groups gave significant differences in I.Q. Shuey suggests that this may be due to the smallness of the numbers tested in some of these categories, but pooling all the middle classes together and testing this single group is illegitimate for just the reasons that Grinder and others eliminated the upper and lower classes from consideration in the first place. The Chi-square values obtained for each of their three groups by these authors were 5·54, 5·25, and 3·93 for the low-middle, middle-middle, and high-middle comparisons respectively, with significance at the 5 per cent level requiring a Chi-square value of 5·99, the basis of their comparison being the number who scored above or below the Jamaican norm for this particular test. When these three Chi-square values are summed, which is a legitimate procedure, we obtain a value of 14·72 which for five degrees of freedom gives a probability of about 0·02. In other words, the evidence is certainly suggestive but hardly very convincing that there may be a positive relationship between I.Q. and racial classification.

One important corollary of the hypothesis that there exist significant racial differences in I.Q., is that there should be a significant positive correlation between I.Q. score and relative racial contribution in the hybrids. As we have mentioned Shuey herself found such a correlation in 12 out of 18 tests, although the differences were not large and no very precise determination of the amount of mixing could be made. In those cases where some estimates of amount of admixture have been made there is no consistent tendency for these with higher White ancestry to record a higher I.Q. score. This is illustrated in a study of Witty and Jenkins (1936) whose results are shown in *Table* 6.29. This study is rather special in that it only involves children with an I.Q. of 125 or greater, and therefore it is possible to argue that it holds little relevance for the population in general. Similar results of a more general nature have been found in studies of Amerindian tribes in the U.S.A. who have varying degrees of White genetic admixture and assimilation of White culture. Within any one tribe there is zero correlation on both verbal and non-verbal tests with amount of White admixture (Anastasi, 1958).

There is one further survey of intelligence in racial hybrids too recent to be included in Shuey's book which is of some importance, and which we shall therefore examine fairly closely. This survey starts from the premise that if

Table 6.29.—PROPORTION OF WHITE ANCESTRY AMONG NEGRO SCHOOL CHILDREN WITH I.Q.S OF 125 OR HIGHER (*From* Anastasi, 1958, *after* Witty and Jenkins, 1936.)

DEGREE OF WHITE MIXTURE	PERCENTAGE IN NEGRO POPULATION	PERCENTAGE AMONG GIFTED NEGRO CHILDREN	
		I.Q. > 125 (N = 63)	I.Q. > 140 (N = 28)
No White ancestry*	28·3	22·2	21·4
More Negro than White	31·7	46·1	42·8
About equal	25·2	15·9	21·4
More White than Negro	14·8	15·9	14·3

* Less than one-sixteenth of mother's and father's ancestry reported to be White.

racial differences in intelligence are determined by additive genetic factors which are not sex-linked, then test scores for children of interracial crosses might be independent of maternal race. If, on the other hand, differences in intelligence between the races are mainly environmental then the mother's race should have an effect on children's performance, since she is the primary socializing agent during the pre-school years. Willerman, Naylor, and Myrianthopoulos (1970) report a study of 88 children from interracial marriages whose I.Q.s were tested at 4 years of age. The authors assumed that the mean intelligence of the parents does not differ with either maternal or paternal race combination and their data show that there were no significant differences between either of the parental groups with regard to the length of received

Table 6.30.—The I.Q. Scores of 4-year-old Children categorized by Race of Mother, Sex of Child, and Marital Status (*Results show mean, standard deviation, and numbers on which they are based.*) (*From* Willerman and others, 1970.)

Race of Mother	Sex of Child	Marital Status	I.Q. Score
White	Male	Unmarried	94·7 ± 12·1 (7)
White	Male	Married	100·8 ± 18·3 (20)
White	Female	Unmarried	100·3 ± 15·7 (15)
White	Female	Married	103·8 ± 18·0 (19)
Mean			100·9 ± 16·8 (61)
Negro	Male	Unmarried	67·5 ± 23·3 (2)
Negro	Male	Married	88·4 ± 11·0 (9)
Negro	Female	Unmarried	88·6 ± 13·7 (5)
Negro	Female	Married	105·1 ± 14·1 (11)
Mean			93·7 ± 16·9 (27)
Overall Mean			98·7 ± 16·8 (88)

education. The data presented in *Table* 6.30 show that in fact there are important differences in I.Q. score between the offspring with mothers of different races. The data cannot be accepted at face value since application of regression procedure showed that there were differences of 8·36 ± 3·75 I.Q. points for race, 8·14 ± 3·52 for sex, and 8·30 ± 3·73 for marital status of mother. But it turns out that if sex is ignored, maternal race is significant only among the children of unmarried mothers ($P < 0{\cdot}05$). Alternatively if marital status is ignored, maternal race is significant only among males ($P < 0{\cdot}05$).

In a parallel control study, the mean I.Q. for children from intraracial matings, matched for hospital of birth, socio-economic and marital status against the current sample of interracial offspring, was 104·3 for children of White matings and 97·4 for children of Negro matings. The mean I.Q. of the interracial matings was 98·7.

The interpretation of the results of this survey can only be tentative because of the relative smallness of the sample, but it does appear 'that environmental factors are playing an important role in the lower intellectual performance of Negro children' (from the abstract to Willerman and others' paper).

One of the strongest arguments against there being significant differences in intelligence between the races has been that throughout man's evolutionary history natural selection will have operated to increase intelligence. This, after all, is the trait above all others which differentiates man from other animals and which has guaranteed his successful conquest and domination of this planet. Jinks and Fulker (1970) show on theoretical grounds that the genes controlling I.Q. have indeed probably been subject to strong natural selection. Why, then, should natural selection have been less successful in some races than in others? Certainly the immediate requirements of, for example, an Australian Aborigine to survive in the arid desert were very different from, say, those of the first Europeans living in a cool temperate climate, but probably each environment was as demanding as the other in intelligence. In the two examples quoted, for instance, we have on the one hand a shortage of food and on the other the difficulties and dangers associated with catching it. However, the likely differences between the races in certain sensorimotor abilities would appear to rob this argument of much of its force, and we might agree that arguments along these lines cannot really be used one way or the other in this particular context.

All in all the weight of the evidence at the present time does seem to favour the suggestion that there may be some small differences between the races in intelligence levels. But deficiencies in our methodology of testing and in our understanding of all facets of intelligence are almost certainly serving to exaggerate these differences in all but the most carefully designed surveys and experiments. And most of the work on which claims of 10, 15, or 20 I.Q. point differences between the races are made does not fall into this latter category. The magnitude of any differences between the races therefore remains unknown but probably not very large.

If it were in fact true that the average I.Q. of the Negro in the United States was 85 compared with a mean I.Q. for the Whites of 100, then if the races interbred entirely at random, the average I.Q. of the resulting population would be 98·46, assuming that the Negroes at present constitute 11 per cent of the United States population. The principal result of this 1·5 per cent reduction in average intelligence would be that the frequency of people with very high I.Q.s would be markedly reduced. In particular it has been pointed out that the number of people with I.Q.s of 160 and over would be halved. This possible decline in the number of highly gifted people has been stated by some to be an example of the sort of genetic catastrophe which would result from racial mixing.

For what it is worth we can set out the calculations to determine just what the magnitude of the decline in the number of moderately intelligent people would be after mixing of races with different mean I.Q.s.

If a population has a mean μ_1 variance σ_1^2 and another has a mean μ_2 variance σ_2^2 (σ_1^2 and σ_2^2 being about the individual means μ_1, μ_2 respectively) and if the two populations are mixed so that the fractions of the total are p_1, p_2 respectively ($p_1 + p_2 = 1$) then the mean overall of the new population

$$= p_1\mu_1 + p_2\mu_2,$$

and the variance overall (about the overall mean)

$$= p_1\sigma_1^2 + p_2\sigma_2^2 + p_1p_2(\mu_1 - \mu_2)^2.$$

Now the probability that a randomly chosen individual has an I.Q. greater than a given figure for the combined population, $X \geqslant x$, say

$$= \text{Prob. } (X \geqslant x \text{ in 2nd population})p_1 + \text{Prob. } (X \geqslant x \text{ in 1st population})p_2$$

i.e., a weighted average of the individual cases.

On the assumption that the mean I.Q. in Whites is 100 with a standard deviation of 15, the probability that an individual has an I.Q. in excess of 130 is approximately 2·3 per cent and in excess of 145 is approximately 0·13 per cent (*see Fig.* 6.27). The mean I.Q. of American Negroes is commonly reported to

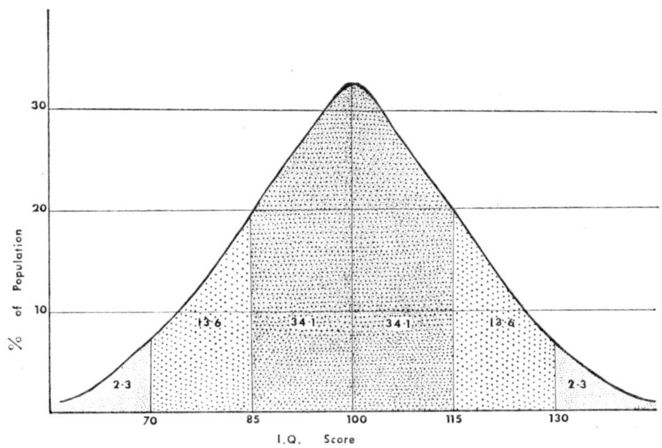

Fig. 6.27.—The normal distribution of I.Q. for a population whose mean is 100. The standard deviation is about 15 points and the proportions of the total population within 1, 2, and 3 standard deviations of the mean are shown.

be up to 20 points below that of Whites and the variance of their I.Q. is usually also reported as being a little less than in Whites. We can therefore calculate for various possibilities of I.Q. differences, variance differences, and levels of admixture, the outcome of racial mixing in the United States situation. The results of these calculations are shown in *Table* 6.31.

Table 6.31.—THE NUMBER OF INDIVIDUALS PER 1,000 POPULATION HAVING AN I.Q AND S.D. 15 I.Q. POINTS HYBRIDIZES WITH A POPULATION OF LOWER I.Q. AND

MEAN I.Q. OF HYBRIDIZING POPULATION	80					85				
Level of admixture (per cent)	2	4	6	8	10	2	4	6	8	10
Mean I.Q. of resulting population					98					98·5
Standard Deviation of hybridizing population										
10										
11	22·34	21·89	21·43	20·98	20·52	22·34	21·89	21·43	20·98	20·5
12	The population with lower I.Q. makes an effectively nil contribution to variance in excess of 130 I.Q. points. We therefore merely observe diluting effects									
13						Diluting effects				
14										
15						22·34	21·89	21·43	20·98	20·5

Under the most extreme assumptions, i.e., a mean I.Q. of 80 among American Negroes and total amalgamation with the Whites, we find that the genetic contribution of the Negroes would make a negligible contribution to the likelihood of individuals having I.Q.s in excess of 130. We therefore would see a reduction in the occurrence of such individuals from about 2,280 per 100,000 to about 2,052, or in other words perhaps a 10 per cent reduction in the numbers of individuals with an I.Q. greater than 130—the level of a very good university graduate. Less extreme and more probable assumptions show lower levels of reduction of mean I.Q. in the resultant population.

It must be remembered, though, that non-random mating with respect to intelligence would in all probability still occur, even if mating with respect to race was entirely random. This in itself would tend to maintain the number of high level I.Q.s segregating in the population. Indeed with the increase in size of the gene pool the phenotypic variance for I.Q. might increase. We are in fact dealing with a situation in which two populations are merging in both of which the genes (for high I.Q.) are at a low frequency. This is just the situation in terms of Muller's analysis (*see Table* 5.10) in which the variance of the trait concerned would be expected to increase. In other words a greater proportion of people with high I.Q. might be produced. Finally it must be remembered that the association between very high I.Q. and achievement is not always very great. It has been shown for academics, for instance (Gibson and Light, 1967; Hudson, 1960), that a certain I.Q. level is necessary for initial success in

EXCESS OF 130 IN A HYBRID POPULATION WHEN A POPULATION OF MEAN I.Q. 100
EDUCED VARIANCE. (*On the assumption of random intermarriage with respect to I.Q.*)

90					95					100				
2	4	6	8	10	2	4	6	8	10	2	4	6	8	10
				99					99·5					100
					22·34	21·89	21·43	20·98	20·52	22·40	22·09	21·48	21·08	20·56
22·34	21·89	21·43	20·98	20·52						22·47	22·10	21·57	21·26	20·87
Diluting effects					22·34	21·89	21·43	21·00	20·53	22·50	22·18	21·68	21·47	21·14
22·34	21·89	21·43	20·98	20·53	22·34	21·90	21·45	21·01	20·56	22·59	22·24	22·03	21·78	21·50
22·34	21·89	21·43	20·99	20·54	22·35	21·91	21·46	21·03	20·59	22·74	22·46	22·13	21·98	21·72
22·34	21·89	21·44	21·00	20·56	22·36	21·93	21·49	21·06	20·62	22·80	22·80	22·80	22·80	22·80

university selection and the ability to graduate successfully. Above this minimum or threshold level, however, there is no great correlation with eventual academic success. Presumably other qualities such as motivation, industry, imagination, personal relationships, and so on, play as much if not a greater part than sheer intelligence in academic success.

6.5. Race Crossing and Evolutionary Fitness

6.5.a. *Dental and Palate Characteristics*

We are considering these particular traits under the section on Evolutionary Fitness because in fact some of them are of importance for survival and reproduction. Nevertheless most previous investigations of either the anthropology or the genetics of different characters of the teeth have not emphasized this aspect. We have already considered some of these characters as evidence of complex genetic control (absence of third molars) or useful indicators of racial affinity (crown patterns, cusp number, shovel-shaped incisors, etc.). What we are here concerned with are those characters which may differ between different races but, more importantly, could lead to incompatibilities in racial hybrids. Thus from this point of view cusp number is almost completely irrelevant but tooth size is not, particularly if it is associated with different sized jaws in the races hybridizing. The whole field is one of considerable genetic complexity and Lasker (1950) lists the following characters which are known to vary, and are probably under genetic control in man.

Development:
- Order of eruption,
- Age at eruption,
- Age at shedding of deciduous teeth.

Number:
- Supernumerary teeth,
- Congenitally missing and peg teeth,
- Crown pattern and cusp number,
- Supernumerary cusps,
- Cusplets,
- Wrinkles.

Incisor Pattern:
- Shovel shape,
- Lingual ridges,
- Dental tubercle.

Enamel Line:
- Extensions,
- Modules.

Root Pattern:
- Supernumerary roots,
- 'Fused' roots,
- Bent roots,
- 'Taurodontism'.

Measurements:
- Crown dimensions,
- Root length.

Maxillary and Mandibular Hyperostoses:
- Palatine torus,
- Mandibular torus,
- Alveolar hyperostoses.

Tooth Spacing:
- Diastemata,
- Trema.

Occlusion:
- Bite and jut,
- Shape of dental arches,
- Prodontism.

To this list we might add the occurrence of defective enamel and dentine; for example, it is known that brown enamel is inherited as a sex-linked character. It is also known that resistance to the development of dental caries is under partial genetic control.

Faced with this amount of genetic variability it is little wonder that not a great deal of progress has been made in systematic genetic studies of these traits and their behaviour in racial crosses. We will therefore examine a few scattered reports of the dental characteristics of racial hybrids and draw from them what conclusions we can. One of the best collections of data on variability in teeth is the volume *Dental Anthropology* edited by Brothwell (1963). This volume has a paper by K. Hanihara on the deciduous dentition of Japanese-American hybrids, including both Whites and Negroes, which can at least serve as a pointer for the general consequences of hybridization in dental traits. This

Table 6.32.—FREQUENCIES OF CARABELLI'S CUSP IN THE UPPER SECOND MOLARS IN JAPANESE/WHITE AND JAPANESE/NEGRO HYBRIDS COMPARED WITH THEIR PARENTAL POPULATION (*From* Hanihara, 1963.)

RACIAL GROUP	*N*	NO CUSP	CARABELLI'S PIT	CARABELLI'S CUSP
Japanese	185	32·4	55·7	11·9
Japanese/White	71	14·1	62·0	23·9
White	56	5·4	58·9	35·7
Japanese/Negro	41	17·1	63·4	19·5
Negro	51	19·6	68·6	11·8

Table 6.33.—FREQUENCIES OF SHOVEL-SHAPED TEETH IN THE DECIDUOUS CENTRAL INCISORS OF JAPANESE/WHITE AND JAPANESE/NEGRO HYBRIDS COMPARED WITH THEIR PARENTAL POPULATIONS (*From* Hanihara, 1963.)

RACIAL GROUPS	*N*	0	1	2	3
Japanese	124	0·0	23·4	76·6	0·0
Japanese/White	65	7·7	55·4	36·9	0·0
White	20	50·0	50·0	0·0	0·0
Japanese/Negro	35	0·0	42·9	57·1	0·0
Negro	10	80·0	10·0	10·0	0·0

paper is a review of a number of years of work and is based on 79 Japanese/White hybrids, 46 Japanese/Negro hybrids, 209 Japanese, 61 Whites, and 52 Negroes.

Tables 6.32–6.34 from Hanihara summarize his overall results. In general he says that the hybrid groups show an intermediate pattern of distribution between the racial groups of their parents, although some characters do show a closer resemblance to one of the parents. In most of the characters examined, the differences between Japanese and Americans are highly significant, while the hybrids show a smaller degree of difference from the parental populations. It is interesting to note that the Japanese/White and Japanese/Negro hybrids differ from one another in their degree of morphological dissimilarity with their parental group. The former are generally equally dissimilar to both the Japanese and the American Whites, while the latter show closer resemblance to the Japanese than to the American Negroes. So in fact in this study, one of the few really comprehensive studies on this topic, we find exactly the same behaviour as for all other physical characters. Certainly some components are not intermediate and since these may be the result of individual genes we would not expect them to be. For instance both hybrid groups are closer to the Japanese as regards the cusp development of the deciduous upper first molars, but are nearer to the Americans when considering Carabelli's cusp in the deciduous

Table 6.34.—FREQUENCIES OF DIFFERENT TYPES OF CUSP DEVELOPMENT IN THE DECIDUOUS UPPER FIRST MOLARS OF JAPANESE/WHITE AND JAPANESE/NEGRO HYBRIDS COMPARED WITH THEIR PARENTAL POPULATIONS. (*From* Hanihara, 1963.)

RACIAL GROUPS	*N*	2 CUSPS	3 M (2 + metacone)	3 H (2 + hypocone)	4 CUSPS
Japanese	184	15·2	60·3	4·3	20·1
Japanese/White	70	17·1	68·6	4·3	10·0
White	55	60·0	23·6	16·4	0·0
Japanese/Negro	42	11·9	45·2	4·8	38·1
Negro	50	28·0	36·0	16·0	20·0

upper second molars. Nevertheless the overall effect is one of intermediate phenotypes.

On the basis of the data which he has obtained Hanihara is able to make some convincing estimates of the frequencies of some of the genes controlling these characters in the different races. Carabelli's cusp, for instance, appears to be under the control of a simple Mendelian dominant gene with incomplete penetrance, and if this is correct Hanihara estimates that the frequency of this gene among Whites is 19·8 per cent and among Japanese is 6·1 per cent. Therefore the frequency of the gene in the Japanese/White hybrids is theoretically estimated to be 13 per cent and that of the individuals carrying the Carabelli's cusp to be 24 per cent. These estimates are very close to the figures obtained by direct observation, and the same is broadly true for Japanese/Negro hybrids.

C. G. Turner (1969) investigated the dentition of three small North American native groups—Aleuts, Koniag Eskimos, and Hopi Indians. He agreed with Hanihara's suggestion of a single gene with 2 alleles controlling the expression of Carabelli's cusp, although he suggested co-dominant alleles. Turner suggested a similar mode of inheritance for maxillary lateral incisor shovelling and for the inheritance of the hypocone. Using a mixture of prehistoric fossil material for these three groups, which were assumed to be completely without European admixture, dental impressions from living peoples who clearly show admixture, and data from various Europeans, Turner was able to estimate the

Table 6.35.—ESTIMATES OF EUROPEAN ADMIXTURE IN SOME NORTH AMERICAN NATIVES BASED ON DENTAL TRAIT ALLELE FREQUENCIES (*From* C. G. Turner, 1969.)

People	Maxillary Lateral Incisor Shovelling C	m	Carabelli's Trait S	m	Molar 2 Hypocone H	m	Mean Admixture from Dental Data	Mean Admixture from Blood Groups
		per cent		per cent		per cent		
European	0·537		0·084		0·351			
Living Koniag	0·483	75	0·402	60	0·593	48	61	47
Prehistoric Koniag	0·316		0·879		0·820			
European	0·537		0·084	*	0·351			
Living Aleut	0·133	6	0·814	—	0·444	26	16	28
Prehistoric Aleut	0·107		0·784		0·477			
European	0·537	*	0·084		0·351	*		
Living Hopi	0·128	—	0·698	6·5	0·747	—	6·5	0
Prehistoric Hopi	0·250		0·741		0·707			

* No estimate of admixture could be made in these cases, since the hybrids were not intermediate between the parental populations.

amount of admixture in these three peoples and obtained values which agreed fairly well with those based on blood group frequencies (*see Table* 6.15). This information can therefore be regarded as supplementary to Hanihara's in that it concerns populations with some history of hybridization. Neither in this paper nor in others concerned with the dentition or general biology of these peoples is there any suggestion of harmful dental effects of European admixture. Note furthermore that in only 3/9 comparisons did the mixed populations fall outside the range specified by the parental populations.

Clearly then this work, so far as it goes, does not indicate that there is any greater risk of incompatibilities among dental characters than among all the others we have so far examined. Other workers, however, have suggested that this may be one area in which interracial hybrids are likely to have an increased probability of physical disharmony or deformity *vis-a-vis* their parental types, perhaps more with reference to the size and shape of the jaws rather than with the teeth themselves.

Fleming (1939), for instance, when studying 110 offspring the result of matings between Negro males and White females, found that 11 of them, or 10 per cent, had sufficient disharmony of the jaws to cause significant pre- or post-normal occlusion. He described, for example, a 12-year-old child who inherited a well-arched jaw from her Negro father and a badly arched jaw from her White mother. She had a small V-shaped palate with small upper teeth and a lower jaw that was widely arched and had large long teeth. The result was that her lower teeth slipped up outside her upper teeth which 'interfered with speech'. Among 49 backcross individuals resulting from the matings of Anglo-Negro women to Negro men, 10 were said to show disharmony of the jaws. Two individuals had marked pre-normal occlusion and the rest post-normal. Among 15 individuals, the result of backcrossing Negro-White males to White females, one person showed disharmony of the jaws. This person had a large well-arched palate with strong wide teeth. The lower jaw was very small, V-shaped, and had small irregular teeth. The result was a marked disharmony of the jaws.

Shapiro (1929) reports that the Norfolk Island hybrids experience considerable dental problems. They are subject to a very early loss of their teeth, a loss which is complete when they are still comparatively young. This trait may, however, be inherited in a fairly simple manner. Edward Young, one of the Bounty mutineers, had very bad teeth and although only 22 years old had already lost several incisors—just those teeth which are removed first among Norfolk Islanders.

Other studies have not suggested that dental problems invariably arise in hybrids. Little (1943), for example, observed the presence or absence of dental caries in more than 200 Anglo-Negro boys and girls and compared his findings with those for a similar number of English children broadly matched for age and background. The relative absence of carious teeth was appreciably greater amongst the 'hybrid' children. Nearly 1 in 3 English children was classified as possessing 'poor' teeth as compared with slightly more than 1 in 10 of the Anglo-Negro children. Little's findings are shown in *Table* 6.36.

Table 6.36.—THE CONDITION OF THE TEETH IN SOME HYBRID CHILDREN (*After* Little, 1943.)

		EXCELLENT: No apparent sign of caries, and absence of extractions	FAIR: 1 or 2 extractions or teeth showing signs of disease	POOR: A large number of extractions or decayed teeth
Anglo-Negro	Boys (108)	50·9	36·9	13·0
	Girls (98)	57·1	31·6	11·3
English	Boys (113)	29·2	43·4	27·4
	Girls (110)	37·9	33·7	28·4

It is a pity that no data were or are available for a comparative series of Negro children as to the usual condition of their teeth. Little mentions that 'overcrowding' of the palate is a feature commonly associated with White-Negro mixture, he simply notes that among the children he surveyed 'About twelve "hybrid" subjects possessed palates showing greater or lesser degrees of the anomaly compared with five of the "English" children'.

It is perhaps important to put this incidence of dental anomaly into perspective. Mills (1963) has reviewed the incidence of malocclusion in primates, including man, and he quotes estimates of the incidence of malocclusion in man varying from between about 20 per cent to more than 90 per cent. This wide variation is probably chiefly due to a difference of definition of normal and abnormal occlusion. However, *Table* 6.37, which gives the results of a survey of malocclusion as classified by one authority, shows that, despite the smallness of the sample sizes, world-wide variation of this trait is very marked. However, there seems little racial association with high or low values. Mills suggests that this high degree of variation may be due to one or all of three possible reasons. Firstly the admixture of populations which takes place in industrialized countries. This is probably responsible for the malocclusions which are caused by discrepancies in the parts of the skull, and for the differences in crowding and spacing of the teeth which are caused by discrepancies in size of teeth and jaws.

Secondly, he mentions the increased likelihood of an individual with a severe malocclusion surviving under present-day conditions of child care and dental treatment, i.e., he invokes the idea of relaxed selection, and thirdly he suggests that some environmental factors may also be influencing malocclusion rates.

Since we are at the moment considering palate characteristics it is appropriate to investigate the position with respect to cleft palate. This is in most respects a fairly typical congenital malformation, of the sort which will be considered in s. 6.5.*c*. But because of the suggestion that there is in fact an increased incidence of dental problems and a higher proportion of ill-adapted individuals so far as the palate is concerned, it is appropriate to consider it now. Cleft palate is more often than not associated with hare-lip and we will consider these two defects together as cleft lip and palate.

Table 6.37.—INCIDENCE OF MALOCCLUSION IN VARIOUS RACES. (*From* Mills, 1963.)

RACE	TOTAL NO. OF SKULLS EXAMINED	PERCENTAGE VARYING
A. Mongoloids		
Ancient	16	6·25
Modern	35	8·6
Eskimo	14	50
Japanese	12	66·67
Chinese (Borneo)	14	91·4
B. Australoids		
Tasmanian	11	27·3
Australian	14	28·6
C. Negroids		
S. Africa	10	0·0
E. Africa	33	21·2
W. Africa	76	32·9
D. Caucasoid		
Whites	62	32·26
Whites	41	33·3
Whites	42	34·2
'Hindu'	39	35·9

So far as broad racial differences are concerned, Mongoloids have a higher incidence than Caucasians, who have, in turn, a higher incidence than Negroes (Fraser, 1970). Racial differences in the incidence of cleft palate unassociated with cleft lip are less obvious than those of both traits together.

The incidence of cleft lip and palate in Caucasians varies between about 0·8 and 1·8/1,000 births, with a highest value of 1·78/1,000 in Iceland and among the lowest figures about 0·76/1,000 among Caucasians in Hawaii (Morton and others, 1967).

The mean figure for Caucasians is about 1/1,000 births. Japanese in Japan have a mean frequency of approximately 1·7/1,000, a figure which is also found among Japanese in Hawaii. One of the largest series of births among Negroes in which the incidence of hare-lip and cleft palate is recorded comprises 10,025 births in Pretoria (Stevenson and others, 1966, two papers). The overall incidence here is 0·5/1,000 births. It seems fairly certain, therefore, that there are genuine racial differences in the incidence of cleft palate. What of its occurrence in racial hybrids? There are perhaps four major areas where comparisons are possible. Firstly, a composite series from Central and South America including Mexico, Panama, Columbia, Brazil, and Chile, in which Negroes, Caucasians, and American Indians are involved. Secondly, there is Hawaii where again the three major races are involved. Thirdly, there is Malaysia in which Europeans, a number of Asian peoples, and their hybrids 'Eurasians' are found. Finally there is South Africa, in which Whites, Bantu, and Coloured can be compared. It must be remembered, though, that there is a significant Asian or Mongoloid component in this last group. The figures are shown in *Table* 6.38 and clearly there is no increase in incidence of this trait in racial

Table 6.38.—THE INCIDENCE OF HARE-LIP AND CLEFT PALATE IN DIFFERENT RACIAL GROUPS AND THEIR HYBRIDS (*Compiled from* Stevenson and others, 1966.)

	CAUCASIANS	MULATTOES	NEGROES	MESTIZOES	MONGOLOIDS
Latin America	11/10,510 (0·10 per cent)	8/7,559 (0·11 per cent)	2/2,138 (0·09 per cent)	81/74,186 (0·11 per cent)	1/815 (0·12 per cent)
South Africa	19/11,285 (0·17 per cent)	1/3,091* (0·03 per cent)	5/10,025 (0·05 per cent)		
Malaysia	1/254 (0·4 per cent)			0/261†	20/7,703‡ (0·26 per cent)
Total	0·14 per cent	0·08 per cent	0·06 per cent	0·11 per cent	0·25 per cent

* Includes significant Mongoloid component.
† Eurasians
‡ Malaysians.

Table 6.39.—THE INCIDENCE OF HARE-LIP AND CLEFT PALATE IN HAWAII. (*From* Morton and others, 1967.)

CAUCASIAN	HAWAIIAN	CHINESE	FILIPINO	JAPANESE	PART HAWAIIAN
45/40,221 (0·11 per cent)	2/1,162 (0·17 per cent)	14/6,478 (0·21 per cent)	29/11,801 (0·25 per cent)	109/45,013 (0·24 per cent)	16/9,126 (0·18 per cent)

crosses. Nor is there any tendency in Hawaii for there to be an increased frequency of cleft palate and hare-lip in part-Hawaiians or in any other racial crosses (Morton and others, 1967) as shown in *Table* 6.39.

It is clear, then, that the incidence of hare-lip and cleft palate is normally intermediate in racial hybrids. We have in fact classed together the three different deformities of hare-lip, hare-lip with cleft palate, and cleft palate. The first two, however, are very similar in etiology and certainly show the racial differences we have described. Cleft palate alone is slightly different in origin. It seems to occur at a frequency of about one-quarter that of the other two combined, and does not show any marked racial variation.

6.5.*b*. *Fertility Differences*

We are concerned in this section with natural fertility differences, i.e., that fertility which exists or has existed in the absence of deliberate birth control. It is recognized that the relevance of this parameter is likely to decline in the future as birth control becomes both more effective and more widespread. Accurate statistics in this field, with which we might include perinatal mortality, are extraordinarily difficult to obtain for a number of reasons. Firstly populations have to be found either where birth control is not practised or where control is so limited that it does not effectively limit fertility. The second difficulty lies in the collection of data for, with the exception of a small number of strict religious isolates in advanced countries, those populations which did not or do not practice birth control are either pre-modern or inhabit countries where demographic statistics are not developed or are of poor quality.

Thirdly, for comparative purposes all measures of fertility need to be standardized for such factors as age of marriage and duration of marriage, both of which are more significant for the expression of fertility than chronological age as such. Finally there appear to be a number of almost cyclical variations in both the level and pattern of fertility in some populations which do not appear to be related to any of the usual social factors, but which are unlikely to be of genetic origin.

The development of effective statistical gathering services is likely to go hand in hand with the development of modern medical services, so that genetically mediated perinatal mortality becomes considerably modified. Clearly, then, the figures for mortality at all ages among populations living today are more a reflection of efficiency of the medical and social services available to that country than of any real genetic differences. There are probably two areas in this general field where reasonably objective statistics can be obtained: firstly, the incidence of congenital malformations; and secondly, the frequency of sterility among married couples. With decreasing reliability one might consider susceptibility to disease, longevity, and perinatal mortality. The incidence of congenital malformations we shall consider in a separate section.

As we have already mentioned, data on fertility levels are difficult to obtain and are in any case becoming increasingly irrelevant as actual fertility becomes less and less a reflection of parental fertility. *Table* 6.40 shows some examples of age-specific female legitimate fertility rates from populations presumed not to practise or have practised contraception. This table indicates that there are possibly differences between the races in intrinsic fertility, both considered overall and at different ages. From these figures there is a suggestion that the Europeans have perhaps a slightly higher level of fertility than non-Europeans.

We can return to the micro-races of Wales in Great Britain to show indeed that these differences are likely to be real ones. Ashley (1969) compared the

Table 6.40.—AGE-SPECIFIC FEMALE LEGITIMATE FERTILITY RATES (*rates per* 1,000) (*From* Henry, 1 61.)

POPULATION	MEAN NO. OF CHILDREN/COMPLETED FAMILY OF WOMAN MARRIED AT 20	INDICES OF MEAN FERTILITY RATES (Base 100: mean rate 20–29)		
		20–29	30–39	40–49
Hutterites, marriages from 1921–30	10·9	100	81	27
Canada, marriages from 1700–30	10·8	100	89	26
Hutterites, marriages before 1921	9·8	100	86	25
Geneva women, wives of men born 1600–49	9·4	100	71	15·5
Tunisian Europeans 1840–59	9·15	100	81	22·5
Normandy marriages 1760–90	8·95	100	78	19·5
Normandy marriages 1674–1742	8·3	100	76	17
Norway marriages 1874–76	8·1	100	81	28·5
Iran (villages) marriages 1940–50	7·5	100	76	19·5
Geneva women, wives of men born before 1600	7·5	100	80	19
Taiwan women born about 1900	6·95	100	81	17
India (Bengal villages) 1945–46	6·2	100	80	22
Guinea 1954–55	6·2	100	67	16

fertility of women in areas of a high frequency of Welsh-speaking, broadly speaking West Wales, an intermediate frequency of Welsh-speaking, Mid and South Wales, and an area of little Welsh-speaking, East Wales and urban South Wales. For these three areas Ashley obtained corrected fertility indices (the number of births/1000 married women/year) of 123, 129, and 139 respectively. When he considered Wales as a whole, he found that the ratio of observed to expected numbers of pregnancies for those women whose maiden names and married names were both Welsh was 95, whereas for those women living in Wales, neither of whose names were Welsh, the ratio of observed to expected numbers of births was 110. This difference in fertility between Welsh and non-Welsh women is not due to differences in social class, nor to different urban/rural distribution and probably not due to a later age of marriage in

Welsh women. Since the Welsh show a high degree of assortative mating amongst themselves, and because the phenomenon is apparently common throughout Wales, the difference is thought to be genetic in origin. Clearly if genetic differences in fertility exist at this level, then they are likely to exist between *local* and *major* races of mankind.

Having suggested, then, that racial differences in fertility exist, we have to determine what effects these might have in racial hybrids. The data from Wales again provide some indications on this point. Ashley (1969) presents data on the obstetric history of 593 women over the age of 25 years whose married and maiden names were known, who gave birth in the Maternity Department at Morriston Hospital, Swansea, during 1962. *Table* 6.41 shows the number of

Table 6.41.—NUMBER OF PREGNANCIES IN WOMEN IN WALES
(*After* Ashley, 1969.)

MATING	NO. OBSERVED	NO. EXPECTED	OBSERVED/ EXPECTED	MEAN NO. PREGNANCIES
Both married and maiden names Welsh (i.e., Intraracial mating)	429	453·8	95	2·58
Maiden name Welsh, married name non-Welsh	377	394·7	96	2·79
Maiden name non-Welsh, married name Welsh	372	373·4	99·5	2·75
Total interracial matings	749	767·1	98	2·77
Both married and maiden name non-Welsh	466	423·9	110	3·04

pregnancies in these women that are the result of Welsh/Welsh marriages, Non-Welsh/Non-Welsh marriages, or 'interracial' marriages. The indication of Welshness was taken to be the surname, which has been shown (Ashley and Davies, 1966) to be an accurate genetic marker in Wales. Overall there is quite a difference in the mean number of pregnancies observed in the Welsh as opposed to the non-Welsh marriages. The interracial marriages were in every respect intermediate between them. There is a suggestion that the difference in fertility is greater in the wives than in the husbands, though this does not reach the level of statistical significance.

As an indication that these data concerning hybridization among very small poorly distinguished *micro-races* in Wales are typical, there are some figures from Brazil which show that the Mulattoes, hybrids between two of the world's major races the Whites and Negroes, also have intermediate fertility.

The proportion of women aged 50 years and over who have had at least one child is 85 per cent Whites; 84 per cent Mulattoes; 81 per cent Negroes; and the total number of births per 1,000 women aged 50 years and over who have had at least one child is White 775, Mulatto 767, Negro 723. Note that these figures are in line with our tentative suggestion, based on Henry's data, that it is the White race which is intrinsically the most fertile.

Fertility, then, like most other characteristics, appears in hybrids to be intermediate between the contributing parental populations. We have already introduced some data in Chapter 4, s. 4.3, suggesting that there may be some heterotic effects for fertility in racial hybrids; Morton and others (1967), on the other hand, in their large-scale study of interracial crossing in Hawaii, find no

Table 6.42.—RACE NATIVITY OF WOMEN: STANDARDIZED DISTRIBUTION OF WOMEN 15–44 YEARS OLD, EVER MARRIED BY NUMBER OF CHILDREN EVER BORN FOR U.S.A., 1950. (*From* Bogue, 1959.)

SUBJECT	NATIVE WHITE	FOREIGN WHITE	NEGRO	OTHER RACES URBAN (largely Asiatic)	OTHER RACES RURAL (largely Indian)
Number of women ever married	21,810,110	1,033,810	2,766,150	42,480	51,210
No children ever born	21·9	22·4	30·8	21·2	12·9
1 child	27·1	29·9	22·1	25·5	17·5
2 children	25·7	25·5	16·0	21·0	16·3
3 children	12·7	12·0	10·0	16·4	13·1
4 children	6·0	4·9	6·6	4·8	12·3
5 and 6 children	3·3	4·3	7·6	8·4	14·7
7–9 children	1·7	1·3	4·9	2·0	10·5
>10 children	0·5	0·6	1·9	0·6	2·6
Children ever born/1,000 women ever married (standardized)	1,829	1,707	2,119	2,047	3,304

real evidence to support the idea of heterosis for this character. They do, in fact, have some evidence for an association of both maternal hybridity and hybridity of the child with increased birth order (which is a rough measure of the fertility of married women), although the effect is a small one. But they say that this association 'is no critical evidence of heterosis, and might reflect only difference in age at marriage or other cultural factors' and, in general: 'The most that can be said on the evidence is that first and later generation hybrids in man suffer no appreciable loss of fertility and may indeed have increased fertility, compared with the parental races'.

The alternative way of looking at this whole problem is to consider the reverse of fertility—namely, the incidence of sterility. Again, of course, it is very difficult to collect reliable information, and in practice we can only get an estimate of fundamental sterility by considering, for example, married women aged 45 years or over who have never produced children. Even these statistics are unavailable, however, and although we know that the percentage of women 45–49 years old who have never borne children varies between 26·8 per cent (U.S.A., 1950), 23·0 per cent (France, 1946), 15·4 per cent (Moçambique, 1940), and 14·9 per cent (Brazil, 1940), we do not know the proportion of unmarried women in these figures. Alternatively the figures available for the U.S.A. (*Table* 6.42) are for standardized distributions of women 15–44 years old, and of course we do not know the proportion of women with incompleted families among them. There are nevertheless some suggestive differences between the proportions of sterile women in the different races in this table.

Henry (1961) has collected some figures on the age-specific percentages of sterility in married women and these are shown in *Table* 6.43. All the different

Table 6.43.—AGE-SPECIFIC PERCENTAGES OF STERILITY IN MARRIED WOMEN (*From* Henry, 1961.)

POPULATIONS	20 YEARS	25 YEARS	30 YEARS	35 YEARS	40 YEARS
England (mid-19th century)	3	7	12	19	32
Canada	3	4	6	10	22
Bourgeosie of Geneva (wives of men born before 1650)	3	8	13	21	39
Europeans of Tunis	3	7	10	16	28
Crulai (Normandy)	2	3	7	15	36
Average (European population)	3	6	10	16	31
Rural Japan (pre-war)	4	10	19	33	53

populations show a similar low figure for women aged 20 years of about 3 per cent, which we may take to be the fundamental incidence of sterility. The series for Japan is the only one for a non-European population. The rise in sterility rates as the women grow older is considerably steeper than that in the European samples, but it is impossible to say whether this is a general phenomenon. In any case it is not likely to be of any significance in interracial crosses today.

We conclude, then, that while there are likely to be some racial differences in the incidence of sterility there is no suggestion that they are of any great magnitude or likely to be of any consequence in social terms for racial hybrids.

There is one further feature which might be considered in this section on fertility, and that is the age at which girls first menstruate. This, at least in the abstract, is a measure of when fertility is first attained, and it is of some interest to determine whether there is any racial variation for this trait. It is known, of course, that there is a great deal of variation around the world, although it is clear that most of this is environmental—partly nutritional and partly climatic (Tanner, 1962). Laska-Mierzejewska (1966) investigated the age at first menstruation of Cuban girls and obtained the following results:

White girls	12·42 ± 0·0326	s.d. 1·42
Black girls	12·33 ± 0·0832	s.d. 1·22
Mulatto girls	12·61 ± 0·0640	s.d. 1·52

In detail it was found that there were more Negro girls aged 9½ to 10½ years menstruating than White girls, and by the age of 15½ years all Negro girls were menstruating, whereas not until they were 18½ years were all White girls menstruating. It was concluded that the Negroes showed a tendency for earlier menstruation than the Whites. This difference was only 0·08 years and not significant statistically; but it was probably indicative of a real biological difference because the living conditions of Negroes, being rather worse than the Whites, would cause a delay in the onset of menstruation. The Mulatto girls begin to menstruate rather later than the others. Observations regarding the development of the breasts, pubic hair, height, and body-weight suggested that in these traits the Mulatto girls were more similar to the White girls than the Black girls, and therefore for these characters the White gene contribution is pre-eminent, which would mean that they would begin menstruating closer to the age at which the White girls do than the Black girls. The social and economic conditions under which Mulatto girls live are identical to those of the Negroes and therefore this has an additional delaying effect. The net effect is that they begin menstruating rather later than either group.

There are a few other studies to substantiate the conclusion reached by Laska-Mierzejewska, firstly of racial differences in the age of female sexual maturity and secondly of the predominance of the White genome controlling this character in racial hybrids.

The racial differences are reviewed by Tanner (1962, 1964) and Garn (1962) and a little further information on racial hybrids is given by Laska-Mierzejewska (1967) and Morton and others (1967). Some of this information will be discussed elsewhere (Chapter 7). Da Costa Martins (1971), in his studies in Moçambique, found that the age at menarche for White girls was 12·15 ± 0·27 years, for Negro girls 13·03 ± 0·78 years, for Indians 12·39 ± 0·10 years, and for Mixed girls 12·57 ± 0·10. The hybrid girls were in this case intermediate although the Negroes are significantly later in beginning menstruating than the Whites.

We have also considered in Chapter 2 some other racial variation in fertility and *Table* 2.9, for instance, shows some of the marked differences in the continuation of fertility past the age of 50 years in different peoples. Some, at least, of this variation is likely to be genetic in origin.

6.5.c. *Race Crossing and the Frequency of Congenital Malformations*

Congenital malformations are those defects which are present at birth and which may be due to genetic causes or embryological disturbances. We are primarily concerned with those due at least in part to genetic causation, although the nature of this genetic causation may be either fairly simple or very complex (Carter, 1965). The definition of congenital defect is in fact arbitrary, and diagnosis depends on the rigour with which examination is conducted, as well as the age and condition of the foetus or baby. Many malformations, in fact, show a continuous variation in their degree of expression (for instance hare-lip and cleft palate, which we have already considered) from virtually normal to very severe. As well as some defects being missed in different surveys, there will be differences between physicians and paediatricians as to where the line can be drawn between malformed and not malformed. Certainly, in some cases, examination of children 8–10 months of age revealed the presence of some malformations including cleft palate, polydactyly, and ichthyosis which had not been diagnosed at birth (Neel, 1958). All these considerations, together with others mentioned by other workers in this field, and in particular by Morton and others (1967), suggest that comparisons between different countries and surveys regarding the incidence of congenital malformations are rather hazardous. Fortunately there exists the report *Congenital Malformations* by Dr. A. C. Stevenson and co-workers (1966, two papers) from Oxford, who analyse data from a series of consecutive births in 24 centres in 16 countries around the world. This study is on the results of 421,781 pregnancies and, so far as was possible, the diagnosis and recording of the defects was uniform in each of these centres. Using these data as relatively reliable reference points, we can also consider the results of such investigations as those by Saldanha (1960, 1961, 1962a), Neel (1958), McKeown and Record (1960), Böök (1951), and others reviewed by Lilienfeld (1970).

The number of different congenital malformations known to occur in man with detectable frequency is quite large. Stevenson and others group theirs into 13 major classes primarily on the system or organs affected, together with other categories for 'Miscellaneous' and 'Minor', viz.:—

A, Downs syndrome;
B, Neural tube defects;
C, Other malformations of the central nervous system;
D, Malformations of the heart and great vessels;
E, Defects of the alimentary canal and associated organs;
F, Defects of the diaphragm;
G, Hare-lip and cleft palate;
H, Talipes;
I, Dysplasia, subluxation, and luxation of the hip;
J, Polydactyly and other digital deformities;
K, Other local skeletal malformations; L, Urinogenital malformations.

This system of classification with its numerous subdivisions is about the most complex employed, and it is therefore the results of this report which we shall examine first in order that we might be able to consider a wide range of malformations. The overall figures for the incidence of *all* congenital malformations are presented first in *Table* 6.44.

Table 6.44.—CONGENITAL MALFORMATIONS IN DIFFERENT RACES AND THEIR HYBRIDS (*Compiled from* Stevenson, Johnston, Stewart, and Golding, 1966.)

	CAUCASIANS	MULATTOES	NEGROES	MESTIZOES	MONGOLOIDS
Latin America	363/33601 =0·0108	131/7450 =0·0176	67/3645 =0·0184	1094/71237 =0·01536	21/1295 =0·0162
S. Africa	252/11176 =0·0225	26/3050 =0·0085	129/10025 =0·0129		
Malaysia	8/246 =0·032			1/260 =0·0038	79/7001* =0·011284 334/40128† =0·008323
Total	=0·0138	=0·0150	=0·0143	=0·00953	=0·0090

* Malays; † Chinese.

The results from this compilation clearly show that there are few differences between the world's major races with respect to their overall incidence of congenital malformations.

It is difficult to generalize regarding the position of racial hybrids. The sample from Malaysia is much too small to read anything into the very low frequency of malformations among hybrids there. In Latin America we know that the distinction into Caucasian and Mulatto or Negro is arbitrary, since all races show greater or lesser degrees of admixture. We might note nevertheless that the two hybrid populations are intermediate between their respective parental populations. In South Africa, where perhaps the racial distinctions are more meaningful (*but see* Chapter 8), the hybrids show a lower frequency than their parental races. These figures from South Africa are in fact significantly heterogeneous. For reasons which are discussed in Chapter 3 we might expect a decreased frequency of malformations in racial hybrids and we might, therefore, claim from the South African data that such an expectation is upheld. It would perhaps be more in keeping with the methodological problems associated with obtaining reliable comparative data in this field merely to note that it appears very unlikely that there is any increase in the incidence of congenital malformations among racial hybrids, whichever are the contributing populations. This general conclusion is firmly supported by two important large surveys within racially mixed populations, those of Brazil and Hawaii.

Table 6.45 from Saldanha (1962a) shows that in Sao Paulo the overall malformation frequency among Whites, Negroes, and their hybrids is almost identical. Indeed the only significant difference for any racial group was for

Table 6.45.—THE FREQUENCY OF CONGENITAL MALFORMATIONS AMONG DIFFERENT RACIAL GROUPS IN BRAZIL. (*From* Saldanha, 1962a.)

MALFORMATION	WHITES (15,532) No.	/10³	NEGROES (3,286) No.	/10³	MULATTOES (3,804) No.	/10³
Achondroplasia	2	0·13				
Ano-microphthalmos	1	0·06				
Anencephaly	8	0·51	1	0·30	1	0·26
Anal atresia	6	0·39				
Congenital heart disease	40	2·57	5	1·52	9	2·37
Hare-lip and/or cleft palate	22	1·42	1	0·30	2	0·53
Hydrocephaly	16	1·03	1	0·30	4	1·05
Lobster claw (ectrodactyly)	2	0·13	2	0·61	1	0·26
Microcephaly	6	0·39				
Mongolism	15	0·97	2	0·61	1	0·26
Polydactyly	38	2·45	23	7·00	20	5·26
Spina bifida	1	0·06			1	0·26
Total of these specific malformations	157	10·11	35	10·65	39	10·25
Frequency of all malformations		31·87		36·21		31·54

polydactyly. When the rates for all congenital malformations are considered together it turns out from Saldanha's study that the Mulattoes show a rate lower than Whites or Negroes, and from Stevenson and co-workers' that they show a rate below the midpoint of the two parental populations.

With regard to Hawaii, Morton and others (1967) conclude a very thorough investigation of the distribution of congenital malformations in different races with the words 'we know of no critical evidence of any effect of hybridity or recombination on the frequency of congenital malformations'. The effects they did find when the defects were obtained by any of 5 possible sources open to them, viz., Birth and Death certificates, records of the Bureau of Children, and two categories of hospital, are shown in *Table* 6.46.

The genetics of most of these congenital malformations is still not well understood, and it is likely that what we have so far discussed are a heterogeneous group with respect to their genetic causation. This is therefore an argument in favour of examining each malformation separately. However, in view of the labour involved in carrying out surveys of congenital malformations in sufficient detail to demonstrate possible racial differences, it is difficult to consider many of the different types of congenital malformation from the point of view of racial crossing. On the basis of investigation of a few of them, such as

Table 6.46.—ESTIMATES OF RACIAL EFFECTS ON CONGENITAL MALFORMATIONS. (*From* Morton and others, 1967.)

RACIAL VARIABLE	MAJOR MALFORMATION	MINOR DEFECTS
Pacific mother	—	—
Hawaiian mother	−0·0186*	—
Chinese mother	—	—
Filipino mother	—	—
Puerto Rican mother	0·0317†	—
Korean mother	—	—
Major maternal hybridity	−0·0065	—
Minor maternal hybridity	−0·0009	—
Major recombination	0·0035	—
Minor recombination	−0·0069	—
Major hybridity of child	−0·0031	—
Minor hybridity of child	0·0011	0·1146

* $P < 0{\cdot}01$; † $P < 0{\cdot}001$; — not significant in stepwise regression.

anencephaly, cystic fibrosis, and others which we shall examine, it seems fairly certain that there are marked differences in racial frequencies of many congenital malformations. Nevertheless the conclusion that their overall frequencies are about equal seems fairly secure. It further seems that there is little effect on their incidence due to racial crossing.

Anencephaly

Anencephaly is the most severe and the most easily recognized congenital malformation of the human central nervous system. The skull has no bony roof, the cerebral hemispheres are much reduced and the rest of the brain and spinal cord may also be greatly affected. The condition is of course lethal. Searle (1959) investigated the incidence of anencephaly among births in the largest maternity hospital in Singapore (Kandang Kerbau) which caters for all races living in Singapore. The W.H.O. study also utilized data collected in this hospital and figures extracted from both of these investigations are presented as *Table* 6.47. Other useful surveys are by Penrose (1957) and Saldanha (1962a) in a mixed population from southern Brazil.

Searle found marked differences between the races, and the important point from our present point of view is that the Eurasians, who, Searle states, are mainly the result of marriages between Europeans and Chinese, have an incidence unquestionably intermediate between the parental races.

Saldanha (1962a) reports on a series of 22,781 consecutive live births which occurred in the clinic Hospital of the University of São Paulo from January, 1944, to December, 1960. The incidence of anencephaly among Whites was 8/15,532 or 0·51 per thousand, among Negroes 1/3,286 or 0·3 per thousand, and among Mulattoes 1/3,804 or 0·26 per thousand.

The relative infrequency of anencephaly among Negroids is confirmed by Penrose's (1957) data and by Stevenson, Johnston, Golding, and Stewart

Table 6.47.—INCIDENCE OF ANENCEPHALY AT THE KANDANG KERBAU HOSPITAL, SINGAPORE

MATERNAL RACE	1953–56 (Searle, 1959) ESTIMATES			1963 (Stevenson, Johnson, Stewart, and Golding, 1966) ACTUAL NUMBER		
	No. Births	No. Anencephalics	Frequency /10³	No. Births	No. Anencephalics	Frequency /10³
Chinese	72,660	45	0·62	31,729	23	0·72
Indians: Sikh	1,230	8	6·49 } 1·20	3·119	2	0·64
Indians: Other Indians	12,990	9	0·69 } 1·20			
Malays	2,458	3	1·22	4,904	3	0·61
Eurasians	690	1	1·46			
Europeans	930	2	2·14			

(1966), who quote an incidence of 0·41 per 1,000 among 14,444 children born in Makerere Hospital, Uganda, between 1956 and 1958.

Cystic fibrosis

Cystic fibrosis is a defect which is genetically fairly simple yet shows rather interesting racial distributions. It is due to an autosomal recessive gene which is very rare in non-European populations and yet shows an incidence among Europeans of between 1/4,000 and 1/2,000 births. A study of 242,354 births during a 15-year period in Hawaii shows this low frequency for the gene for cystic fibrosis in non-Caucasians and the disproportionately low incidence in hybrids (Morton, 1970). *Fig.* 6.28 shows the differing incidence of cystic fibrosis in racial hybrids with different degrees of Caucasian ancestry.

Other Malformations

Other congenital malformations clearly show some of the consequences of outcrossing for characters controlled by recessive genes. For example, talipes equinovarus (a form of clubfoot) is low in Orientals and very high in Hawaiians. Other Polynesian populations also have a high incidence of talipes equinovarus. These differences are largely additive in outcrosses, but there is a significant recessive component to the Hawaiian effect (*Table* 6.48). Similarly pyloric

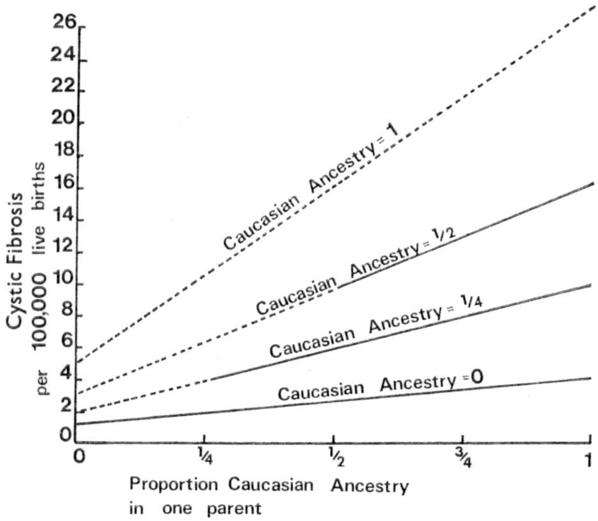

Fig. 6.28.—Incidence of cystic fibrosis per 100,000 live births in unmixed and hybrid matings. On the abscissa the proportion of Caucasian ancestry may vary from 0 to 1 for one of the parents. For the other parent the proportion of Caucasian ancestry is fixed at one value. Calculated on the assumption of a single locus. (*From* Morton, 1970.)

Table 6.48.—INCIDENCE OF TALIPES EQUINOVARUS PER 100,000 BIRTHS (*From* Morton, 1970.)

SOURCE	INCIDENCE
Orientals in Hawaii	57
England	124
Caucasians in Hawaii	112
Hawaiian incross	681
Caucasian-Hawaiian F_1	238

stenosis shows a lower incidence in Pacific populations than European ones, with the high incidence being partially recessive (*Table* 6.49).

Table 6.49.—INCIDENCE OF PYLORIC STENOSIS PER 100,000 BIRTHS IN HAWAII (*From* Morton, 1970.)

SOURCE	INCIDENCE
Caucasian	159
Japanese	68
Caucasian-Japanese F_1	54
Hawaiian	5
Filipino	14

Although we have not presented a systematic survey of the very many congenital malformations known to vary in frequency between the races (*see*, for example, Lilienfeld, 1970), it seems clear that in most cases the frequency of whatever malformation is examined is intermediate in racial hybrids. The exceptions are those cases where there is an element of recessiveness in the genetic control of the character concerned. In these cases the frequency in the hybrid population is below the mid-point value of the two parental populations.

6.5.d. *Mortality, the Incidence of Disease, and Overall 'Genetic Load'*

This section, in a sense, ties together the two previous sections and unites some scattered reports of a possibly heterogeneous nature. We begin by examining foetal and neonatal death.

Many of the difficulties of accurately collecting this information have already been touched on when considering fertility statistics. There is, in addition,

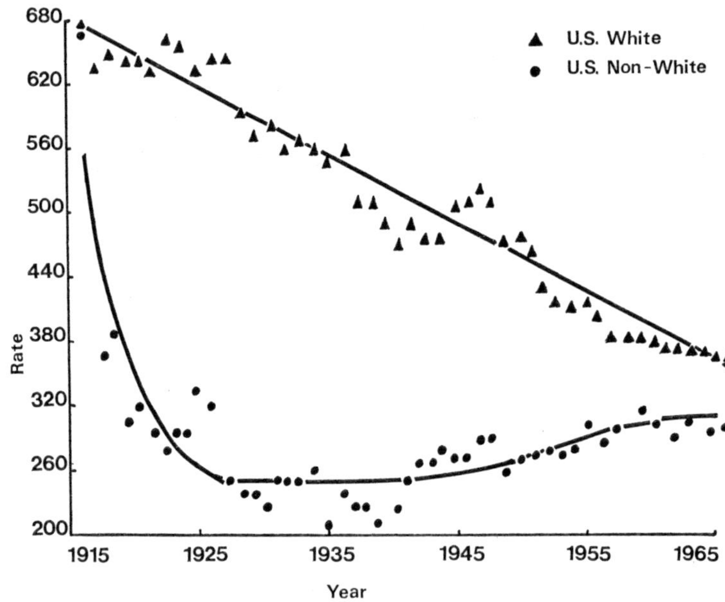

Fig. 6.29.—Infant deaths ascribed to congenital malformations per 100,000 live born. (*From* Morton and others, 1967.)

the profound interaction of the levels of medical service available to the people concerned and the foetal mortality rates. This interaction will normally completely override any possible racial differences. Even where the races concerned inhabit the same countries it is usually the case that it is socio-economic differences not genetic ones which are mainly responsible for observed mortality differences, as for example Aboriginal/White differences in Australian, or Negro/White differences in the U.S.A. In certain circumstances, however, it is possible to obtain statistics indicative of the true biological situation. Mortality due to congenital defect is one such possibility and *Fig.* 6.29 shows

Table 6.50.—ESTIMATES OF RACIAL EFFECTS ON MORTALITY IN HAWAII, 1948–1958 (*From* Morton and others, 1967.)

RACIAL VARIABLE	EARLY FOETAL DEATH	LATE FOETAL DEATH	NEONATAL DEATH	POSTNATAL INFANT DEATH
Pacific mother	−0·0059‡	—	—	—
Hawaiian mother	—	—	—	—
Chinese mother	—	—	—	—
Filipino mother	—	—	—	—
Puerto Rican mother	—	—	—	—
Korean mother	—	—	—	—
Pacific parent	—	−0·0072‡	—	—
Hawaiian parent	−0·0205‡	0·0185‡	—	0·0075†
Chinese parent	−0·0062†	—	—	—
Filipino parent	−0·0097‡	0·0057‡	—	0·0030‡
Puerto Rican parent	−0·0123†	—	—	—
Korean parent	—	—	—	—
Major maternal hybridity	−0·003	−0·0014	−0·0031	−0·0013
Minor maternal hybridity	0·0006	−0·0031	−0·0061*	0·0013
Major recombination	−0·0036	−0·0044	−0·0004	0·0007
Minor recombination	−0·053	0·0012	0·0107†	−0·0017
Major hybridity of child	−0·0014	−0·00005	0·0027*	−0·0009
Minor hybridity of child	−0·0009	−0·0003	0·0017	0·0014
Part Hawaiian mother	—	—	—	—
Part Hawaiian parent	—	—	—	—
Hawaiian incross	—	—	—	0·0167‡
Part-Hawaiian incross	—	—	—	—

— Not significant in stepwise test; * $P < 0{\cdot}05$; † $P < 0{\cdot}01$; ‡ $P < 0{\cdot}001$.

the rate of infant deaths per 100,000 live births in the U.S.A. ascribed to this cause. For Whites there is an approximately linear decrease from 680 to 366 between 1916 and 1963. The non-Whites from continental U.S.A. show a complex pattern of change, but invariably exhibit lower mortality rates than the Whites. The main factors leading to these changes are, firstly, improved medical services reducing the death-rates, and secondly, improved social services improving the recognition rate and the number of children available for treatment. It seems probable that the convergence of the two groups will continue, leaving only a small difference between them which can be explained by genetic factors.

Mortality, of course, is an event which may occur from conception onwards, and we can divide it up into a number of different groups. Early foetal death (abortions and miscarriages) under 20 weeks gestation; late foetal death (still births) of 20 or more weeks gestation; neonatal death (less than 28 days after birth); postneonatal infant death (at least 28 days but less than 1 year after birth); juvenile and adult mortalities. Morton and others use this division in their analysis of mortality among interracial crosses in Hawaii. Their results are shown in *Table* 6.50 and it is obvious that there is very little in the way of any racial or interracial effects on mortality. Indeed, if anything mortality may be lower in interracial matings. We can briefly quote the conclusions that Morton and others come to. With regard to early foetal death they say that 'Pacific and Puerto Rican parents report fewer early foetal deaths than Caucasian parents' and their 'Ancillary and hybridity effects are non-significant. Furthermore since parental hybridity is not associated with an increased early foetal death-rate, these data provide no evidence for chromosomal polytypy or other mechanisms for subvital recombination of coadapted genotypes'. With regard to late foetal death Morton and others say that 'Pacific parents except Hawaiians report fewer late foetal deaths than Caucasoid parents' and that 'Filipinos are intermediate between Hawaiians and other Pacific groups. There is no significant effect of hybridity'. They conclude for neonatal death that 'There is no significant main effect of race, but neonatal death decreases with minor maternal hybridity and increases with minor recombination and major hybridity of child.'

Turning to postneonatal infant death we encounter important effects of environmental mortality factors, particularly from infectious diseases. Hawaiian parents have a higher postneonatal infant death-rate, but because part-Hawaiian incrosses do not have elevated risks (which would be expected if recessive genes were involved) and because infant death among Hawaiians declined from 29 per cent in 1924 to 6 per cent in 1950, it is thought that its origin must be environmental. There is, in any case, 'no significant hybridity effect'.

Other studies with data bearing on this problem include Krieger (1970) in north-eastern Brazil and Bresler (1970) in the U.S.A. Krieger investigated a trihybrid population in north-eastern Brazil. A total of 1,069 families were studied and the individuals were assigned a score on a seven-fold racial classification. The results were analysed by a stepwise multiple regression, just as in Hawaii, and for 4 metrical traits, 6 pathological conditions, and 5 specific congenital abnormalities there was no real evidence of racial effects on incidence. Mortality was divided into foetal death and post-natal death and in neither case was there any significant racial effect. In Krieger's words, 'The evidence presented . . . is consistent in that the relatively small increase in heterozygosity, occurring by interracial crosses, is irrelevant to fitness'.

Bresler's (1970) investigation is concerned only with Caucasians. He interviewed a number of mothers who had given birth in Providence, Rhode Island, regarding their reproductive history. Each was also asked for details of race and place of birth of herself and her husband, and race and national origin of the 4 parents in the P_2 generation and the 8 grandparents of the P_3 generation.

Bresler analyses his data in a number of ways, but *Table* 6.51 shows perhaps his most striking results. With an increasing number of countries in the background there is an increasing level of foetal loss at the F_1 level. Considering the

Table 6.51.—THE RELATIONSHIP BETWEEN THE NUMBER OF COUNTRIES IN THE P_3 GENERATION TO THE RATE OF FOETAL LOSS IN THE F_1 GENERATION AMONG MIGRANTS TO THE U.S.A. (*From* Bresler, 1970.)

NO. COUNTRIES	NO. FAMILIES	PREGNANCIES	LIVE BORN	FOETAL LOSS (per cent)	FAMILIES EVER F.L. (per cent)
1	224	736	680	7·7	19·6
2	273	901	807	10·3	22·0
3	161	471	419	11·0	22·4
4	40	114	99	13·2	30·0
5	9	28	23	17·9	55·6
6	1	2	1	50·0	100·0
Total	708	2,252	2,029	9·9	22·3

data for 4, 5, and 6 countries, a Chi-square test indicated significant heterogeneity. He also found that increased foetal loss is related to greater distances of mates within the P_1 generation. He interprets his findings in terms of a greater mixing of Mendelian gene pools and 'concomitant genetic imbalance between loci'. Despite the fact that this investigation is confined only to Caucasians, its general conclusions are clearly at variance with those derived from interracial hybridizations. The reasons for this are not at all clear.

The other approach to this problem which has been adopted is that of determining the increase in mortality and morbidity which is exposed by inbreeding. We have already considered the ideas behind this sort of approach in Chapter 2 when considering overall genetic variation. There are, in fact, two different quantities which can be determined in this context, which are usually termed the 'random load component A' and the 'inbred load B'. The complex genetic/environmental interactions and the rather subtle statistical difficulties usually tend to make the interpretation of completed surveys rather uncertain. We have already concluded that, on average, each individual may contain up to 4 lethal equivalents and up to 10 sublethal and detrimental equivalents. In fact there is considerable variation in estimates of the load values from different peoples, as shown in *Table* 6.52 from Cavalli-Sforza and Bodmer (1971). However, it must be remembered that the random load component A, measuring prereproductive mortality among non-inbred individuals, encompasses many causes of death that are non-genetic and we have already seen that this makes genetic analysis very difficult. The quantity B would seem to be a measure of real genetic difference, particularly as it is largely a measure of lethal equivalents, i.e., recessive lethals and semi-lethals. The frequency of these rest heavily on mutation rates or selective values operative in the population concerned, which seem unlikely to vary from country to country. The lower frequency of lethals in Japanese as compared to Europeans is thought to be due to the inbreeding history in Japan, where first-cousin marriages are common, indeed often encouraged, whereas in Europe they are rare (especially in Catholic Europe). It should be emphasized, of course, that since interracial crosses are essentially outbred, the level of B, the inbred load, is of no consequence to the

Table 6.52.—DISTRIBUTION OF A AND B VALUES COMPUTED FOR PRE-REPRODUCTIVE MORTALITY IN DIFFERENT POPULATIONS (*From 'The Genetics of Human Populations'* L. L. Cavalli-Sforza and W. F. Bodmer, 1971. Copyright London: W. H. Freeman & Co.)

		B VALUES						
		<0	0·0–0·5	0·5–1·0	1·0–1·5	1·5–2·0	2·0–2·5	>2·5
	0–0·1		1*	3*				
A	0·1–0·2	1*	1*	2*	2*		1†	
VALUES	0·2–0·3			1†		1†		
	>0·3	1†, 1	1†, 1	1†				1†

* Japanese populations; † Caucasian populations; Rest Negroes and Indians.

F_1 hybrids. We are considering, if anything, the possible gene flow of lethal equivalents from populations with high frequency to those with low.

We can also consider the average number of detrimental recessives carried by individuals in different populations, although here, unless we specify the particular conditions involved as in the surveys on congenital malformations, we run into difficulties regarding the criteria used to define genetic morbidity as a disability.

There are some results based on surveys in the same areas using careful objective methods of assessment which suggest that the inbreeding load in different races might vary—particularly between the Whites and Negroes in South America (Freire-Maia, 1963). More recent work suggests that this is in fact not so. Freire-Maia and Azevedo (1971) obtain the results shown in *Table* 6.53 where non-Whites refer to both Mulattoes and Negroes. None of the differences between Whites and non-Whites are significant.

Table 6.53.—ESTIMATES OF NUMBER OF LETHONS IN STATE OF MINAS GERAIS, BRAZIL. (*Data from* Freire-Maia and Azevedo, 1971.)

	ABORTIONS	STILLBIRTHS	INFANTILE-JUVENILE MORTALITY	TOTAL
Whites	0·33 ± 0·48	0·62 ± 0·26	1·91 ± 0·62	2·90 ± 0·85
Non-Whites	1·06 ± 0·63	0·02 ± 0·55	2·65 ± 1·12	3·69 ± 1·45

Table 6.54.—RACIAL VARIATION IN DISEASE—SOME SIMPLE AND COMPLEX DISORDERS WITH AT LEAST SOME GENETIC CAUSATIVE FACTORS (*Partly from* Damon, 1971, *and* McKusick, 1967.)

POPULATION	RELATIVELY HIGH FREQUENCY	RELATIVELY LOW FREQUENCY
Ashkenazic Jews	Tay-Sachs disease Pentosuria Gaucher's disease Familial dysautonomia Stub thumbs Bloom's disease Factor XI (PTA) deficiency Niemann-Pick disease Abetalipoproteinemia Dystonia musculorum deformans Spongy degeneration of brain	Phenylketonuria Pyloric stenosis Tuberculosis
Mediterranean Peoples (Italians, Greeks, Sephardic Jews)	Thalassaemia (mainly β) G6PD-deficiency, Mediterranean type Familial Mediterranean fever	Cystic fibrosis
Africans	Haemoglobinopathies, esp. Hbs, HbC, α and β thalassaemia, persistent HbF G6PD-deficiency, African type Tuberculosis Polydactyly	Cystic fibrosis Haemophilia Phenylketonuria Wilson's disease Congenital hip disease Major C.N.S. malformations
Japanese	Acatalasia Oguchi's disease Dyschromatosis universalis hereditaria	Breast cancer
Chinese	α-Thalasaemia G6PD deficiency, Chinese type Nasopharyngeal cancer Trophoblastic disease	Chronic lymphatic leukemia Prostatic cancer
Armenians	Familial Mediterranean fever	
Irish	Major C.N.S. malformations (anencephaly, encephalocele)	
Polynesians (Hawaiians)	Clubfoot Coronary heart disease Diabetes mellitus	
American Indians	Tuberculosis Congenital dislocation of hip	Duodenal ulcer

In a re-analysis of her earlier data Freire-Maia concludes that the differences detected are probably spurious and mainly environmental and that 'The "total" inbreeding load estimated for the pooled sample showed nonsignificant difference between Whites and Non-Whites' (Freire-Maia, 1971).

We have dealt so far almost entirely with mortality and malformations manifest among embryos and infants. Disease and mortality continue throughout life, although it becomes progressively less and less easy to disentangle the effects of heredity and environment. *Table* 6.54 partly, from McKusick (1967) and Damon (1971), is a tabulation of the incidence of some diseases among different racial groups, particularly those represented in the population of the U.S.A. Clearly the diseases represented are a very heterogenous collection and the main point of this table is really to show that each race has its array of frequently occurring diseases and rare diseases. With regard to those diseases due to simple genetic abnormalities, it is a straightforward matter to determine their behaviour on hybridization. For the remainder it is by no means so simple,

Table 6.55.—FREQUENCIES OF VARIOUS DISEASES IN CAUCASOIDS, NEGROIDS, AND THEIR HYBRIDS IN AFRICA.

CAUSE OF DEATH	RATE PER THOUSAND IN LAGOS, 1963	RATE PER THOUSAND IN SOUTH AFRICA (COLOUREDS), 1962	RATE PER THOUSAND IN SOUTH AFRICA (WHITE), 1962
Malignant neoplasm	25·9	81·1	134·7
Benign neoplasm	2·7	2·0	2·4
Diabetes mellitus	8·0	4·4	9·8
Vascular lesions affecting C.N.S.	37·3	82·8	96·1
Arteriosclerotic and degenerative heart disease	8·8	67·6	205·5
Hypertension with heart disease	4·1	35·2	20·2
Hypertension no heart disease	26·8	6·6	4·9
Pneumonia	245·6	179·4	55·1
Intestinal obstruction and hernia	2·7	3·8	7·4
Gastritis, enteritis, etc.	121·5	341·6	6·0
Non-meningococcal meningitis	14·1	10·1	1·5

although we have seen that for one or two of them their frequencies in racial hybrids are intermediate between the parental populations.

In order to put this particular problem into perspective we might compare the death-rates for various non-infectious non-parasitic diseases among the White population of South Africa and the Coloured population, which is a hybrid population of Whites with the Negroid Bantu. There are unfortunately no data available for the Bantu themselves, either from South Africa or her neighbours Zambia or Rhodesia. To provide some comparison we can use the

Table 6.56.—COMPARISON OF NUMBER OF CASES OBSERVED IN SOUTH AFRICAN CAPE COLOURED POPULATION OF JOHANNESBURG (1953–55) WITH NUMBER EXPECTED IN UNITED STATES POPULATION (*From* Higginson and Oettlé, 1960.)

Site of Cancer	Males			Females		
	Number observed	U.S.* White	U.S.* Negro	Number observed	U.S.* White	U.S.* Negro
All cancers	81	87·5	70·5	78	118·4	107·8
Buccal cavity						
Tongue	2	0·8	0·6	1	0·2	0·3
Salivary gland	1	0·7	0·9	2	0·6	1·4
Mouth, other	1	1·0	0·4	0	0·4	0·3
Adamantinoma	0	0·0	0·0	1	0·0	0·0
Pharynx	2	1·0	1·1	0	0·0	0·0
Other digestive system						
Oesophagus	2	1·9	2·5	1	0·7	0·8
Stomach	12	7·8	9·6	6	5·2	6·9
Colon	2	11·3	7·6	0	0·0	0·0
Rectum	0	0·0	0·0	3	6·1	6·0
Liver	7	1·1†	1·8†	3	1·2	1·1
Biliary passages	1	0·0	0·0	1	1·4	0·7
Pancreas	1	0·0	0·0	0	1·8	2·1
Respiratory system						
Sinuses	1	0·0	0·0	0	0·0	0·0
Larynx	0	1·7	1·3	0	0·0	0·1
Lung and bronchus	18	7·3†	6·9†	0	2·2	2·2
Breast	0	0·0	0·0	11	26·2†	20·1†
Female genital system						
Cervix uteri				29	13·1†	28·4
Corpus uteri				1	3·6	3·7
Unspecific uterine				2	4·5	4·7
Ovary				5	5·8	4·5
Male genital system						
Prostate	10	6·5	10·7			
Urinary system						
Kidney	1	1·7	1·6	0	1·0	1·2
Bladder	4	4·3	1·6	0	2·3	2·4
Skin	3	14·7†	1·3	3	13·4†	2·1
Central nervous system	1	3·3	2·3	0	0·0	0·0
Endocrine system	0	0·4	0·0	1	1·5	1·4
Thyroid	0	0·4	0·0	1	1·5	1·4
Other endocrine	1	0·3	0·3	0	0·0	0·0
Soft tissue	2	0·8	1·6	3	1·2	0·7
Other and unspecified sites—						
metastatic	2	0·0	0·0	2	0·0	0·0
Lymphomas	3	4·0	3·8	2	3·4	2·8
Leukemias	4	3·7	3·3	1	3·1	1·4

* Number expected according to United States rates.

† The calculated figure differs signficantly (3 × S.D.) from the observed figure.

figures for Lagos in Nigeria, which are among the few accurate figures available for Negroids in Africa, recognizing, of course, that they might well differ from those of the southern Bantu. The data from the United Nations *Demographic Yearbook* are presented in *Table* 6.55, and they show some useful comparisons.

It should be emphasized that the figures in this table are only those which show reasonably large differences in death-rate. There are very many other causes of death in this broad category of non-infectious non-parasitic diseases which do not show a great deal of racial variation.

In 7 of the 11 comparisons shown in *Table* 6.55 the rate of occurrence of the particular disease is intermediate in the hybrids between the Negroes and Whites. In 2 cases the hybrids show the lowest incidence and in 2 the highest. In some cases it is fairly clear that it is environmental factors which are probably responsible for the different death-rates. This certainly includes the gastritis group, pneumonia, and probably also arteriosclerotic and degenerative heart disease. In the case of malignant neoplasms and the vascular lesions affecting the central nervous system, it is possible that genetic factors are more important, particularly in view of the fact that benign neoplasm shows little racial variation.

Table 6.57.—CANCER INCIDENCE IN THE JOHANNESBURG RESIDENT BANTU AND CAPE COLOURED PER 100,000 PER ANNUM (ALL SITES). (*From* Higginson and Oettlé, 1960.)

AGE-GROUP	MALES				FEMALES			
	Bantu	Cape Coloured	U.S.* White	U.S.* Negro	Bantu	Cape Coloured	U.S.* White	U.S.* Negro
0–	10	13	18·1	16·2	7	12	16·7	9·3
15–	17	25	25·2	26·9	16	22	27·0	25·7
25–	34	57	48·8	37·4	46	66	85·0	91·5
35–	69	124	131·6	106·1	129	229	253·0	275·5
45–	182	479	389·1	371·9	244	342	567·6	535·4
55–	320	817	990·5	759·9	556	543	959·5	894·5
65–	579	1596	1752·4	1251·2	646	623	1365·3	1167·7
75+	614	1170	2815·3	1545·9	495	1282	1924·0	970·6
Crude rate	64·8	141·4†	146·1†	118·8	74·8	106·7†	151·6†	145·0†

* Number expected according to United States rates.
† The calculated figure differs significantly (3 × S.D.) from the observed figure.

Higginson and Oettlé (1960) report the results of a large-scale cancer survey carried out between 1953 and 1955 in Transvaal, and they suggest that in fact a great deal of the observed differences in cancer incidences between the races in South Africa are due to environmental influences, particularly the fact that the Bantu have a standard of living much lower than 'more civilized communities' [*sic*]. They suggest, therefore, that the average intermediate incidence of cancer in Cape Coloureds versus Bantu and Whites is merely a reflection of their intermediate standard of living. They do show convincingly that there is great variation for different types of cancer in different peoples, and therefore many of the generalizations made about this subject should be treated with caution. Two tables taken from Higginson and Oettlé (*Tables* 6.56 and 6.57) show both of these features—an intermediate level overall; great differences in detail for some cancers.

At least part of the explanation for malignant tumours is a loss of control of regulatory mechanism controlling cell division. It is possible that in certain circumstances such control may be lost in interracial crosses in some species and for some systems, as for example is markedly demonstrated in the Lepidoptera (Ford, 1971) and Poeciliid fish (Gordon and Gordon, 1957). There is no evidence for anything at all similar in man.

CHAPTER 7

SOME RACIAL HYBRIDS

We will be concerned in this chapter to consider, in a rather general way, the biology of some of the world's major hybrid populations, restricting ourselves, as before, principally to racial amalgamations which have taken place during the last 500 years. We will examine some of the biological systems that we have considered in the previous chapters to see if the generalizations which we arrived at hold good. And we will, in general, show the important part that recent racial hybrids form of the world's population today. We will only consider in any detail hybrid populations between the world's major racial stocks, on the assumption that any effects manifest among them will also be present, but to a much lesser extent, in hybrids between two populations more closely related.

7.1. Negroid-Caucasoid Hybrids

Hybrids between these two races live in North and East Africa, in southern Africa, in the West Indies, in North America, and in South America, where there is usually a degree of Indian admixture as well. There are also small but increasing numbers in Britain.

On the African continent live the oldest established hybrids. They have resulted from the mixture of Hamitic Negroes with Caucasoids who entered Africa from the Middle East and spread westwards and southwards as we discussed in Chapter 3. The hybrids can be divided into three groups—Nilotes, Negro-Hamites, and Bantus. The Nilotes are restricted mainly to the Nile Valley, extending southwards from Khartoum to the north-east shores of Lake Victoria. They can be subdivided into three subgroups, but they are all, on the average, very tall, black, dolicocephalics. They have a prominent forehead, thin lips, a high nasal bridge, and nostrils that are not very wide. They have a long face and narrow or medium prognathism.

The Negro-Hamites live in East and East Central Africa. They are rather less tall than the Nilotes, less dolicocephalic, have a less dark skin, and, on average, a wider nose. The Bantu, who occupy most of Central and southern Africa, are a very variable people showing varying degrees of influence of Hamitic admixture. The western Bantu show mostly Negroid features such as long head, broad flat nose, thick lips, and considerable prognathism. Even within them, though, there are a number of tribes showing marked Hamitic influence. The southern Bantu always show some Hamitic influence, mostly restricted but sometimes quite marked, and their features are never as extreme as the 'True Negroes' of West Africa.

Clearly this group of hybrids is very widespread, very variable owing to varying degrees of admixture, and of very long standing. In South Africa, into which the Bantu have only recently migrated, there was initially extensive racial mixing with the Europeans, and the hybrid population which resulted, who also incorporated smaller but important components from Indians, Asians, and the Hottentots, now form the Coloured peoples of South Africa. They can therefore be regarded as secondary hybrids or, at least in part, almost as 'back-cross' peoples, between Bantu, who carry a small but significant Caucasoid element, and the early European colonizers. This rather curious situation has perhaps been insufficiently well understood by the Race Classification Board of South Africa, or for that matter by the successive governments of that country, who insist that the complex genetic situation obtaining among the inhabitants of southern Africa can be reduced to *a priori* simplicity on the basis of a few legal generalities. It is also perhaps insufficiently realized that the majority of the White population in South Africa also carries significant Negro genetic admixture, on average about 6 per cent.

There are a number of detailed studies of these Negroid/Caucasoid crosses (or backcrosses) living in South Africa, none of which provides any evidence of physical disharmony (Fischer, 1913; Lotsy and Goddijn, 1928). Among the other races in South Africa are the Hottentots and Bushmen. Coon, as we have seen, went so far as to suggest that these represented a distinct subspecies of man. Few workers agree with this idea, and yet all recognize the marked differences of the Bushmen. The Hottentots are short, with yellowish loose skin; they have peppercorn hair, are free of body hair, and are markedly steatopygous (have large protruding buttocks). These Hottentots are about as physically different from Europeans as any race, and yet Fischer's study of their European hybrid offspring shows that they are a harmoniously developed people. These hybrid offspring, the Rehoboth Bastaards numbering about 3,000 individuals in all, are taller than their parental stocks and considerably more fertile, but otherwise intermediate between them. Tobias (1955, 1961) has also examined Bushman-European hybrids, without finding any evidence of detrimental effects among the offspring.

Table 7.1 shows that for the cephalic index the Rehoboth Bastaards are significantly less variable than either of their parental races. The values for the parental races are those calculated by Trevor from a series of 74 male Hottentots measured by Schultze (1928) and 74 Dutch measured by Steggerda (1932a). Fischer himself used a very small sample of Hottentots for his comparison and a German series to represent their European ancestry. In fact in three of the four traits shown in *Table* 7.1 the Rehoboth Bastaards are less variable than the Europeans (Fischer, 1913; quoted by Wallis, 1938) and for each of these four

Table 7.1.—VARIABILITY OF MALE REHOBOTH BASTAARDS
(*From* Trevor, 1953.)

	STATURE	CEPHALIC INDEX	FACIAL INDEX	NASAL INDEX
Male Hottentots (73)	6·47 ± 0·54 (73)	2·62 ± 0·22 (74)	5·08 ± 0·42 (74)	7·96 ± 0·82 (74)
Male Rehoboth Bastaards (69)	6·58 ± 0·59 (69)	2·45 ± 0·20 (74)	5·57 ± 0·46 (74)	8·27 ± 0·68 (73)
Male Dutch (70)	7·24 ± 0·61 (70)	3·85 ± 0·33 (70)	6·00 ± 0·66 (41)	7·55 ± 0·84 (40)

The figures shown are standard deviations and sample sizes.

traits the mean values are intermediate between those of the Hottentot and Dutch samples. *Table* 7.2 shows that for three of these characters the parental races themselves differ significantly. Some of the problems of comparison associated with studying these people and the determining of appropriate parental values are discussed by Trevor in his monograph.

We turn then to the West Indies, where Negro slaves from Africa were transported during slavery, and where a considerable degree of hybridization between them and the Europeans has subsequently taken place. There are two important studies worth analysing briefly—firstly, that of Davenport and Steggerda (1929) in Jamaica; and secondly, that of Laska-Mierzejewska (1967) in Cuba.

The study by Davenport and Steggerda included observations on three adult groups: Negroes, Europeans, and 'hybrids between them of all degrees'. Although questions were asked of all participators regarding their pedigrees, and opinions of other persons on colour were obtained, the genetical evidence on the extent of race mixing that we have today was not available in the 1920s when this work was carried out. It is very likely that, just as in the United States, virtually all Negroes contain some admixture, whatever their appearance. The distinction between Blacks and Browns, Negroes and Mulattoes, which these surveys make is therefore likely to be one which oversimplifies the true situation.

Trevor (1953) also indicates a number of methodological defects to the Jamaica study, and for these reasons restricts his analysis to anthropological indices which are less likely to be affected by possible changes during growth. Davenport and Steggerda give the means of 20 indices, and they find that Black and Brown males differ significantly in 3 of these characters and females in 6. The male and female Blacks differ from male and female Whites, on the other hand, by 11 and 12 characters respectively, and the male and female Browns differ from male and female Whites in 11 characters each. While this suggests that the two coloured populations are closer genetically than either are to the White, the fact that in all cases where there are significant differences in mean value between the three races (either males or females) the Browns are intermediate between the Blacks and Whites, is in line with all other evidence we have presented. We have discussed in the previous chapter (6.*2.e*) some of the conclusions of Davenport and Steggerda on the variability and association of characters in the Jamaica Browns, and Castle's refutation of them. *Table* 7.2 shows the relationship of the variability of various of the characters measured in the Jamaican Browns relative to the Blacks and Whites. These results are slightly unusual in that the characters which are more variable in female hybrids are not so in males, and vice-versa. This suggests that the differences which do exist cannot be very fundamental ones.

The Cuban investigation was carried out in 1963 and 1964 on 3,872 individuals, of whom 1,680 were White, 1,027 Negro, and 1,165 Mulatto. Measurements were made of 15 somatic characteristics among males and females of these three races aged between 6½ and 20½ years. We will discuss the results for the 20½-year-olds, on the assumption that these have reached maturity. *Table* 7.3 shows for each character in females the mean measurement and the standard deviation. So far as these means are concerned, in 11/15 cases the Mulattoes are intermediate in absolute value between the Whites and Negroes. In 4 cases the Mulattoes have a lower value than either of the parental races, and in 1 case there is an obvious misprint in Laska-Mierzejewska's tabulation so no reliable comparison can be made. In no case where the Mulattoes have a lower value than both Whites and Blacks do they in fact differ significantly from the lowest of these two values.

With regard to the degree of variability shown by the hybrids, in 6/15 cases they are more variable in the hybrids than either of the parental races, and in 5/15 cases they are less variable than the parental races. In 4/15 cases, therefore, they are intermediate.

The data from the males, which are not separately presented here, show that for 7/15 characters the hybrids are intermediate between Whites and Negroes, in 5/15 cases they are greater than both, and in 3/15 they are less than both. So far as variability is concerned the males follow the females in being more variable in 6 characters, although they are not the same characters as the females, and in some cases where the male hybrids are more variable than the Whites and Negroes the females are less variable than either. In 6/15 comparisons the male hybrids are of intermediate variability and in 3/15 cases, therefore, the

Table 7.2.—COMPARISON OF VARIABILITIES OF SINGLE CHARACTERS BETWEEN HYBRID AND PRESUMED PARENTAL SERIES IN NEGROID/CAUCASOID CROSSES IN TERMS OF THE ORDER OF DIFFERENCE BETWEEN PARENTAL MEANS (*After* Trevor, 1953.)

HYBRID SERIES AND NUMBER OF CHARACTERS COMPARED	CLEARLY SIGNIFICANT DIFFERENCES BETWEEN PARENTAL MEANS						ALL OTHER CASES. NONE OF THEM IN WHICH THE HYBRID SERIES IS APPRECIABLY MORE VARIABLE THAN EITHER PARENTAL SERIES BUT DIFFERENCE BETWEEN MEANS IS SIGNIFICANT
	Hybrid series appreciably more variable than either parental series	Hybrid series appreciably more variable than one but not appreciably different in variability from the other	Hybrid series not appreciably different in variability than either parental series	Hybrid series appreciably more variable than one but appreciably less variable than the other parental series	Hybrid series not appreciably different in variability from one but appreciably less variable then the other parental series	Hybrid series appreciably less variable than either parental series	
Male hybrid American Negroes (12)	Stature. Nasal breadth	—	—	Facial height	Ear breadth	Ear length	Head length. Head breadth. Minimum frontal breadth. Bizygomatic breadth. Nasal height. Mouth width. Cephalic Index
Female Jamaican Browns (17)	Relative sitting height. Relative span	Relative arm length. Relative kneeling height. Zygomatico-pupillary index. Ear index	Total arm length. Upper arm length index	Nasal index	—	—	Total arm length. Forearm length index. Brachial index. Hand index. Foot index. Cephalic index. Head length, head height index. Head breadth—head height index. Parieto-frontal index. Facial index

Male Jamaican Browns (17)	Zygomatico-pupillary index. Nasal index. Ear index	Relative kneeling height	—	Relative sitting height	Relative span	—	Relative arm length. Total arm length–upper arm length index. Total arm length–forearm length index. Brachial index. Hand index. Parieto-frontal index. Foot index. Facial index. Cephalic index. Head length–head height index. Head breadth–head height index
Male Rehoboth Bastaards (4)	—	—	—	Stature. Nasal index	—	Cephalic index	Facial index
Female Cuban Mulattoes (12)	Width of shoulders. Hip circumference	Weight. Stature. Neck circumference	Sitting height	—	Arm width. Waist circumference	Pelvis width. Chest circumference	Chest width. Width of hips
Male Cuban Mulattoes (12)	—	Weight	Stature	—	—	Arm width	Sitting height. Shoulder width. Chest width. Pelvis width. Hip width. Neck width Chest circumference. Waist circumference. Hip circumference

hybrids manifest lower variability than the parental races. Again we can say that since there is this lack of correspondence between males and females with

Table 7.3.—ANTHROPOMETRIC CHARACTERISTICS OF BLACK, WHITE, AND HYBRID CUBANS (FEMALE) (For each character the mean (m) and variance (s) are shown.)

CHARACTER		WHITE	MULATTOES	BLACK
Weight	m	51.8 ± 1.16	54.24 ± 1.78	58.84 ± 1.68
	s	7.54	10.10	9.53
Stature	m	155.24 ± 0.86	156.72 ± 1.12	158.72 ± 1.12
	s	5.54	6.32	6.34
Height in sitting	m	82.05 ± 0.48	81.27 ± 0.56	80.88 ± 0.59
position	s	3.09	3.19	3.33
Width of arm	m	156.56 ± 1.04	162.36 ± 1.19	166.48 ± 1.57
	s	6.72	6.75	8.91
Width of shoulders	m	33.97 ± 0.24	34.78 ± 0.36	35.15 ± 0.32
	s	1.59	2.04	1.84
Width of chest	m	23.95 ± 0.26	23.90 ± 0.28	24.18 ± 0.29
	s	1.71	1.57	1.65
Width of pelvis	m	26.73 ± 0.29	26.59 ± 0.26	25.87 ± 0.31
(bi-ilio cristalis)	s	1.88	1.46	1.77
Width of hips	m	30.64 ± 0.25	30.71 ± 0.28	30.75 ± 0.35
	s	1.65	1.57	1.98
Circumference of	m	31.6 ± 0.17	31.5 ± 0.28	32.4 ± 0.26
neck	s	1.10	1.6	1.5
Chest	m	75.2 ± 0.86	74.1 ± 0.94	77.5 ± 1.24
circumference	s	5.60	5.30	7.00
Circumference of	m	64.6 ± 0.90	66.0 ± 1.07	67.8 ± 1.29
waist	s	5.80	6.10	7.30
Circumference of	m	92.6 ± 0.92	93.5 ± 1.44	95.0 ± 1.18
hips	s	6.00	8.10	6.70
Subcutaneous fat	m	16.2 ± 0.77	?16.6 ± 1.06?	16.2 ± 0.92
of triceps	s	5.00	6.00	5.20
Subcutaneous fat	m	18.8 ± 1.14	18.6 ± 0.94	20.2 ± 1.17
under shoulder blade	s	7.40	5.30	6.60
Subcutaneous fat	m	27.9 ± 1.17	26.1 ± 1.50	27.4 ± 1.48
on stomach	s	7.60	8.50	8.40

regard to excess variability in the hybrids, the characters concerned, and the nature of their variability, cannot be regarded as being fundamental for the fitness of the peoples involved.

We have already discussed in Chapter 6 the attainment of the age of fertility among Cubans. It is worth while to consider the more complete data of Laska-Mierzejewska on the development of secondary sexual characteristics. In terms of Tanner's grades of development of the secondary sexual characteristics, there are found in Cuba the normally significant differences between the Negroes and Whites in the ages at which the stages are reached—the Negroes invariably maturing earlier. In all cases, except one, where there are these significant differences, the Mulattoes are intermediate between the parental races. When there are no significant differences between the Whites and Negroes the hybrids are not intermediate, development in fact mostly being completed earlier.

The racial composition of Cuba at the 1954 census was Whites 73·5 per cent, Negroes 11·8 per cent, Mulattoes 13·4 per cent. In general the Negroes and Mulattoes belong to the lower socio-economic groups, and this factor clearly has its effect on the rate of development of some characters, particularly sexual maturity. It can be regarded as of importance that the stresses imposed by these adverse conditions are not made manifest in the racial hybrids in any marked way. The data on this point are shown in *Table* 7.4.

The greatest differences between the races in fact involve differences in the width of the shoulders, and in indices constructed using this parameter. For instance the index 'width of shoulder/width of hip' shows important differences between all three racial groups. This is a reflection of the fact that the young Negro tends to have a trapezoidal trunk, while the young White has a rectangular trunk. This might possibly be one of those areas where developmental differences might lead to inequalities or imbalance in hybrids.

The figures bearing on this point are shown in *Table* 7.5. In the majority of cases the Mulattoes are intermediate between the two races, although in a few cases, particularly among women, the index exceeds that in the Negroes, who always have a higher index than the Whites. In only 1 case does the mean value of the hybrids drop below that of the Whites (and that at age $7\frac{1}{2}$ years) so that the possibility of odd combinations among hybrids is convincingly eliminated. The male Mulattoes show some evidence of increased variability for this index, although also interspersed with ages at which variability is reduced, but among older females variability is on the whole reduced.

Although Laska-Mierzejewska demonstrates a number of important differences between the peoples of Cuba in her large-scale study, in no cases are any biological disharmonies mentioned or are the Mulattoes said to be at any sort of disadvantage.

Probably the most numerous and well-studied population wherein are demonstrated the effects of Negroid/Caucasoid mixture is the American Negro. We have already mentioned that the amount of White admixture in this population

Table 7.4.—DEVELOPMENT OF SECONDARY SEXUAL CHARACTERS IN NEGROES, WHITES, AND THEIR HYBRIDS IN CUBA
(*From* Laska-Mierzejewska, 1967.)

	STAGE 2		STAGE 3		STAGE 4		STAGE 5	
	Mean Age	Difference	Mean Age	Difference	Mean Age	Difference	Mean Age	Difference
	Women—Development of the Breasts							
Whites	10·41		11·57		13·32		15·77	
Mulattoes	10·31	1·28†	11·39	0·73†	13·14	0·36	15·21	0·65*
Negroes	9·13		10·84		12·96		15·12	
	Women—Growth of Pubic Hair							
Whites	11·13		11·80		12·48		14·11	
Mulattoes	10·95	0·76†	11·83	0·49†	12·47	0·71†	13·91	0·43*
Negroes	10·37		11·31		11·77		13·68	
	Men—Growth of Genitals							
Whites	12·62		13·63		15·42		18·80	
Mulattoes	12·31	0·46†	13·34	1·00†	14·94	1·0†	17·98	1·75†
Negroes	12·16		12·63		14·42		17·05	
	Men—Growth of Pubic Hair							
Whites	12·34		13·60		14·81		17·96	
Mulattoes	12·07	0·17	13·41	0·17	14·52	0·21	18·28	0·36
Negroes	12·17		13·43		14·60		17·60	

* Significant at the 0·05 level.
† Significant at the 0·01 level.

Table 7.5.—THE INDEX $\frac{\text{WIDTH OF SHOULDERS}}{\text{WIDTH OF HIPS}} \times 100$ AMONG YOUNG CUBANS (*From* Laska-Mierzejewska, 1967.)

AGE	FEMALES			MALES		
	Whites	Mulattoes	Blacks	Whites	Mulattoes	Blacks
6·5 m	124·2	127·7	126·0	126·5	126·7	130·5
s.d.	6·6	5·6	5·5	6·0	7·0	6·5
7·5 m	124·8	118·4	124·0	125·7	127·8	131·3
s.d.	5·5	6·3	7·4	6·1	5·7	6·1
8·5 m	123·1	125·4	126·1	123·0	127·8	129·1
s.d.	6·2	6·5	6·2	6·4	6·0	6·8
9·5 m	122·6	125·7	126·1	124·0	127·0	128·8
s.d.	6·2	7·0	6·8	6·7	5·3	5·9
10·5 m	119·8	122·8	125·1	123·3	125·4	127·1
s.d.	6·6	6·6	7·6	8·0	6·9	7·8
11·5 m	119·4	122·0	122·9	123·7	126·1	127·7
s.d.	7·0	7·2	7·4	7·2	6·7	8·1
12·5 m	117·4	117·4	120·6	121·3	127·2	128·9
s.d.	6·4	7·2	6·4	7·1	9·2	8·6
13·5 m	114·7	116·5	115·6	121·9	125·5	126·3
s.d.	6·6	7·7	6·0	6·0	6·8	6·7
14·5 m	113·4	116·3	117·3	119·5	122·5	124·9
s.d.	6·9	6·5	8·1	6·8	7·1	8·7
15·5 m	112·8	116·6	114·6	122·1	125·6	126·9
s.d.	4·9	5·9	6·0	5·0	7·7	5·1
16·5 m	113·1	113·5	116·3	122·1	124·3	127·2
s.d.	7·9	5·5	5·8	7·4	6·3	8·3
17·5 m	110·6	113·9	116·7	122·3	127·5	126·9
s.d.	6·1	5·5	6·2	6·6	6·8	6·2
18·5 m	111·8	113·0	115·5	122·1	124·9	127·7
s.d.	5·8	6·4	7·9	6·9	7·5	6·9
19·5 m	110·4	116·3	115·0	123·6	126·3	127·2
s.d.	6·1	6·3	6·7	6·4	6·2	6·4
20·5 m	111·0	113·3	115·1	123·0	127·1	126·7
s.d.	4·7	6·0	7·7	6·3	7·0	6·3

is far from uniform, but in addition to this we must remember that the racial composition of the originally transported slaves to the U.S.A. was far from uniform. Slaves were taken to the Americas from virtually the whole of the western coastal areas of Africa, with a small contribution from East Africa. The racial composition of the slaves therefore included the True Negroes, eastern, western, and perhaps some southern Bantu, and Hamitic Negroes. *Table* 7.6

Table 7.6.—The Origin of North American Negro Slaves. (The data are averages for Virginia (1710–1769), South Carolina (1733–1807), and for the British slave trade (1690–1807).) (*Adapted from* Reed, 1969b.)

Present Area	Peoples involved	Approximate Proportion from Region
Senegal and Gambia	Mainly Bambara and Malinke from interior	0·133
Sierra Leone	Sierra Leone, Guinea, Portuguese Guinea Peoples + Bambara and Malinke	0·055
Ivory Coast, Liberia	Various	0·114
Ghana	About ¾ Akan from south, rest from north	0·159
Togo, Dahomey, Nigeria west of Benin river	Peoples of Togo, southern Dahomey, and western Nigeria	0·043
Nigeria (east of Benin river) to 1° S. (Gabon)	About ¾ Ibo, rest Ibibio and Cameroonians	0·233
1° S. to south-west Africa Gabon, Congo, Angola	Very varied, including many from inland	0·245
Mozambique and Madagascar		0·016

shows some estimates quoted by T. E. Reed (1969b) on the African origins of slaves imported into the North American mainland. The distribution of geographic areas is approximate, but the variety of peoples involved is very apparent.

The importation of Negro slaves into what was to become the U.S.A. really started about 1700 and continued at a high rate for most of the eighteenth century, until 1808 when importation became illegal. The total number of slaves brought into the United States was probably less than 400,000 (Potter, 1965). The number of Negroes in the U.S.A. today is in excess of 20 million, a rate of increase which seems to belie any marked disadvantages in these Negroes.

Racial mixing between the Whites and Negroes obviously began shortly after the first introduction, probably before 1700, and has continued at varying levels since. Nevertheless, as we saw in Chapter 5, the levels of Caucasian genes among American Negroes is very heterogenous and this shows how impossible it is, in fact, to talk of them as if they were one uniform race. This important qualification must be remembered in all that follows.

One of the most detailed anthropometric studies on the American Negro is that of Herskovitz (1930). A total of 1,893 Negroes of both sexes was studied and careful genealogies for 1,551 of them were constructed. These showed that 22 per cent claimed to be entirely Negro, while 26·5 per cent claimed some Indian ancestry. Later work on blood groups and biochemical characteristics, which we have reviewed in Chapters 5 and 8, suggests that the White admixture is rather more pervasive than this, and also casts considerable doubt on the high degree of Indian ancestry which was claimed. A series of males from Howard University measured by Herskovitz himself, has been generally accepted as giving valid indications of the effects of admixture on physical characters, and we have looked at some of the results of this survey in Chapter 6. The means

Table 7.7.—THE MEANS AND STANDARD DEVIATIONS OF MALE AMERICAN NEGROES AND OLD AMERICAN WHITES

CHARACTER		MEANS AND STANDARD DEVIATION				
		N	NNW	NW	NWW	W
Stature	m	170·4	170·4	171·7	172·2	174·3
	s.d.	5·65	6·77	5·93	7·05	5·80
Head length	m	198·0	196·9	196·1	195·9	197·1
	s.d.	6·54	7·06	6·40	6·98	6·04
Head breadth	m	151·9	151·7	151·4	153·2	153·9
	s.d.	5·23	6·44	4·78	6·76	5·20
Minimum frontal	m	107·2	106·6	106·0	105·7	105·9
breadth	s.d.	4·19	5·15	4·26	4·40	4·24
Facial height	m	123·9	122·7	122·6	120·1	119·3
	s.d.	5·59	6·20	6·03	5·93	6·72
Nasal height	m	53·1	53·3	53·7	53·7	53·5
	s.d.	3·16	3·71	3·24	3·45	3·36
Nasal breadth	m	43·4	41·3	40·0	37·5	36·1
	s.d.	2·81	3·44	3·01	3·86	2·51
Ear length	m	59·8	59·7	60·9	60·4	66·9
	s.d.	4·57	4·00	3·79	3·45	5·72
Ear breadth	m	33·8	33·2	33·1	32·7	37·9
	s.d.	3·21	2·58	2·67	2·06	2·52
Mouth width	m	55·0	52·9	52·4	52·1	53·7
	s.d.	3·77	4·23	3·21	3·46	3·78
Cephalic Index	m	76·9	77·1	77·2	78·3	77·9
	s.d.	3·22	3·77	2·88	3·58	3·01
Bizygomatic breadth	m	140·7	138·4	139·0	137·3	138·6
	s.d.	5·40	6·87	6·00	7·18	4·82

and standard deviations for all the characters measured of the different groups in this series are given in *Table* 7.7, which is constructed from Trevor's invaluable analysis. The White group used for comparison by Trevor is, as we have mentioned, that of the 'old Americans' studied by Hrdlicka (1925).

The data of *Table* 7.7 divide the characters into a number of different groups. Firstly, nasal height and head length do not vary in any systematic way between the races. Stature, nasal breadth, and facial height show a regular gradation from the Negroes at one extreme to the Whites at the other, a situation illustrated diagrammatically in *Fig.* 6.13. For head breadth, minimum frontal breadth, ear length, and cephalic index the same gradation would be present if one or two pairs of means which differ insignificantly from one another were transposed. In the case of ear breadth and mouth width, the four Negro groups show a gradation according to expectation, but the White group deviates. Trevor suggests that the fact the White group was part of an independent investigation may have been responsible for the discrepancy in these two cases.

The conclusion is clear then, that for all these physical characteristics, not just those of the face, the hybrid population tends to approach the population from which it has received its greatest contribution.

So far as variability is concerned Herskovitz himself suggested that, excluding the Whites, the Negroes were the least variable, followed by the NW, NWW, and NNW. Trevor in his monograph suggests that it is the NW series which is least variable, followed by the Negroes, Whites, NWW, and the NNW. He points out that the NW, being equal mixtures of Negro and White, would be expected to be less variable than the NNW and NWW which each represent several grades of admixture. In *Table* 7.7 we see that in three characters, stature, minimum frontal breadth, and nasal breadth, the three intermediate hybrid groups are all more variable than Whites and Negroes. In one, ear length, the intermediate hybrids are less variable than the presumed non-hybrids.

In spite of many outrageous statements to the contrary (*see*, for example, Haller, 1970) we can identify no important physical features in which the American Negroes, hybridized to a greater or lesser extent with the Whites in North America, are in any way inferior to the Whites. Nor, indeed, does it appear that they are much more variable for any important traits than the Whites.

Turning to specific genetic defects, we have referred several times previously to some traits which serve to protect against falciparum malaria, and which are prevalent among Africans, but which also possess a number of definite dis-

Table 7.8.—EXPECTED HAEMOGLOBIN GENOTYPES IN AMERICAN NEGROES (*From* Rucknagel and Laros, 1969.)

Haemoglobin genotype	S/A	C/A	β-thal	S/S	S/C	S/β-thal	C/C
Expected frequency	1/10	1/50	1/125	1/400	1/1,000	1/2,500	1/10,000

advantages in non-malarial environments. *Table* 7.8 shows the expected frequencies of various haemoglobin genotypes among American Negroes based on average values over the whole population.

It is well known that most of these traits have important effects on viability and fertility, and therefore a reduction in their average frequency among

American Negroes, compared with the frequencies which are found in Africa, can be seen as an important advantage of admixture. The simple fact of admixture will, as we saw in Chapter 5, give an intermediate mean gene frequency in the resultant population, but this will result, if the gene is recessive, in a marked reduction in the incidence of the trait. In other words the number of homozygotes is reduced. We have to remember that there is admixture both between the Whites and Negroes and among the different Negro groups themselves. This is important because the proportion of sicklers among the Africans from whom slaves were taken ranged from 0 per cent to frequently well over 30 per cent, and even in some tribes more than 40 per cent (Tobias, 1966). In the American Negro the frequency of sickle-cell trait varies from 6 per cent in Washington D.C. to 16 per cent in the Gulla Negroes of Charleston, South Carolina (Rucknagel and Laros, 1969). This may be a reflection of the original heterogeneity, or the varying amounts of White admixture, or both.

The second phenomenon contributing to the reduced frequency of haemoglobinopathies in American Negroes is selective elimination, possibly because of some slight disadvantage as heterozygotes, but mostly through the rather severe effects in the homozygotes. The reality and magnitude of these changes have been demonstrated in a rather general way by Allison (1964), Workman and others (1963), and Workman (1968), and we have discussed some of this in Chapter 2. We also mentioned in that chapter the striking differentiation of sickle-cell frequencies between Surinam and Curaçao owing to the selective pressure being present in Surinam but absent in Curaçao. There seems little doubt that selective pressures can alter gene frequencies in this way, although how general the phenomenon is has not yet been agreed on (Morton, 1955).

There is little doubt that the Negro population in the U.S.A. shows many biological defects compared with the Whites. Their neonatal mortality is higher, expectation of life at birth is lower, death-rates throughout life are higher, and general physical condition often rather poorer. This latter disadvantage is indicated by the lower proportion of high medical grade Negroes who have been inducted into the U.S. army from the time of the First World War onwards. There is also little doubt that the American Negro, starting from his position of slavery and the incredible prejudice against him in the nineteenth century, is at a considerable sociological disadvantage. To exactly what extent the latter is responsible for the preceding list of biological defects is a matter for considerable debate, although it must certainly be a large contributory factor.

For our final examination of Negroid/Caucasoid crosses we will briefly consider some from Britain where, although some environmental disadvantage on average undoubtedly exists for the Negroes and their hybrids, it is much less pronounced than in the United States. Geographically and socially the Negro in Britain is much more integrated than in the United States. The other important difference, which it is useful to investigate, although it is not likely to have many important biological effects, is that most of the interracial matings in Britain, at least in the past, are between Negro men and White women, whereas most of the gene flow into the American Negro population has been as a result of matings between White men and Negro women. The possible difference this might make is in changing the social environment of the children (Willerman and others, 1970). For the importance of this see s. 6.*4.e*.

The survey which we will particularly discuss is that of Little (1943). He examined a total of 460 male and female English and Anglo-Negro children, mainly in the ports of Liverpool and Cardiff. Negro and hybrid populations have been established in these ports, and also in lesser numbers in London and other ports, since the First World War, and their dockland environment and socio-economic circumstances are similar to those of the White children which

Table 7.9.—MEANS AND STANDARD DEVIATIONS (WITH STANDARD ERRORS) FOR VARIOUS INDICES OF ANGLO-NEGRO AND ENGLISH MALE CHILDREN (*From* Little, 1943.)

INDEX	TOTAL ANGLO-NEGRO SERIES	FIRST GENERATION CROSS	ENGLISH
100 B/L	78·0 ± 0·27 (113)	77·7 ± 0·47 (41)	78·7 ± 0·28 (113)
100 B^1/J	84·4 ± 0·32 (108)	83·7 ± 0·46 (38)	83·7 ± 0·28 (112)
100 GH/J	83·4 ± 0·49 (109)	82·8 ± 0·73 (40)	82·3 ± 0·51 (112)
100 G^1H/J	47·6 ± 0·34 (109)	48·5 ± 0·56 (40)	48·2 ± 0·32 (112)
100 NB/NH	79·0 ± 0·51 (114)	78·7 ± 0·82 (41)	65·4 ± 0·73 (113)
100 EW/EL	62·3 ± 0·44 (111)	60·7 ± 0·60 (41)	59·4 ± 0·42 (112)
$100 \frac{\text{Wt. (lb.)}}{\text{Age (mth.)}}$	48·5 (109)	48·4 (37)	46·2 (116)
100 B/L	2·91 ± 0·19	3·02 ± 0·33	2·92 ± 0·20
100 B^1/J	3·30 ± 0·22	2·78 ± 0·32	2·98 ± 0·20
100 GH/J	5·14 ± 0·35	4·60 ± 0·51	5·41 ± 0·36
100 G^1H/J	3·56 ± 0·24	3·52 ± 0·40	3·36 ± 0·23
100 NB/NH	5·80 ± 0·38	5·20 ± 0·58	7·81 ± 0·52
100 EW/EL	4·62 ± 0·31	3·77 ± 0·42	4·41 ± 0·30

L, Maximum length of head; B, Maximum breadth of head; B^1, Minimum forehead width; GH, Total facial height from nasion to gnathion; G^1H, Upper facial height from nasion to prosthion; J, Maximum interzygomatic breadth; NH, Nasal height; NB, Nasal breadth; EW, Ear width; EL, Ear length.

Little investigated. The age of the children he examined ranged between 5 and 15 years and at no age did his sample size exceed 20 children. Despite the smallness of the samples a number of useful points emerge.

In summary Little found that 'the hybrid children are slightly taller, slightly longer in the head, longer and broader in the face, and greater in the external and internal orbital diameters' and 'Relative to age, the Anglo-Negro children are heavier than the English in the case of both sexes'. So there are certainly no disadvantageous effects. With respect to variability of the characters, the smallness of the sample size precluded any direct comparison at any age, but for the character indices, in which the variation due to age becomes statistically of less importance, comparisons are possible. For both boys and girls the hybrids are more variable than the English for three indices, and less so for three. When the first-generation hybrids are considered separately from later generation hybrids, they are more variable than the English in only two characters for both boys and girls. The results of these comparisons are shown in *Table* 7.9.

We have already discussed elsewhere in this book the growth rates and condition of the teeth of these hybrid children. Little considers a number of

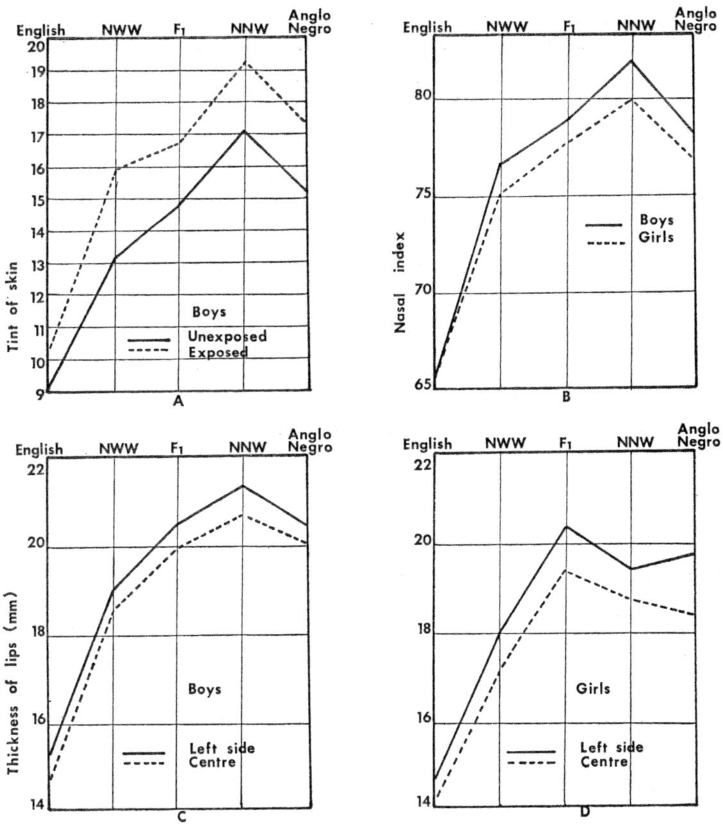

Fig. 7.1.—Differences between 'mean' values of exposed and unexposed skin colour (von Luschan's scale) (A), between means of the nasal index (B), and between mean thicknesses of lips (C, D) for the English, subdivisions of the Anglo-Negro, and the total Anglo-Negro series.

In each of these figures, which are taken from Little (1943), NWW and NNW refer to second and later generation hybrids with respectively more White than Negro and more Negro than White ancestry. The F_1 group are first generation in which, of course, the racial contributions are equal.

other characters in more detail, dividing his sample of Anglo-Negro children into NWW (nearer White), first-generation cross (F_1), and NNW (nearer Negro). This sample shows that the thickness of the lips, nasal index, and skin colour all show gradations from one group to the other with increasing proportion of White or Negro ancestry, just as we have found for many other similar characters in other hybrid populations. The expression of these characters in these three groups is shown in *Figs.* 7.1 A–D. Eye colour and condition of teeth were also classified by Little into these more detailed groups and they show similar behaviour. In other words, whether or not these traits are due to one or a few major genes, the modifying effects of genetic background are clearly seen.

Little does not report any disadvantageous consequences from racial mixing—on the contrary he suggests a number which are positively advantageous, and this would seem to be a reasonably general assessment for all Negroid/Caucasoid hybrids.

7.2. Mongoloid-Caucasoid Hybrids

There are two very widespread parts of the world in which hybrids between these two major races are common, South-East Asia and Latin America, with other small numbers of hybrids living in North America and various parts of Asia. These interracial matings, as well as involving very distantly related members of the Mongoloid race, also involve very different Caucasoid peoples.

There are a number of well-reported instances of European/Amerindian crosses in North America, some of which we have already referred to, including the Half-blood Sioux investigated by Boas (1895), whose results were later analysed by Sullivan (1920) and Trevor (1953). In general Trevor concluded, on the basis of 9 metrical characters, that the Half-blood Sioux were closer to the pure Sioux than they were to Hrdlicka's Old Americans, which he used as a standard for comparison in lieu of a true European parental population. In the case of stature, head breadth, bizygomatic breadth, physiognomical facial height, nasal breadth, relative sitting height, and cephalic index, the Half-blood Indians were intermediate between the Sioux and the Americans. The mean span of the half-bloods was greater than the Sioux or Americans, although not significantly different from the Sioux. Head length was less than both other groups, although again not significantly different from the Sioux. So far as variability for various metrical characters was concerned, the Half-blood Sioux appeared to be, on average, marginally more variable than either the Pure-blood Sioux or the Old Americans, although the mean difference was small, particularly between the Pure and Half-blood Sioux. There were about as many characters in which the Sioux were more variable as those in which the Half-bloods were more variable.

The Ojibwa or Chippewa are the largest Algonkian tribe of Indians living in the Lake Superior region of North America. Jenks (1916), who studied crosses between these peoples and Whites, stated that amalgamation has been continuous and cumulative since 1660, with the result that among the 25,000 or so living at the time Jenks studied them it was difficult for him to find any Full-bloods. Most of the hybridization took place with French and Scottish traders of the Hudson's Bay Company active in this area for over 200 years. Only three metrical characters were measured by Jenks and the hybrids were intermediate in value for each of these, and also intermediate in variability.

Other European/Amerindian crosses were studied by Gates (1928) in Canada. His study was mainly of family pedigrees, with no measurement of anthropometric characters. He does conclude, though, that 'The Indian half-breed of nothern Ontario are a hardy race of hunters and trappers and woodsmen well adapted to the wild pioneering conditions under which they live; they appear to have the hardiness of the native Indians combined with greater initiative and enterprise than the pure Indian would ever show. They push the fringe of civilization farther north than it would otherwise extend'. It would be unwise, however, to attach too much importance to subjective statements such as this.

Moving south into Mexico and central America there is the general Mestizo population and, in particular, the well-studied Maya-Spanish hybrids of Yucatecan. It is estimated that the number of Mestizos in central America is about 15 million, a number which is greater than the number of unmixed Indians living in this area. In Mexico the Mestizos are said to be taller and more fertile than the original parental stock and 'are in every way a thoroughly vigorous group biologically' (Montague, 1964). A comprehensive monograph on Maya-Spanish crosses was published in 1931 by G. D. Williams, based on fieldwork carried out in 1927 in the Mexican state of Yucatecan. Racial crossing has continued for nearly 400 years, with the Indians being in a numerical superiority throughout most of this period.

Williams himself did not measure either any pure Maya Indians or any Spaniards, and Trevor, in his analysis of these people, utilized a series of Mayas measured by Steggerda (1932b) and measurements on Spanish university students originally communicated to Williams by the original investigator Barras, as standards of comparison. One of the problems, in fact, is to find sufficient numbers of 'pure' Mayas, and with respect to the 77 Mayas who Steggerda measured, he estimated that none possessed more than one-eighth to three-sixteenths of White blood and that all of these were 'relatively pure Maya'. It is hardly surprising, then, that for all characters for which reliable comparisons are possible, the absolute differences between the Maya and Yucatecan means are markedly smaller than those between the Yucatecan and the Spanish. The mean absolute values for the Yucatecans are still mostly intermediate. The variability for the characters is also mostly intermediate, the relationships being shown in *Table* 7.10.

Our general conclusion for this category of Mongoloid/Caucasoid hybrids, therefore, is that they are numerous, long-established, are no more variable than their parental races, show slight tendencies to exceed their parental races in a few traits, and on the whole are fit and successful human populations.

There are also many hybrids between Asiatic Mongoloids and Whites, especially in Indo-China between the Indo-Chinese and French, resulting from the long French domination of that area. Other well-known hybrid populations include Japanese/Europeans both in Japan and elsewhere, and a whole variety in Hawaii involving all the many races living on those islands.

A sample of Japanese/European hybrids examined by R. Ruggles Gates (1958b) are of interest, because they can be compared with some Japanese/Negro hybrids who were conceived under similar circumstances, viz., the post-war

Table 7.10.—The Relationship between Male Yucatecans and Their Parental Races

Character	Mean	Standard Deviation
Stature	Intermediate	Intermediate
Head length	Intermediate	Intermediate
Head breadth	Greatest	Greatest
Bizygomatic breadth	Intermediate	Least
Bigonial breadth	Intermediate	Least
Nasal breadth	Intermediate	Intermediate
Ear length	Least	Intermediate
Ear breadth	Least	Least
Cephalic index	Greatest	Intermediate
Ear index	Intermediate	Intermediate

American occupation of Japan, and have since been raised under identical, rather favourable, institution conditions. The characters on which information are available are mainly those of major gene effects, such as eye colour, hair colour, and form, etc. This is partly due to the age at which Gates examined these children (mainly between 2½ and 6½ years) and partly due to the smallness of the samples. A summary of Gates's results is given in *Table* 7.11.

For each of the three characters presented in *Table* 7.11 the relatively greater prominence of the Negroid features in the hybrids is apparent. Gates states that 'The children all appeared in excellent health, except that some suffered severely from skin diseases. . . . They show a great range of physiognomy, some essentially European, some very Negroid, a few essentially Japanese'.

Gates remarks that there is probably considerable independent segregation between the various racial characters concerned. Eye colours of the Japanese/Whites, with but two exceptions, range from 1 to 5 of the scale he employs, and skin colours are mostly near 8 (White), but the degree of development of the eyefolds does not correspond with the depth of hair colour or the amount of depression of the root of the nose. 'Similarly the cephalic index varies independently of ear size'.

Gates concludes that the F_1 hybrids from White fathers are relatively uniform and intermediate in such features as skin colour and hair form. He suggests that the variation in eye colour and hair colour will be largely determined by variability of the features in the European fathers, whereas variation in eyefolds and nasal depression will result mainly from variability of these features in the Japanese mothers.

In the children of Negro fathers and Japanese mothers (and therefore presumably all other Negroid/Mongoloid crosses in the same direction), skin

Table 7.11.—A Comparison of Japanese/European and Japanese/Negro Hybrids (*Based on data of* Gates, 1958b.)

Eye Colour	No.	1	1·5	2	3	3·5	4	4·5	5	6	7	8
Japanese/White ♂	45	1	2	12	17	–	3	1	6	1	1	1
Japanese/Negro ♂	15	–	–	11	2	–	2	–	–	–	–	–
Japanese/White ♀	24	1	5	11	1	6	–	–	–	–	–	–
Japanese/Negro ♀	12	–	10	2	–	–	–	–	–	–	–	–

Hair Colour		Black	Dark Brown	Medium	Light Brown
Japanese/White ♂	43	7	9	21	8
Japanese/Negro ♂	15	14	0	0	1
Japanese/White ♀	22	3	9	7	3
Japanese/Negro ♀	17	10	6	1	0

Hair Form		Straight	Wavy	Curly	Kinky
Japanese/White ♂	45	44	1	0	0
Japanese/Negro ♂	15	1	2	12	0
Japanese/White ♀	25	21	4	0	0
Japanese/Negro ♀	17	0	1	9	7

colour, hair form, lip thickness, and ear size and lobe type are mostly determined by the Negro father. In addition to the difference shown in *Table* 7.11, the ears are smaller in both dimensions than in the sons of White fathers. Skin colour, of course, is also normally darker too and the depression of the root of the nose is much more pronounced. Finally the lips are normally thicker. The presence of the Mongolian eyefold is not normally affected by the race of the father, the inner fold being normally fully present and the top fold mostly half present, although occasionally it too is fully represented.

We have already mentioned the lack of harmful effects among the teeth of Mongoloid/Caucasoid hybrids (*see* Chapter 6, s. 6.5.*a*) and the absence of any complicating factors in pregnancy (6.2.*c*). The youthfullness and small size of this sample really precludes any very firm conclusions on anthropometric characters, so we turn to a rather larger study on Franco-Vietnamese to consider this aspect of the biology of Mongoloid/Caucasoid hybrids. The anthropology of these peoples was the subject of a searching monograph in 1967 based on a long history of investigations, particularly by French anthropologists. In no cases have there been any reports of disadvantageous effects of race mixing. Indeed Hiernaux and Heintz (1967) suggest that first-generation hybrids show some evidence of hybrid vigour in height and weight during development. The graphs showing the growth of the Franco-Vietnamese were presented in Chapter 6 (*Figs.* 6.7 and 6.8). Hiernaux and Heintz state that at each age the boys show a marked tendency to exceed the mid-parental value of the two parental races, and since they are F_1 hybrids, this is some evidence of hybrid vigour. This phenomenon is apparent for both height and weight and is present in girls, although to a slightly lesser extent.

With regard to relative weight, the position of the hybrids *vis-a-vis* the two parental races changes somewhat during development, but is still intermediate throughout. As we saw in s. 6.2, Europeans and South-East Asians are rather markedly separated in terms of height and weight. This constant intermediary of relative weight during development assumes importance as being a typical example of what almost invariably occurs in racial hybrids. The Franco-Vietnamese are therefore vigorous well-adapted hybrids on all scores.

Of course Asiatic Mongoloids are a very variable peoples and it would be inappropriate to generalize on the basis of French/Vietnamese hybrids as to the normal outcome of Caucasoid/Mongoloid crosses. A number of other hybrid populations are known of Indo-Chinese and Chinese/European marriages, Japanese/European marriages, Malay/European marriages in both directions, and rarely, if ever, are any detrimental effects reported on by those investigating the offspring. Morton, Chung, and Mi, for instance, include a considerable proportion of Caucasoid/Mongoloid crosses or backcrosses among those studied in their thorough survey of interracial crossing in Hawaii. There are in fact a total of 10,767 individuals in their study who are the offspring of such marriages, and these represent 6 per cent of the total number of 179,327 offspring analysed in that definitive study. Their overall conclusions, which we have mentioned elsewhere, that 'human populations do not represent coadapted genetic combinations which are disrupted by outcrossing' rest to a significant extent, therefore, on a wide variety of Caucasoid/Mongoloid race crosses.

7.3. Negroid-Mongoloid Hybrids

There are no very large populations in existence who are the offspring of matings between these two major races, primarily because they have had little geographical contact in the past. Small numbers can be found in North America, Hawaii, Japan, South Africa, and rather larger numbers in South America. We have already briefly examined some Negro/Japanese crosses studied by Ruggles Gates, and the largest other study of Negroid/Mongoloid crosses was also carried out by him. This was a study of a number of families in Cuba who have a Chinese father and a Negroid or Spanish mother (Gates, 1956). The numbers involved are not sufficiently large to come to any very general conclusions, but a number of important facts do become clear, which we can discuss.

A summary of the results which Gates obtained has been compiled in *Table* 7.12. Gates concluded that 'The F_1 children were often surprisingly Mongoloid in appearance due to the relative dominance of the eyefolds, the straight hair, and the flat cheeks of the father'. In a number of these families some of the offspring had eyes which were darker than either parent, over a third of the children in the families involved showing this phenomenon. Only one son in one family had eyes lighter than either parent. Gates suggests this may be due to a cumulative effect of different genes for iris pigmentation in the two parent races.

Table 7.12.—THE EXPRESSION OF SOME SUPERFICIAL CHARACTERISTICS IN CHINESE × CUBAN CROSSES (*Compiled from data of* Gates, 1956.)

HAIR COLOUR		HAIR FORM				EYE COLOUR	SKIN COLOUR
Black	Brown	Straight	Wavy	Curly	Kinky		
6	0	6	0	0	0	2·5	7·5
6	1	0	3	1	3	2·5	5·5
27	4	16	7	2	6	2–2·5	7–7·5

With regard to hair form these crosses confirm our expectation from the genetic situation outlined in Chapter 6, namely that the order of expression is straight, wavy, curly, kinky. For this character, then, the Mongoloid features would show preferential expression. The eyefold is another 'dominant' character, and the data which Gates presents, shows that 22 mixed offspring had the full epicanthic and median eyefolds and 40 mixed offspring had partial eyefolds.

Certain other racial traits, such as depression of the nasal root and the thickness of the lips, are also discussed by Gates. Both races possess depressed noses, although in Mongoloids the depressed nasal root is associated with the eyefolds, while in the Negro race it is not. Gates suggests that this depressed nose has been retained independently in these two races as a primitive character and that the high narrow nose of the Caucasians is a later specialization.

Gates has very little to say on other aspects of the biology of these people, but mentions no disadvantageous characters appearing in any of the individuals he studied, whether they were the result of Chinese/Negro or Chinese/Mulatto crosses, and in the absence of any more large bodies of data that is as far as we can go.

7.4. CAUCASOID-MONGOLOID-NEGROID HYBRIDS

We have examined hitherto in this chapter populations which can arbitrarily be defined as 'biracial'. Even so, with the wide variety of peoples included in the Negroes of America, the complex situation of South Africa, and the progress along the path to complete panmixia among twelve or fifteen races in Hawaii, the term 'biracial' can hardly be taken too literally. More, perhaps, for the sake of semantic completeness than because we are considering any new biological situation, we shall now consider some triracial hybrids.

The population more than all others which can be considered to be triracial is that of Brazil. Of course by no means the whole of Brazil's population contains components of all three races; indeed, as we have seen in Chapter 5, the country shows significant heterogeneity with respect to racial mixing. It is in particular the north-east states in which contributions from all races are most frequent, and we shall therefore concentrate on peoples from this area.

The 1970 population of Brazil is estimated at 85 million, but the last census for which the racial mixture of the whole population is available is that of 1950, when the total population was about 52 million. Of this number close to 62 per cent were classified as White, nearly 27 per cent as Brown (which includes all categories of hybrids), almost 11 per cent as Negro, and less than 1 per cent each for Yellow (mainly Japanese) and Indian.

The Indians were the first settlers in Brazil. The commonly accepted date for the entry of the Indians into South America was for a long time 15,000 B.P. As we discussed in Chapter 3, this figure may well have to be increased on the basis of recent archaeological evidence. The first modern European settlers (Portuguese) arrived in 1501, although the first Portuguese women did not arrive until 1551. Initially Indians were enslaved and of course under these conditions racial mixing began almost at once.

The number of Indians living in Brazil has never accurately been known and never will be. In 1798 it was estimated that there were 250,000 'domesticated' Indians. In 1900 the proportion of Indians who were integrated or partially integrated was less than a third of the whole and so it might be estimated that the original Indian population of Brazil was in excess of 500,000.

The first Negroes to be introduced came from Spain and landed at São Domingues in 1502. The first traffic in the slave trade which lasted until 1850 probably began in the first half of the sixteenth century. The total number of Negroes transported in this period of more than three hundred years is not known exactly, but is almost certainly more than three million. By 1800 Negroes and Mulattoes formed more than 70 per cent of Brazil's population. We have already discussed in s. 7.*1* the very varied origins of the Negroes transported to America. In addition to the areas of Africa detailed in *Table* 7.6 from which slaves were taken to the Americas, there is evidence that large numbers of people from the Sudan were transported to Brazil.

The Portuguese themselves, who were the earliest Europeans in Brazil, have a very mixed ancestry. Portugal was conquered successively by Romans, Germanic tribes, and Arabs (mainly Caucasoids, but possibly with some Negro admixture—*see* Chapter 3). In addition, thousands of Negro slaves were introduced during the sixteenth century directly into Portugal, and by 1557 one-eighth of the population of Lisbon was Negro.

Since the end of the slave trade, immigration from Europe into Brazil has been on a very large scale together with a significant Japanese contribution. Between 1884 and 1944 it is estimated that the migrants came in the following numbers (Salzano and Freire-Maia, 1970):

Italy	1,413,767
Spain	584,060
Japan	189,143
Germany	173,339

In *Tables* 8.5 and 8.6 are shown the average amounts of racial admixture in North-east Brazil, and the fact that mating is 80 per cent of the way to becoming completely random is there illustrated. Clearly this situation is of enormous interest and significance for geneticists and anthropologists, and it has been studied in great detail by a number of Americans and Brazilians (*see* in particular the book by Salzano and Freire Maia, 1970, and the reviews by Morton, 1964, Krieger, Morton, Azevedo, Freire-Maia, and Yasuda, 1965, Krieger, 1970, and references cited therein). Bearing in mind the history that we have just discussed we can now summarize rather briefly some of the findings of these workers.

The most important results were gathered from a large number of migrants from north-eastern Brazil passing through a hostel in São Paulo on their way to the interior of that state. They comprised 1,068 complete families chosen essentially in order of decreasing cohabitation time, but without selection for the presence or absence of children. The samples were interviewed, racial assignments made (on the basis of morphology and pigmentation), reproductive histories determined, and blood and saliva samples taken during the 24 hours they were in the hostel. There are a number of fairly obvious disadvantages in relying on a migrant population in this way but, as Morton (1964) points out, the logistic problems associated with large population surveys in remote rural areas of north-eastern Brazil rendered it almost essential.

This population was extensively studied with respect to a number of different characters (Morton, 1964) and these various characters were analysed by Krieger (1970) for admixture effects. Krieger studied a total of 4 metrical traits, 6 pathological conditions, 5 specific abnormalities, abnormalities in general, and mortality, divided into foetal death and postnatal death.

Now the sample involved in this survey, a total of 8,646 pregnancies in 1,068 families, was classified into seven different classes according to racial phenotypic appearance. But since the proportion of Indian admixture was approximately constant among the eight groups (*see Table* 8.5), only the proportion of Negro admixture was used in determining racial indicators. Furthermore, in the case of mortality, no information of the racial phenotype of the dead individual was available. The racial indicators used were therefore as follows:

N_z = Proportion of Negro admixture of the zygote and equal to $\frac{1}{2}(N_p + N_m)$ or the mean of maternal and paternal admixture,
N_z^2 = Zygote's probability of Negro endozygosity,
H_z^2 = Zygote's probability of racial endozygosity and equal to $N_z^2 + I_z^2 + C_z^2$,
H_p = Paternal endozygosity,
H_m = Maternal endozygosity,
where H_m and $H_p = N_i^2 + I_i^2 + C_i^2$ and i refer to paternal or maternal proportion of admixture respectively.

Therefore with 18 characters to study, and their variation with 5 racial indicators, there are a total of 90 possible comparisons, on each of which it is possible to carry out an appropriate regression analysis. Krieger found, in fact, that only 5 of these comparisons turned out to be significant at the 5 per cent level, a number which one would expect on the basis of chance alone. Therefore it seems that there is no racial effect at all (in a genetic sense) in this population.

The traits which Krieger analysed, and those that gave a significant racial admixture effect, are as follows:

	Significant Effects for
Height	N_z
Weight	N_z
Haematocrit	
Diastolic blood-pressure	
Pathology in general	H_z (negative)
Hepathomegaly	
Splenomegaly	
Smallpox	
Diffuse goitre	
Nodular goitre	
Abnormalities in general	H_p
Polydactyly	
Digital	H_m
Prehelicine fistula	
Auricular appendage	
Strabismus	
Foetal death	
Post-natal death	

From this listing we see that for all the comparisons with zygotes homozygosity only 1 out of 18 was significant at the 5 per cent level (and that was negative). Thus the analysis of this trihybrid population lends no support to theories of either heterosis or coadaptation with respect to interracial marriages.

There are no other large-scale studies on trihybrid populations. Witkop (1954) briefly examines some of the 28 known triracial isolates in the United States, but these are usually fairly small in number and quite inbred. For instance one such isolate in Maryland consisted of about 5,000 individuals sharing but 14 family names who have been isolated for about 250 years. Other work on these American hybrids has been carried out by Pollitzer and co-workers (Pollitzer, Menegaz, and Herion, 1966), Pollitzer, 1969; Pollitzer and Brown, 1969; and Beale (1957) considers the significance of these groups for genetical research. These triracial isolates have been carefully studied from many points of view, yet on the whole they have not contributed too greatly to racial studies largely because of their smallness. We might briefly consider some of the findings of Pollitzer and others (1966) on a small isolate of about 4,000 individuals living in north-eastern North Carolina. The population can be traced at least as far back as the American Revolution, and the surnames now common in the isolate appear in the censuses from 1790 onwards in people listed as 'free white' and increasingly in those listed as those 'free persons of colour'. After 1850 the designations 'white', 'black', and 'mulatto' were adopted and the common surnames appear in each of these three categories.

Members of the isolate have not wished to marry Negroes, and Whites have not wished to marry them, so a high degree of consanguinity exists within the group. In 1957 a segment of the isolate broke away considering itself to be Indian, with the remainder classified as Negro. Only those who appeared to be healthy and non-Negroid were invited to become members. However, the two groups are not physically dissimilar and the degrees of admixture in each segment of the isolate seem similar:—

	English White (per cent)	Charleston Negro (per cent)	Cherokee Indian (per cent)
Non-Indian segment	35	33·5	31·5
Indian segment	17·5	41·5	41

Pollitzer and others suggest that this situation in fact confirmed, on a very small scale, the suggestion of Stern (1960) that the interaction of miscegenation and the 'passing' of persons from the Negro to the White group will lead to similar gene frequencies in Negroes and Whites, except for those few genes that control their phenotypic appearance.

With regard to the few comparisons which can be made of their morphological features the members of the isolate appear, as expected, to reflect the genetic contributions of three races involved. *Table* 7.13 shows three comparisons of this isolate with Negroes, Whites, and Sioux Indians. In nose width and face height the isolates are intermediate between White and Negro, and in face width they show the effects of their Indian ancestry.

The Martiniquais have some Chinese admixture, as do the Cubans, and we consider these in Chapter 6, and the 'mixed' category in Da Costa Martins' (1971) study in Mozambique is also significantly triracial. The South African coloured people are another example of an emerging race with important contributions from the three major racial stocks, and for some data on these peoples *see Tables* 6.50, 6.51, and 6.52. The island of Madagascar (now the Malagasy Republic) would prove a valuable area to investigate, since this has

Table 7.13.—FACIAL MORPHOLOGY OF THE MALE ADULT NON-INDIAN SEGMENT OF THE TRIRACIAL ISOLATE COMPARED WITH THAT OF NEGROES, WHITES, AND INDIAN.

	White*	Indian†	Isolate‡	Negro§
Nose width	36·1	39·9	39	43·4
Face height	119·3		123	123·9
Face width	138·0	149·1	143	140·7

* Old Americans (Hrdlicka, 1925).
† Sioux Indians (Sullivan, 1920).
‡ Pollitzer and others, 1966.
§ Unmixed Negroes (Herskovitz, 1930).

been a meeting point for early Caucasoid, later African, and recent Mongoloid immigrants and triracial mating has proceeded considerably. There are as yet few physical anthropological or genetic studies on this island.

It would seem overall, though, from the evidence available from Brazil and to a lesser extent from elsewhere, that the situation where three races are mixing is little different biologically from that where there are two.

CHAPTER 8

FUTURE TRENDS IN RACIAL INTEGRATION

8.*1*. Introduction

We have already discussed in Chapter 3 some of man's racial history and the considerable degree of racial amalgamation which has probably occurred in the past. In Chapter 7 we considered some of the many peoples around the world who are the product of racial amalgamation and gained some idea of the rate at which these hybrid populations are reproducing themselves. Our task in this chapter is to set this in the context of the whole species. Our main thesis so far has been that racial amalgamation has few if any harmful effects for the resulting populations. The most important consequence is, in fact, the genetic mixing and redistribution which occurs. This, as we suggested in Chapter 5, may be of considerable evolutionary significance, markedly increasing the amount of genetic variability released into an enlarged human gene pool, and therefore altering the potential for rapid human evolution in the next few generations.

Our final task should therefore be to consider the amount of hybridization which has occurred in the recent past, and to compare this with the amount of mixing taking place today. This will give us some basis for making estimates of the likely amount of amalgamation in the future. Having done this we can, on the basis of what we have learnt so far, discuss the fate of man's genetic and phenotypic variability. In other words we shall investigate to what extent and in what direction man is likely to change as a result of future racial mixing. Finally we might try to determine what are likely to be the general consequences of these changes for *Homo sapiens* within the foreseeable future.

It is less than 500 years since the continent of America was rediscovered by the western world, and it is since that time that human migration and population growth have really achieved enormous significance. The world's total population has grown in size from about 500 million to more than 3,500 million, and in the course of this growth there has been enormous redistribution of people around the globe. The date of Colombus's voyage to America can therefore be regarded as something of a milestone in man's history.

The first colonizers to leave Europe for America were from Spain, and it has been estimated that about three million Spaniards emigrated to Latin America during the first 150 years after its discovery. The colonization of North America got under way shortly after, and migration increased continuously until well into the twentieth century. The net number of permanent settlers, mainly in the U.S.A., was probably between 35 and 40 millions between 1820 and 1935. During this period about 4 million people migrated from Europe to Australia and more than 1 million to South Africa. Europeans have also moved eastwards and millions of Russians have moved from European Russia into Asia at a rate considerably increased since the Russian Revolution.

In addition to these movements of Europeans, there have been considerable migrations of non-Europeans, forced and otherwise, into new areas. The total number of Negroes imported into the New World will never be accurately known, but is variously estimated at between 14 and 15 millions from the time the Spaniards began the slave trade in the sixteenth century until the last shipments to Brazil in 1850. Although the Chinese, Arabs, and Hindus had been expanding and travelling long before the advent of the Europeans, the rate and extent of their migrations were enormously increased during the nineteenth and twentieth centuries as a result of conditions made possible by European hegemony. The Chinese, for example, increased their rate of migration to the Pacific Islands, South-East Asia, North America, and other parts of the world. Indians were transported to Africa, Fiji, and South-East Asia for cheap labour, and more recently the Japanese too have begun migrating.

The number and types of racial hybrids resulting from all these migrations cannot be accurately enumerated, but in many countries of the world the population is almost entirely hybrid, and many other distinct peoples are also known to carry important bi-racial components in their gene pool.

This is the case for the American Negro for instance, because in this case, for census purposes, all individuals of pure or mixed Negro ancestry are listed as Negro. Some of the other difficulties confronted in enumerating the number of racial hybrids are illustrated by the fact that in many Latin American countries where Indian and White mixtures are well established, the generally accepted criterion of racial status is largely cultural. Thus Mestizos who have accepted Spanish customs, dress, language, etc., are considered Spanish; whereas those adopting native customs are classified as Indian.

Although the world-wide number of recent racial hybrids is not known (and of course in some senses is a problem impossible of any very precise formulation) it is certain that the majority live in North and South America. In South America it is estimated that probably 20 per cent of the total population are Mestizos (Indian/White hybrids), this proportion varying from only 2–3 per cent in Argentina, for example, to more than 90 per cent in Paraguay. In Mexico and central America the Mestizos may form more than 30 per cent of the population, and there are small numbers (less than 1 per cent of the population) in Canada and the U.S.A.

The Mulattoes (Negro/White hybrids) are the other major group of hybrids, and the problem of estimating their number is even more difficult since the line between mixed and pure Negro is so hard to establish. Indeed, as we have elsewhere indicated, one can consider virtually all the Negroes of the U.S.A. as hybrids. In Brazil, too, most of the Negroes contain some White admixture (*see* pp. 335, 351) and therefore should be classified as hybrids as well as the Mulattoes. Other crosses such as Negro/Indian, White/Negro/Indian, and varying mixtures including Chinese, Japanese, and Hawaiian are also found in relatively small numbers.

All in all then, it is estimated that between 25 per cent and 30 per cent of the total population of North and South America can be classified as racial hybrids. In other parts of the world there are equivalent numbers of hybrids, especially in South Africa, Malaya, Indonesia, Oceania (particularly Hawaii), India, and other areas of South-East Asia. There are many millions of hybrids stemming partly from European settlement in these areas, and partly from the vast movements of the Asian population which the Europeans themselves instigated. On a world-wide basis it would appear that more than 5 per cent of the world's population are the outcome of racial hybridizations within the last 500 years. Of course, if we go back a little farther in time the proportion of the world's population who can be said to be racial hybrids is markedly increased, as we saw in Chapter 3. Indeed, if we go back far enough it can truthfully be said that we are all in some sense racial hybrids. For the purposes of this chapter we will consider only the last 500 years. Taking this as our starting point, then, we can examine whether this fraction is increasing, as a result either of the increased fertilities of the hybrids who already exist or an increasing proportion of interracial marriages.

8.2. Total Amount of Genetic Admixture

We have already considered in Chapter 5 some of the means of assessing genetic admixture between the races. We can here gather together some of the information which has been obtained using these methods on the extent to which some races have already mixed their gene pools.

Among the earliest estimates of population admixture was that of Glass and Li (1953), who used 4 genetic variables, the Rhesus, ABO, and MN blood groups and PTC tasting ability, to assess the amount of White admixture in Negroes from Baltimore, U.S.A. By 1965 Krieger and others had 17 different genetic systems at their disposal to investigate racial mixing in north-eastern Brazil. Later discoveries, largely through the exploitation of starch gel and polyacrylamide gel electrophoresis, have considerably increased the number of systems available to research workers in this field (*see*, for example, Harris, 1970, Appendix 2; and Cavalli-Sforza and Bodmer, 1971, Chapter 11).

In addition to this reliance on single gene differences, various physical features have been used in the past to assess racial mixing. These are mostly too subjective to be used much today. But the merging of human skin colour, which, as we have seen, can be measured accurately and objectively by reflectance spectrophotometry, has also been used to assess the overall degree of genetic admixture of two peoples with different skin colour. This is of some importance because it in fact allows a check on the amount of genetic amalgamation between races using what is probably the most important sociological character difference of them all. Harrison and others (1967) have estimated the degree of racial integration in southern Brazil by examining the skin coloration of the Negro and White populations of Porto Alegre. Essentially they compared the skin coloration of three racial groups, West African Negroes of either Yoruba and Ibo tribes (Barnicot, 1958) or Kuangali Okavangu Bantu (Weiner and others, 1964), Brazilian Negroes, and Brazilian Whites from Porto Alegre. The estimated degree of White admixture in the Brazilian Negroes in the Porto Alegre populations, on the basis of skin colour, varied between 38 per cent and 57 per cent. The results obtained from three analyses of single-gene differences in exactly the same populations have a mean value of 46 per cent White admixture (Tondo and Salzano, 1962; Salzano and Hirschfeld, 1965). Clearly this method of assessing skin coloration is of considerable utility and importance in supplementing the data supplied by single-gene frequencies, and deserves to be more widely utilized in a number of different situations.

The most thoroughly analysed population from the point of view of racial mixing is the Negro population of the United States. We may consider some of the analyses of this population, initially, to demonstrate some of the methodological problems involved in this field. The total amount of White admixture is given by the formula

$$\mathrm{M} = \frac{q_{\mathrm{N}} - q_{\mathrm{AF}}}{q_{\mathrm{W}} - q_{\mathrm{AF}}}$$

where q_{N}, q_{W}, and q_{AF} are the frequencies of an allele in a Negro, White, and African sample respectively. Reed (1969b) sets out the four main criteria which have to be met for critical estimation of M, and shows how difficult it is to meet all these criteria. They are as follows:

'1. The exact ethnic compositions of the two ancestral populations, African Negro and Caucasian, are known.

2. No change in gene frequency (for the gene in question) between ancestral and modern populations, either of African Negroes or of American Caucasians, has occurred.

3. Interbreeding of the two ancestral populations is the only factor affecting gene frequency in U.S. Negroes—that is there has been no selection, mutation, or genetic drift.
4. Adequate samples (that is, samples that are unbiased, from the correct populations, with small standard error) of the modern descendants of the ancestral African Negroes and U.S. Caucasians, and of modern U.S. Negroes, are available.'

In the article from which this quotation is taken, Reed brings together a number of published estimates of the proportion of White genes in American Negroes and part of his results are given here (*Table* 8.1). The purpose of doing

Table 8.1.—ESTIMATES OF THE PROPORTION (M), IN AMERICAN NEGROES, OF GENES THAT ARE OF CAUCASIAN ORIGIN (*From* Reed, 1969b.)

REGION AND LOCALITY	GENES	M
Non-Southern:		
Baltimore	R^0	0·306
Baltimore	R^0	0·216*
Five areas	R^0, R^1, Jk^b, T, S	~0·20
Cleveland and Baltimore	Gm^1, Gm^5	0·310
Various	R^0, R^1, R^2, r, M, S, Jk^b, k, Fy^b	0·232–0·261
Chicago	AK^2	0·13
Washington D.C., Baltimore, and New York City	R^0, R^1, Fy^a	0·20–0·24
Oakland, Calif.	Gm^1, $Gm^{1,2}$, Gm^5	0·273 ± 0·037
Southern:		
Evans and Bullock counties, Ga.	R^0, R^1	0·104
Evans and Bullock counties, Ga.	Gm^1, $Gm^{1,5}$	0·073
	Gc^1	0·010
Charleston, S.C.	R^0, R^1, Fy^a	0·04–0·08
James Island, S.C., and Evans and Bullock counties Ga.	R^0, R^1, Fy^a	0·09–0·12

* A revised estimate of the first one (Glass, 1955).

this is to show the variety of genetic systems which have been used and the variety of results which have been obtained by using them.

One feature which does emerge fairly clearly from these figures is the much lower amount of Caucasian admixture in southern Negroes as compared with non-southern Negroes. In an approach to determining a more critical value of M using the Duffy Fy^a gene, Reed confirms this difference.

The Fy^a gene, Reed suggests, may be an almost ideal 'Caucasian gene' for determining the amount of gene flow between Whites and Negroes. The mean frequency of Fy^a in the area from which most of the slaves were taken is probably not more than 0·02, whereas the frequency for U.S. Caucasians is about 0·43. Reed himself has shown (1968, 1969b) that there is no evidence for natural selection, such as infant growth and viability nor adult growth and fertility, associated with the Duffy phenotypes. His estimate of 21·95 ± 0·93 per cent admixture for a large sample of Californian Negroes may therefore be used as a best estimate of M for a non-southern area.

In North America, of course, most of the gene flow in the past has been in one direction, resulting mainly from irregular unions between White males and Negro females, and therefore the amount of Negro admixture in the White gene pool is very low, less than 5 per cent according to Glass and Li (1953). The North American Indian has also contributed little to either the White or the Negro gene pool. It is possible, in the light of data presented later in this chapter, that both these circumstances are changing at the present time, since the frequency of interracial marriages are showing marked increases in some areas.

Some gene flow is probably always taking place, even when minority groups are involved who try to maintain isolation. The Ashkenazic and Sephardic Jews, for instance, who have lived in various parts of Europe and the Near East, tend to exhibit frequencies for many genes similar to those of the surrounding populations among whom they live. It is possible that selection, operating on both populations in the same environment is responsible for this, although it is improbable that all genes would respond at the same rate to selection as would be necessary for selection to mimic gene flow.

8.*3*. Amount of Gene Flow per Generation

The true measure of the rate of biological integration is that of how much gene flow is taking place at each generation. There are two different approaches which can be and have been adopted to give this information, and we have already examined in Chapter 5 the algebraic method of Glass and Li which rests on knowledge of the original gene frequencies of the two contributing races.

The second approach is much more straightforward, and simply involves measuring the difference in gene frequencies between younger and older members of the same race. These are taken to be representatives of different generations, and if different rates of gene transfer prevailed at the time of their conception, this would become apparent. By sampling the very youngest and very oldest in a population one can span three generations by this method.

Salzano and others (1968) adopted this approach in their study in southern Brazil. They first used the differential frequencies of the sickle-cell gene to estimate the total White admixture in Negroes, Dark Mulattoes, and Light Mulattoes, which turned out to be 39 per cent, 58 per cent, and 75 per cent respectively. They then considered the changes in White admixture among the Negroids (non-Whites) in general at different age levels, to determine the changing rates of gene flow. Among those aged 60 or more the White admixture was 37 per cent, between 30 and 59 it was 50 per cent, and between 0 and 29 it was 67 per cent. The estimate of total White admixture among those aged more than 60 is almost half that obtained in the 0–29 age-group. If we take the conventional estimate of 30 years as the generation interval, we have an observed

White admixture of more than 10 per cent per generation. On the other hand calculations by the method of Glass and Li previously outlined, suggest a mean figure of 7 per cent gene flow per generation. It would therefore appear that the rate of gene flow per generation is increasing. This is confirmed by the fact that the admixture among 0–9-year-olds is 77 per cent, rather greater than that for 10–29-year-olds. In nother-east Brazil, too, Kreiger and others (1965) show that the amount of Negro and Indian admixture is 0·17 and 0·11 respectively among parents, whereas among children it is 0·18 and 0·12. The amount of White admixture among Negro children also shows a 4 per cent increase on average over that in their parents in this part of Brazil.

8.4. Frequency of Interracial Mating

The frequency of racial crossing is clearly one of the most important factors governing the speed of genetic integration. In sociological terms, in fact, the amount of crossing between two races is in one sense an absolute measure of their integration. If no crossing occurs, separation is complete; if marriage is equally likely between as within two races, integration is complete. Progress should then be rapid towards the formation of a hybrid race and within a short period of time the problem of racial isolation completely disappears.

We can state numerically that if p is the frequency in a population of members of race A, and q is the frequency of members of race B, then if racial characteristics are having no effect at all on the choice of mate, there will be p^2, q^2, and $2pq$ matings within race A, within race B, and between races A and B respectively. This immediately establishes a hybrid race AB and this must also take part in the marriage pattern in a fully random manner until the original two races have completely disappeared; we have examined this in more detail in Chapter 5, s. 5.3.*a*. This notional concept of a completely random mating population rests on the assumption that a male chosen at random will select his spouse from a mixed universe of women, every one of whom he has an equal probability of marrying. This is unlikely ever to be the case, in fact, for reasons not directly associated with race. There are, for instance, a number of other constraints on truly random mating such as age, kinship, religion, caste, polyandry, polygamy, distance and others, which we discuss in Chapter 2, s. 2.3.*b*. A number of studies (Hajnal, 1963; Cavalli-Sforza and others, 1966; Yasuda and Morton, 1967) have shown that the probability of a consanguineous marriage is a function of the age limitations imposed upon the choice of a spouse, of the distance over which spouses are commonly drawn, and of the abundance of relatives of the appropriate kind and age within this common distance.

In addition, as we discussed in s. 2.3.*b*, Boyce and others (1967) have shown that, at least in the past, the likelihood of marriage between two individuals is a decreasing exponential function dependent on the distance between their homes. This type of curve, illustrated in *Fig.* 2.29, is, as we mentioned, very frequently found in measurements of gene flow in both animals and plants (Bateman, 1950; Wallace, 1966). Distance is likely still to be an important consideration in many human societies today, and therefore the possibility of complete random mating between two races is dependent on their complete geographical integration. We know, of course, for historical and sociological reasons that this is likely to be a very long time materializing in the U.S.A. and elsewhere, and we must make due allowance for this. We must also remember that there are likely to be strong biological pressures tending to preserve racial isolates once they have become established, because of this neighbourhood effect in the choosing of mates. In the eastern United States, for instance, there are upwards of fifty bi and tri-racial isolates which have remained largely completely isolated since their establishment and are only now, in some cases, beginning to break up.

In addition to this possibility of mating based on kinship and geographical proximity, there is the possibility of mating based on phenotypic similarity, i.e., simple physical similarity. Spuhler (1967, 1968) has collected data from a number of reports showing that mates are positively and significantly correlated with respect to a wide variety of bodily dimensions, some of the results obtained being shown in *Table* 8.2. It is true that many of the correlations are not large, but they are nevertheless present and significant. Many of these characters shown to be correlated between mates are positively associated with racial characteristics, or are what we might term 'secondary racial characters', and therefore mating would very likely be non-random with respect to race, even if the primary racial characteristics themselves were exerting no influence on choice of mate. One of the most important traits for which there is assortative mating is intelligence. Kiser (1968), among others, shows that the association between husband and wife for educational attainment is very strong, and Burt and Howard (1956) found a correlation between mates of 0·39 for I.Q. If therefore we suggest that a random mating model does hold with respect to any gene, genotype, or phenotype, it must be because the particular character in question is not linked or correlated with any of the above-mentioned limitations of

Table 8.2.—HOMOGAMY FOR PHYSICAL CHARACTERISTICS. NUMBER OF STUDIES BY CORRELATION COEFFICIENTS (*From* Spuhler, 1968.)

CHARACTERISTIC	CORRELATION COEFFICIENTS						
	0	0–0·1	0·1–0·2	0·2–0·3	0·3–0·4	0·4	Total
Weight	–	1	2	3	1	–	7
Stature	1	6	8	7	4	1	27
Chest circumference	–	2	5	–	–	–	7
Sitting height	1	0	3	3	–	–	7
Head circumference	2	3	1	2	–	–	8
Cephalic index	2	12	5	3	–	–	22
Facial index	4	7	3	–	1	–	15
Nasal index	3	2	1	2	–	–	8
Hair colour	–	–	2	2	1	–	5
Eye colour	1	1	1	1	–	1	5

choice of mate. Clearly this ideal of purely random mating is not likely to be attained in many cases, even without the intervention of prejudice.

Finally it must be remembered that in many countries and states, intermarriage between the races is forbidden by law. This is the case in 19 of the states of U.S.A. and in South Africa for instance. Attempts to treat the U.S.A. as a whole in measuring racial integration are clearly vitiated by such circumstances.

It is nevertheless a useful exercise to consider the approach to random mating in multiracial situations, since this provides a uniform frame of reference which can be applied to each different situation.

One of the most interesting approaches is that of Heer (1967), who uses published statistics of racial intermarriage to build a model of the approach to racial amalgamation in the U.S.A., based on the following four assumptions: (1) the generations are non-overlapping and identical in life span; (2) all Americans can be classified as White or Negro; (3) all children of a Negro-White marriage will be socially defined as Negro; (4) the number of children per married couple are identical for White, Negro, and mixed couples.

The consequence of these assumptions is that the racial composition of the United States population becomes a simple function of the proportion of Whites who marry Negroes:

$$W(g+1) = Wg \times [1 - A(g)]$$

where $W(g)$ = the proportion of the United States population which is White in generation g; $A(g)$ = the proportion of Whites who marry Negroes.

In other words, the proportion of the total population which is White in a given generation is equal to the proportion White in the previous generation, multiplied by the proportion of Whites in the previous generation who married within their own group. Now if marriage took place without regard to race, the proportion of Whites who would marry Negroes is simply the proportion of all brides and grooms who are Negro. The product of the ratio of actual to expected proportions of the races intermingling, and the proportion of all brides and grooms who are Negro, is of course equal to the actual proportion of Whites who marry Negroes, i.e.,

$$A(g) = R(g) \times [1 - W(g)],$$

where $R(g)$ is this ratio of actual to expected proportion of Whites marrying Negroes. $R(g)$ is a product of two additional concepts: $R(0)$ which is the ratio of actual to expected interracial marriages at the beginning of the period over which the model is applied, and $Y(g)$ which is a function of the number of generations since the beginning of the application of the model.

The full model is therefore

$$W(g+1) = W\{g \times [1 - Y(g)]\} \times R(0) \times [1 - W(g)]$$

where $R(0)$ and $Y(g)$ can be varied.

Data on the present rate of interracial crosses are available from four states of the U.S.A. These are given in *Table* 8.3. The value of $R(0)$ we choose is

Table 8.3.—Actual and Expected Percentages of Whites marrying Negroes in California, Hawaii, Michigan, and Nebraska in Recent Years (*After* Heer, 1967. Copyright, University of Chicago Press.)

Year	California		Hawaii		Michigan		Nebraska	
	A*	B†	A*	B†	A*	B†	A*	B†
1953					0·07	0·008		
1954					0·06	0·008		
1955	0·14	0·024			0·08	0·009		
1956			0·13	0·232	0·09	0·010		
1957	0·17	0·024	0·20	0·333	0·07	0·008		
1958	0·17	0·024	0·10	0·147	0·09	0·010		
1959	0·21	0·029	0·22	0·310	0·10	0·011		
1960			0·24	0·444	0·10	0·011		
1961			0·29	0·426	0·11	0·012	0·00	0·000
1962			0·24	0·393	0·12	0·013	0·00	0·000
1963			0·27	0·370	0·15	0·017	0·01	0·003
1964			0·38	0·458			0·02	0·006

A* Actual percentage of Whites marrying Negroes.

B† Ratio of actual percentage to expected percentage, where the expected percentage is calculated on the basis of random mating.

clearly of importance, but in the absence of more substantial data than those from *Table* 8.3, it is also to a large extent arbitrary. The values we take for $Y(g)$ can also vary, again from rather arbitrarily chosen starting-points. If, for instance, we set $Y(g) = 1$, we are assuming that the willingness of Whites to marry Negroes will not increase at all as time passes. But as the proportion of White genes in Negroes increases, and therefore the physical and social barriers decrease, it is possible that interracial marriage may become more likely. One possible increase in $Y(g)$ is therefore found by comparing the proportion of Negroes among all brides now, with this proportion at generation 0, and since Negroes at present form about 11 per cent of the population

$$Y(g) = \frac{1 - W(g)}{0{\cdot}11}.$$

Other more rapidly increasing values of $Y(g)$ can be determined from the short-term trends for increasing interracial marriage apparent in the statistics for California, Hawaii, Nebraska, and Michigan since 1950, shown in *Table* 8.3. Such values can then be substituted in the general model.

Burma (1963) gives further evidence of the rate of interracial marriages in California, concentrating particularly on Los Angeles County during the years 1948–59. There are a number of limitations on the data which Burma describes, yet they can be regarded as giving a valid indication of an underlying trend. Furthermore the data are slightly different to Heer's in that interracial marriage is used to mean cases in which an Anglo or Mexican married either a Negro,

Filipino, Japanese, Chinese, or Indian, or member of another non-European racial group.

Until November, 1948, marriage of Whites with individuals of any other race was illegal, and in 1959 the Californian State legislature forbade asking the race of applicants for marriage licences. Burma's data are therefore complete in one sense, in that they cover all the years in which such statistics could be collected in California. The figures, on the other hand, are minimum ones in that marriages between American Indians and any other group, marriages between two ethnic groups where neither is White, and Anglo-Mexican marriages, are all excluded.

It is clear that during the middle years of the study there began a marked increase in the frequency of interracial marriages which became very pronounced in the last three years, so that by 1959 the interracial marriage rate was nearly three times what it was ten years earlier.

The figures of *Table* 8.4 use data gathered by the Los Angeles Marriage

Table 8.4.—INTERRACIAL MARRIAGE IN LOS ANGELES COUNTY, 1948–59
(*From* Burma, 1963.)

YEAR	TOTAL MARRIAGES	INTERRACIAL MARRIAGES	INTERRACIAL MARRIAGES PER 1,000 MARRIAGES
1948 (2 months)	5,376	28	5·2
1949	31,779	187	5·8
1950	31,915	168	5·3
1951	29,459	187	6·4
1952	30,178	171	5·7
1953	31,980	197	6·2
1954	32,095	238	7·4
1955	33,996	264	7·8
1956	36,365	268	7·4
1957	38,333	346	9·1
1958	37,700	465	12·3
1959	39,300	631*	16·1*

* In the 8½ months of 1959 in which race could be recorded, there were 447 interracial marriages which were prorated to 631 for twelve months and 16·1 per 1000 per year.

Bureau, and therefore a racial classification determined by them. Twenty-eight per cent of all interracial marriages in 1958/9 were between Negroes and Whites, with the next highest category being 21·8 per cent between Filipinos and Whites; Japanese/White marriages comprised the next highest group at 20 per cent of the roll and then Hawaiian/White marriages at 11·4 per cent and Chinese/White at 8·4 per cent. The Indian/White intermarriages comprised 7·8 per cent and finally Korean/White marriages were 1·6 per cent. Other interracial marriages together were only 3·0 per cent of the total.

Returning to Heer's model, *Fig.* 8.1 shows three curves calculated by him using three different assumptions. His first assumption, that the propensity for intermarriage will not increase at all with time, is clearly incompatible with the genetic data, presented in Chapter 5, showing how the frequency of interracial mating has markedly increased in the past. The second two curves are both based on what Heer terms 'liberal' assumptions. In the first case he assumes a continuation of the trend for increased intermarriage that is observed in the data for the state of Michigan between 1953 and 1963 (*Table* 8.3). In the

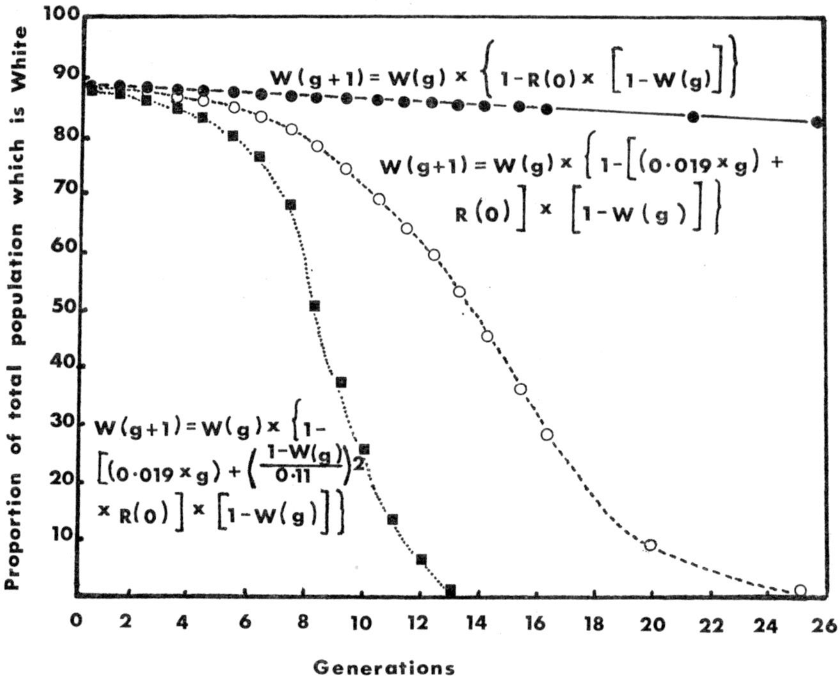

Fig. 8.1.—Three curves of predicted racial change in the U.S.A. derived from Heer's model with different assumptions regarding future rates of intermarriage. (*From* Heer, 1967, © University of Chicago Press.)

second case he assumes progressive additional relaxation of barriers as the Negroes gradually accumulate more and more White genes, and in particular become lighter in colour. On these two assumptions Heer's model predicts complete racial integration within 25 and 11 generations respectively, approximately 700 and 300 years.

Burma's data suggest that, at least in some areas of the U.S.A., interracial marriages may be increasing at a rate greater than Heer allowed for. We have, furthermore, various predictions based on rates of gene flow in the past based on genetic not demographic data, and in some cases these too lead us to suspect that Heer's predictions on the rate of approach to panmixia may need to be modified. The situation may, in some cases, be in more of a state of flux than previously imagined, with progress toward racial fusion likely to occur more rapidly than hitherto supposed.

There is a great contrast in the frequencies of interracial marriage in different states of the U.S.A., but the difference between mainland U.S.A. and the position in South America is even greater. In north-eastern Brazil, for instance, seven distinct racial types can be recognized, each containing various combinations of Negro, Indian, and Caucasian genes, and Krieger and others (1965) give details of the genetic admixture and the degree of intermarriage between these racial types. The actual number of matings and interracial crosses was determined by simple enumeration of published statistics, and it appears that observed interracial marriages amount to 80 per cent of those expected. This is almost a situation of panmixia—very different from the position in the U.S.A. This situation is not uniform of course; in central Brazil the observed proportion of interracial marriages forms only 40 per cent of those expected, and in the south it is lower still (Salzano and others, 1968), which shows how difficult it is to generalize about any aspect of this subject. It is nevertheless instructive to examine this situation in north-eastern Brazil a little more closely. *Table* 8.5

Table 8.5.—RACIAL ADMIXTURE IN NORTH-EAST BRAZIL

RACIAL CLASS	NEGRO ADMIXTURE	INDIAN ADMIXTURE	CAUCASIAN ADMIXTURE
0 White	0·176	0·116	0·708
1 Light Mestizo	0·236	0·128	0·638
2 Dark Mestizo	0·319	0·186	0·495
3 Light Mulatto	0·316	0·118	0·566
4 Medium Mulatto	0·400	0·116	0·484
6 Dark Mulatto	0·490	0·094	0·416
8 Negro	0·668	0·053	0·279
Total	0·301	0·113	0·586

shows the amount of racial admixture in the different racial types, and *Table* 8.6 shows the situation with regard to mating frequencies. If mating is not

Table 8.6.—RACIAL-MATING-TYPE FREQUENCIES IN NORTH-EASTERN BRAZIL (*After* Krieger and others, 1965.)

		Wives by Racial Type 0	1	2	3	4	6	8	TOTAL HUSBANDS	EXPECTED NUMBER ON MAIN DIAGONAL
Husbands	0	*269*	24	4	95	41	14	5	452	183·27
by	1	8	*4*	0	7	3	1	2	25	13·88
racial	2	2	1	*1*	5	1	2	0	12	0·19
type	3	94	18	7	*82*	39	10	1	251	62·81
	4	39	4	3	44	*46*	11	11	158	24·38
	6	14	5	1	23	36	*19*	7	95	7·33
	8	5	3	1	10	8	25	*18*	70	2·90
TOTAL WIVES		431	59	17	266	164	82	44	1063	294·76

random we expect a greater number of matings to take place between partners of the same racial type, and this is indeed shown in *Table* 8.6 when we compare the observed numbers of matings on the main diagonal with the numbers expected under random mating, which are shown in the last column. Now the total frequency of matings in the main diagonal is 439, which is much closer to that predicted by completely random mating (294·76) than completely assortative mating (1,063).

The expression

$$\frac{439 - 294{\cdot}8}{1{,}063 - 294{\cdot}8} \sim 0{\cdot}2$$

gives an approximate measure of the deviation from random mating, which turns out to be about 20 per cent, as we have already stated.

It has been estimated for the whole of Brazil (Salzano and Freire-Maia, 1970), that between the years 1872 and 1940 (census dates in Brazil) about

Table 8.7.—The Estimated Extent and Rate of Passing, United States, 1861–1960 (*From* Stuckert, 1964.)

Period	Estimated Number of Persons Passing	Annual Mean	Annual Rate
1861–1890	90,000	3,030	0·68
1891–1910	101,300	5,065	0·68
1911–1930	183,200	9,160	0·93
1931–1940	42,700	4,270	0·36
1941–1950	155,500	15,550	1·21
1951–1960	96,900	9,690	0·64

Table 8.8.—Percentage of Live Births reported as of Mixed Racial Ancestry in Hawaii (*From* Glick, 1970. Copyright, University of Chicago Press.)

Race of Mothers	July, 1931–Dec. 31, 1950		Calendar Years 1960–69	
	Total Live Births	Percentage Mixed	Total Live Births	Percentage Mixed
Hawaiian	11,939	63·4	2,288	82·3
Part-Hawaiian	32,115	100·0	34,638	100·0
Caucasian	38,675	15·2	55,272	19·4
Chinese	12,502	27·1	6,474	43·9
Japanese	72,389	8·6	35,058	25·2
Korean	2,967	48·3	1,556	77·3
Filipino	18,904	8·8	20,196	35·0
Puerto Rican	5,242	41·0	2,651	68·3
Negro			1,299	8·8
Samoan			1,975	44·9
All Others	730	60·5	2,584	40·9
Total	195,463	31·2	163,995	36·8

four million Mulattoes passed into the White population. The annual number of people passing during this period would be of the order of 59,000. In the United States by comparison it has been calculated (Eckard, 1947) that 26,000 people previously Negro passed into the White population within the decade 1930–40. The annual rate of passing was therefore on this estimate about 2,600, a rate considerably less than that obtaining in Brazil. Stuckert (1964) suggests that the rate was a little greater than this during this period, although still well below that obtaining in Brazil. Stuckert's figures are shown in *Table* 8.7.

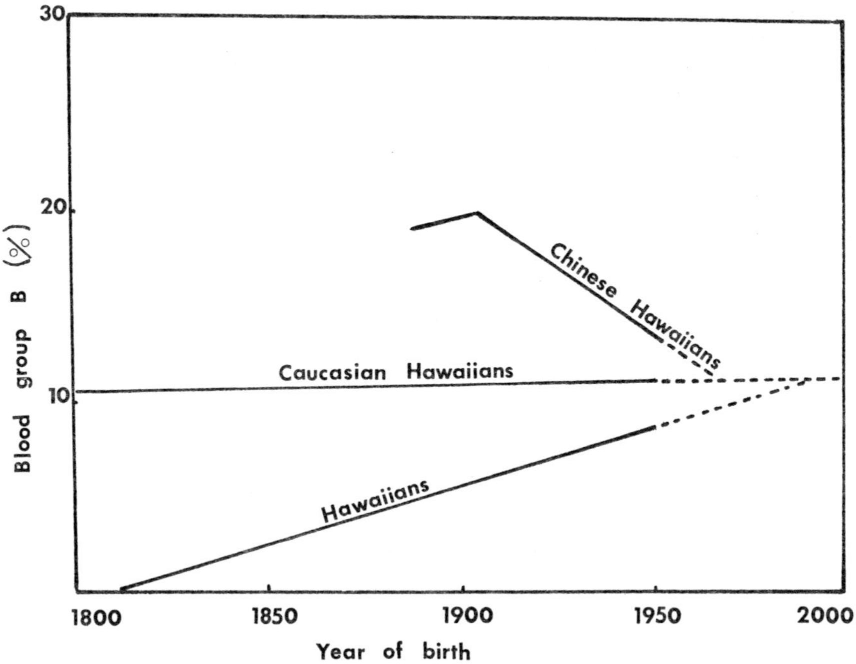

Fig. 8.2.—The trend towards complete amalgamation of Chinese, Caucasians, and Hawaiians shown by changing frequencies of blood group B. (*After* Morton, Chung, and Mi, 1967.)

The other area in the world in which interracial mating patterns are well known is Hawaii. There are here 12 racial groups involved: Caucasian, Hawaiian, Chinese, Filipino, Japanese, Puerto Rican, Korean, Samoan, Negro, Indian (an ambiguous category including Americans, Asians, and East Indians), Guamanian, and other Pacific Islanders, although the first 7 groups are predominant. The designation used by the census is always into either one of these seven major types or 'part-Hawaiian', but on the birth certificate is recorded all the race components, thus allowing very detailed analysis of the speed and effects of integration between all these various racial groups. From

Table 8.9.—INTERRACIAL MARRIAGES AS PERCENTAGE OF ALL MARRIAGES, HAWAII, 1912–1969 (*From* Glick, 1970. Copyright, University of Chicago Press.)

	Total No. of Marriages						Percentage Out-marriages					
	1912–16	1920–30	1930–40	1940–49	1950–59	1960–69	1912–16	1920–30	1930–40	1940–49	1950–59	1960–69
Hawaiian												
Grooms	1,018	1,972	1,459	1,137	955	741	19·4	33·3	55·2	66·3	78·9	85·8
Brides	1,366	2,748	1,751	1,682	1,091	1,063	39·9	52·1	62·7	77·2	81·5	90·1
Part-Hawaiian												
Grooms	455	2,164	3,567	5,142	7,502	10,385	52·1	38·8	41·0	36·9	41·3	52·1
Brides	645	3,133	5,001	8,147	10,583	11,487	66·2	57·7	57·9	64·2	58·4	56·7
Caucasian												
Grooms	2,297	7,301	9,921	19,887	17,848	28,337	17·3	24·3	22·4	33·8	37·4	28·2
Brides	2,151	6,408	8,893	14,670	13,360	25,354	11·7	13·8	10·7	10·2	16·4	19·8
Chinese												
Grooms	480	1,617	1,956	2,865	2,641	2,407	41·7	24·3	28·0	31·2	43·6	58·2
Brides	297	1,442	1,969	3,176	2,717	2,616	5·7	15·7	28·5	38·0	45·2	61·5
Japanese												
Grooms	6,395	8,261	10,114	14,680	15,509	13,354	0·5	2·7	4·3	4·3	8·7	19·6
Brides	6,376	8,301	10,331	16,909	17,498	14,924	0·2	3·1	6·3	16·9	19·1	28·0
Korean												
Grooms	267	443	412	687	632	587	26·4	17·6	23·5	49·0	70·3	75·1
Brides	196	384	516	1,055	736	817	0·0	4·9	39·0	66·7	74·5	82·3
Filipino												
Grooms	684	3,355	3,593	5,092	6,108	6,682	21·8	25·6	37·5	42·0	44·5	50·6
Brides	550	2,522	2,339	3,738	5,282	6,344	2·8	1·0	4·0	21·0	35·8	47·9
Puerto Rican												
Grooms	197	706	677	1,138	1,147	1,117	24·4	18·6	29·8	39·5	51·3	69·0
Brides	203	953	873	1,385	1,415	1,140	26·4	39·7	42·8	50·3	60·5	70·3
Negro												
Grooms	—	—	—	—	—	827	—	—	—	—	—	45·9
Brides	—	—	—	—	—	515	—	—	—	—	—	13·2
Samoan												
Grooms	—	—	—	—	—	542	—	—	—	—	—	38·2
Brides	—	—	—	—	—	681	—	—	—	—	—	50·8
TOTAL	11,826	26,090	31,863	51,140	53,344	65,673	11·5	19·2	22·8	28·6	32·8	36·1

these records it has been ascertained that the proportion of children of mixed racial origin during the period 1948–58 was approximately 35 per cent, and that this figure was twice as great as in the parental generation. *Table* 8.8 shows

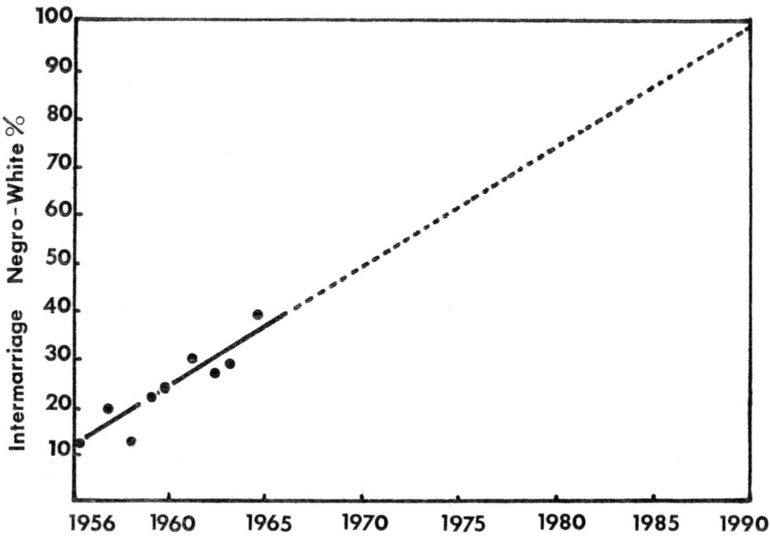

Fig. 8.3.—The proportion of Negro-White marriages in Hawaii. The actual figures are those for 1956–1964. (*From* Dyer, 1969a.)

the increase in the number of births in Hawaii which are of mixed racial ancestry, and this mirrors the increase in the proportion of marriages which are interracial (*Table* 8.9).

There have been a number of different predictions regarding the approach toward complete racial amalgamation in Hawaii. Morton and others (1967), for instance, show that from the evidence of differences in blood group B frequencies, complete amalgamation of Caucasian, Hawaiian, and Chinese will have occurred in Hawaii by around 1980 (*Fig.* 8.2). The changing frequencies of Whites marrying non-Whites are shown in *Fig.* 8.3, and the rather uncertain extrapolation from these figures also suggests that racial amalgamation will be complete before the end of the century. As a final check on the Hawaiian situation we have the total proportion of mixed marriages occurring now, compared with those occurring at various times in the past. We know both from historical and census records that it was only in 1860 that non-Polynesian races began to arrive in Hawaii, and that it was not until about 1890 that they formed any significant part of the Hawaiian population. The changing patterns of intermarriage can therefore be represented as in *Fig.* 8.4. In this case it appears that all marriages will be interracial in about 100 years. This, of course, indicates that complete amalgamation will, in effect, have been achieved much earlier.

All three of these predictions therefore agree in suggesting that racial amalgamation will be irreversibly completed in Hawaii within two generations.

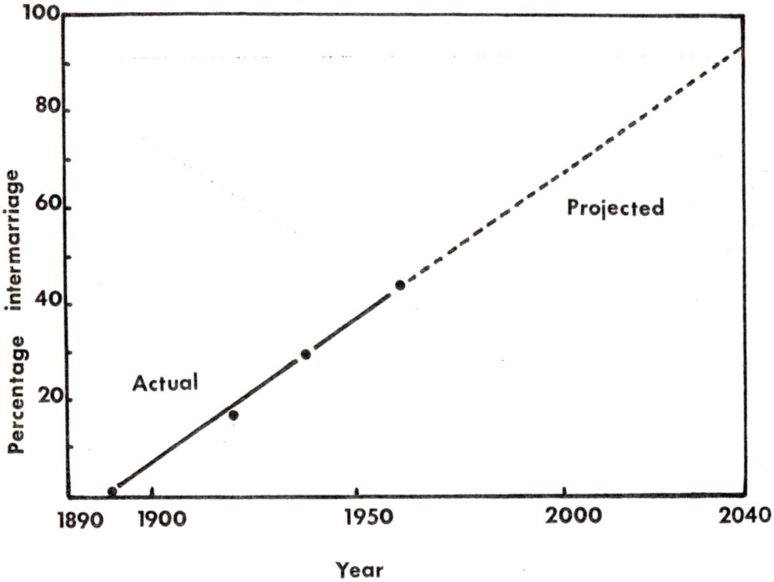

Fig. 8.4.—The overall proportion of mixed marriages occurring in Hawaii. (*From* Dyer, 1969a.)

8.5. Recent Demographic Trends

A final source of information on the progress of racial amalgamation around the world is that of censuses and surveys in those countries which differentiate the various components of their population. This will show us whether hybrid populations are in general increasing in numbers and proportions, and at what rates. A careful reading of many of the census documents and reports on these censuses will also highlight the absurdities which abound in trying to fit man's infinite genetic variety into the neat pigeon holes of demographers and statisticians.

We can consider only a small number of the very many multiracial countries in the world and see what general principles we can distil from them. We might first consider the half-blood Aborigines in Australia; the Aborigines being a minority population and on the whole socially and culturally severely disadvantaged. The situation in Australia is also fundamentally bi-racial.

Table 8.10 shows the changes in the Aboriginal population since the European occupation of Australia. In 1921 mixed-bloods accounted for 18 per cent of Australia's Aboriginal population. In 1933 this proportion had increased to 27 per cent, by 1947 to 37 per cent, 1954 to 44 per cent, and in 1961 49 per cent.

Between 1921 and 1961 the number of mixed-blood Aborigines in Australia increased by 238 per cent, and it is clear that the rate of increase of the mixed-bloods is at a much more rapid rate than full-bloods. Although the projected figure of 87,015 for mixed-bloods is the maximum likely, it seems probable that the mixed-blood Aboriginal population will at least double between 1961 and 1981. There is one other phenomenon which may be increasing the number of mixed-bloods over and above those who can be enumerated; this is the

Table 8.10.—ESTIMATES AND PREDICTIONS OF THE NUMBER OF ABORIGINES IN AUSTRALIA (*From* Lancaster-Jones, 1970.)

	ABORIGINES	MIXED-BLOOD ABORIGINES	TOTAL
1788	251,000		251,000*
1901			66,950
1921	49,593	10,886	60,479
1947	46,505	27,312	73,817
1954	39,580	31,098	70,678
1961	40,081	39,172	79,253
1966 Census	44,224	47,512	79,620
1966 prediction			91,736
1971	49,166	57,902	107,068
1976	54,348	70,961	125,309
1981	60,963	87,015	147,978

* A. R. Radcliffe-Brown's estimates of original population (1930).

phenomenon of 'passing'. Lancaster Jones (1970) shows that a significant number of individuals who described themselves as 'half-caste Aboriginal' in 1954 could not have so described themselves in 1961. He quotes Brereton (1960), who estimates the rate of 'passing' as around 20 per cent in New South Wales over the ten years 1947–57. Jones himself suggests a figure of 1 per cent per annum. If this is so, the actual numbers of mixed-blood Aboriginals will be consistently underestimated.

The impossibility of dealing with this sort of situation statistically, and the difficulty of making demographic predictions in this field, are indicated by the changes in questions on the Australian census form. In 1961 the census instructions asked a person to state whether he was a European, or another named race or half-caste European. In 1966 the instructions were for respondents to state the racial proportions of which they were composed. Those who claimed more than half Aboriginal ancestry were subsequently classified as Aboriginals. Such procedure served to increase the number of full-bloods and decrease the number of mixed-bloods. In 1971 respondents, if of mixed ancestry, were merely asked to indicate the race to which they felt they belonged and this will further increase the difficulties of comparison with past years.

Nevertheless despite these difficulties the general trends are quite striking. By the end of the century Australia's Aboriginal population will have equalled and surpassed the number estimated to have been in Australia at the time of the first European settlement. Genetically, though, it will be a very different population from that which existed 220 years before.

In contrast to Australia is South Africa, where the natives were and still are in considerable numerical superiority. Furthermore interracial mating between the native Bantu and the Whites is now forbidden by law. There nevertheless exists a large and growing coloured population which are mainly the offspring of unions between Whites and Bantu. *Table* 8.11 shows the changes in composition of the South African population this century.

Table 8.11.—THE POPULATION OF SOUTH AFRICA, 1904–60

		WHITES	COLOURED	BANTU	ASIATICS	ALL
1904	Number	1,117,234	444,991	3,490,291	122,311	5,174,827
	Percentage	21·6	8·6	67·4	2·4	
1911	Number	1,276,319	525,466	4,018,878	152,094	5,972,757
	Percentage	21·4	8·8	67·3	2·5	
1921	Number	1,521,343	545,181	4,697,285	163,594	6,927,403
	Percentage	22·0	7·9	67·8	2·4	
1936	Number	2,003,334	769,241	6,595,597	219,691	9,587,863
	Percentage	20·9	8·0	68·8	2·3	
1946	Number	2,372,044	928,062	7,830,559	285,260	11,415,925
	Percentage	20·8	8·1	68·6	2·5	
1951	Number	2,641,689	1,103,016	8,560,083	366,664	12,671,452
	Percentage	20·8	8·7	67·6	2·9	
1960	Number	3,088,492	1,509,258	10,927,922	477,125	16,002,797
	Percentage	19·3	9·4	68·3	3·0	
Percentage Increase, 1904–60		176·4	239·2	213·1	290·1	209·2

It is ironic, if instructive, that in this country of apartheid and White immigration, and in which miscegenation has been forbidden by law since the 1940s, and where before that mixed marriages comprised less than 1 per cent of all marriages involving Europeans, the Coloured population should show a 239 per cent increase between 1904 and 1960, whereas the Whites show only a 176 per cent increase and the Bantu a 213 per cent increase over the same period. The Coloured people in 1960 formed 9·4 per cent of the total population of South Africa whereas in 1904 they comprised only 8·6 per cent. The rate of increase of the Coloured population has, if anything, increased in the last few years with a 36·8 per cent increase in the decade 1951–60 and a marked jump during the same period of the proportion they form of the whole population. It remains to be seen whether this trend will continue in South Africa's special situation.

Finally we might briefly consider the population of the U.S.A. This country has been called the great melting-pot, through its ability to assimilate such a wide variety of basically European cultures. It has been less successful with its non-European inhabitants, but here, as we have seen, things are changing. In this country, too, the Negroes, who as we have noted are markedly if variably a mixed population, show a sustained and increasing rate of population growth.

Despite the considerable White immigration into the U.S.A., the rate of increase of the Negroes has exceeded that of the Whites since 1930 (*see Table* 8.12). This is one important factor which was not built into the model by which

Table 8.12.—POPULATION OF THE U.S.A., 1790–1950.

	WHITE (000's)	PERCENTAGE INCREASE	PERCENTAGE WITH AFRICAN ANCESTRY*	NEGRO (000's)	PERCENTAGE INCREASE	PERCENTAGE WITH NON-AFRICAN ANCESTRY*
1790	3,172		2·0	757†		19·0
1800	4,306	35·8	3·8	1,002	32·4	20·9
1810	5,862	36·1	5·2	1,378‡	37·5	32·7
1820	7,867	34·2	7·9	1,772	28·6	35·0
1830	10,537	33·9	10·8	2,329	31·4	36·2
1840	14,196	34·7	13·7	2,874	23·4	36·2
1850	19,553	37·7	15·2	3,639	22·6	38·2
1860	26,922	37·7	16·7	4,442	22·1	39·1
1870	33,589	24·8	18·0	4,880	9·9	39·7
1880	43,403	29·2	18·3	6,581	34·9	49·4
1890	55,101	27·0	18·8	7,489	13·8	52·1
1900	66,809	21·2	19·5	8,834	18·0	56·6
1910	81,732	22·3	17·4	9,828	11·2	61·6
1920	94,820	16·0	17·7	10,463	6·5	64·8
1930	110,286	16·3	18·5	11,891	13·6	68·0
1940	118,214	7·2	22·8	12,865	8·2	77·4
1950	134,942	14·2	21·0	15,042	16·9	73·0
1960	158,455	17·4	19·5	18,860	25·4	69·9

* These figures from Stuckert (1964).

† The total number of Negroes imported is thought to be about 400,000.

‡ The importation of slaves was made illegal in 1808, but continued at a low level for a few years subsequently.

Heer was attempting to predict the approach to complete random mating. The sheer increase in availability of Negro spouses will certainly increase the frequency of interracial marriages.

In these three different countries in three different circumstances we find sustained increase in the absolute and relative size of the hybrid populations. Once entered into, racial mixing can never be reversed. What we are really witnessing today is part of the continuing evolution of man. Old races amalgamate, new races emerge, peoples migrate to new environments and their habits change.

All of these processes have profound evolutionary consequences, and it is absurd to pick on any one moment of time or any one particular racial distribution as being of more importance in this context than any other. The population of man is growing now at a greater rate than at any previous period in his history; a simultaneous tribute to his genius and his stupidity. It is this fact which should be engaging all man's attention. At the present moment in time we are all the product of a complex history of racial shifts, minglings, and amalgamations—more complex than many will care to admit. In this book we have searched carefully for any evidence of detrimental effects of this racial mixing and have found very little. On the other hand the positive effects were there for all to see, in man's mastery of the earth and his colossal achievements on his way to this mastery.

Complete alteration of all the races existing today seems, on the basis of evidence available to us at the moment, to be absolutely inevitable in the long run. This does not mean dull uniformity. Genetic and social variety is, we have seen, not destroyed by racial mixing—merely rearranged. Various races of man will still exist, hopefully to be described in terms of endearment, not enmity.

CHAPTER 9

GENERAL CONCLUSIONS

IN the previous eight chapters of this book we have considered a wide variety of biological features and their variation in man. This would be not so remarkable were it not for the fact that all this variation has been considered within the context of human races. We have considered everything from tiny differences in the base pair sequences of the DNA in the nucleus of the cell, through to incredibly complex processes of skeletal growth and mineralization, taking in *en route* behavioural and motor differences large and small. We have drawn certain conclusions at each stage, and the general findings of our analysis could be said to be twofold: firstly, hybrid populations are in general intermediate between the contributing parental populations with respect to most traits considered, and do not differ greatly from them in variability; secondly, there are few or no harmful effects resulting from racial mixing, no matter how widely separated the races on genetic or geographical criteria.

But we cannot leave the matter there. Merely to answer a question in the negative, when perhaps the question itself is misconceived, is of course totally inappropriate. We are in the position of having considered a number of questions with respect to racial mixing and generally answered in the negative with respect to the question 'Are there any harmful effects?'. At every stage, though, a number of objections can be raised to merely considering the biology of a certain variable system and then coming to some simplistic answer, after some seemingly objective consideration of the evidence for or against.

These objections centre on two main questions: Have we, despite our discussions in Chapters 1, 2, been too typological in our approach? In other words it may be the case that to deal in terms of cephalic indices or of I.Q. differentials, for instance, vests such measures with a greater degree of significance in terms of racial distinction than they in fact possess. To use anthropological data collected fifty years ago almost inevitably means implicitly accepting the assumptions regarding race held by those collecting the data.

Secondly, when enquiring whether there are any harmful effects, nowhere have we specifically asked 'Harmful to whom?'. The failure to answer this question again carries with it connotations of racial distinctness and purity nowhere justified by the facts. This chapter will attempt to draw some overall conclusions, particularly in the light of these two questions.

9.*1*. Typological Implications

In his book entitled *Race Intelligence and Education* published in 1971, H. J. Eysenck opens Chapter 4 with the following sentences:—

> 'We have already noted that the difficulties in comparing intelligence between groups differing not only in race but also in general cultural background, such as for instance American whites and African negroes, are such that at the moment only exploratory research is possible. There are certain interesting findings which are somewhat peripheral, although certainly not irrelevant; thus it is repeatedly reported that African negro children (and American negro children as well) show highly precocious sensorimotor development, as compared with white norms. Thus most of the African negro babies who were drawn up to a sitting position could keep head erect and back straight from the very first day of life; white babies typically require six to eight weeks to sustain these postures! It is implausible to assert that socio-economic differences or other extrinsic variables (particularly in so far as these disfavour negroes) could have produced this astonishing difference; the observed precocity lasts for about three years, after which time white children overtake the black ones. These findings are important because of a very general law in biology according to which the more prolonged the infancy, the greater in general are the cognitive or intellectual abilities of the species. This law appears to work even within a given species; thus sensorimotor precocity in humans, as shown in so-called "baby-tests" of intelligence is negatively correlated with terminal I.Q.'

Statements of a similar type and presenting similar arguments appear in various places in Chapter 6 of this book, but particularly in ss. 6.*2*.*d*, 6.*4*.*d*, and 6.*4*.*e*. There is nevertheless a great deal to be criticized in the above statement, and many like it both in this and in Eysenck's book. It is typical of many statements in books and articles on this topic, including the present one; indeed it represents a way of thinking which pervades the whole subject of racial biology. In criticizing Eysenck's statement we are therefore also criticizing some of what has already been written in this book, with the hope that what withstands this criticism will be of greater value.

Eysenck begins by noting the difficulties in comparing intelligence between groups differing in race and cultural background. Now intelligence is a continuously varying character, and only in an individual can an absolute value be assigned to it (leaving aside for the moment whether the absolute value so assigned has any biological validity). Groups or races do not have an intelligence, only the individuals within them. Furthermore, although we can certainly say that members of some groups or race show a mean intelligence, this does not really tell us a great deal, unless we also specify the variance around that mean and the shape of the distribution curve. If we state that we are comparing the height of Pygmys and Dinkas, for instance, as we did in Chapter 6, this has some meaning, since these peoples are very different in mean height and part of the normal description of them includes reference to their height. These circumstances never apply to I.Q. The range of I.Q.s in different populations is almost always identical and in the case of American Whites and Blacks for instance there is only a 5 per cent better than chance probability of determining a person's race from his I.Q. alone. Races can't be defined by their I.Q.s!

Within this same first sentence of Eysenck's we read that the difficulties allow only exploratory research to be carried out comparing intelligence in people of different general cultural background, and yet elsewhere in his book (p. 56 and p. 125, Australian edition) we read of culture fair tests which minimize these differences. To admit that these tests are inappropriate for some racial comparisons, and yet valid for others, suggests that considerable caution should be exercised in interpretation of results obtained by using them.

Eysenck then moves on to a consideration of sensorimotor development and its relevance to terminal I.Q. In particular, mention is made of the abilities of Negro babies at birth and the marked difference with Whites. Now this itself is largely an all-or-none trait and the difference between Negroes and Whites is almost absolute. Virtually all Negro babies can sit up and hold their heads up; virtually no White babies can do so (Geber and Dean, 1957). To suggest that this trait is of a similar nature to that of intelligence conveys a totally false impression and is surely incorrect. To talk of 'the observed precocity lasting for about three years' is the clearest possible example of typological thinking. What precocity are we talking about? The ability of Negro babies to hold the head erect, or the more advanced skeletal development of Negro children, or more rapid rates of growth of Negro children, or the changing abilities of Negroes and Whites at any one of a hundred specific tests of sensorimotor development, or the earlier onset of menstruation in Negro girls, or the earlier development of secondary sexual characters in Negro boys and girls, or the earlier eruption of permanent teeth in Negro children? We considered each of these traits in some detail in Chapter 6, particularly in s. 6.*2*, and found that a wide variety of genetic mechanisms is responsible for these differences, each of which is interacting to a greater or lesser extent with the others and with the environment. The times at which these different examples of 'the observed precocity' of Negroes are manifest, vary from the first days of life until at least 18 years of age. To speak in such typological terms about such genetic variety and variability is a very unfortunate reflection of our ingrained way of thinking on this problem.

The final sentence in this quotation attempts to be more specific with regard to sensorimotor precocity, asserting that this precocity, as shown in baby tests of intelligence, is negatively correlated with the terminal I.Q. Now we must first remind ourselves that the whole argument is about the validity or otherwise of any such concept as a terminal I.Q. We are concerned to investigate whether the I.Q. of a Negro reaches the level it does because it has reached some terminal point determined by his genetic complement, or whether the adult I.Q. level is a flexible thing which can, or could have been, modified by exposure to an improved environment. The appearance of this phrase, then, is an assumption of what the preceeding sentences were trying to prove.

One can, in fact, take issue with the assertion contained within this unfortunate sentence, the ignoring typological assumptions it also contains, and on the basis of the evidence presented in s. 6.*4.d.*i and in particular *Table* 6.19 from Goffeney and others (1971) it would appear to be just plain wrong.

More important than all this, though, is the nature of the whole argument running through this paragraph. From the way in which the argument is presented it would appear that although the evidence that we have today is unfavourable, nevertheless the discovery of new evidence might well validate the conclusions as they are presented. The general nature of the argument, however, is totally wrong and could not, under any circumstances, become an acceptable one, no matter what new evidence is discovered. We can summarize the steps in the argument as follows:

1. A general law of biology states that prolonged infancy in a species signifies greater intellectual abilities in that species, compared with species which have a less prolonged infancy.
2. There is a negative correlation of baby intelligence with terminal I.Q. in man.
3. Therefore the law holds in man.
4. Negroes have a shorter infancy than Whites.
5. Therefore (implied, but not stated in this paragraph) Negroes have lesser intellectual abilities than Whites.

Point (1) may be a general law, but it is concerned with large differences in speeds and lengths of development, and is concerned with intergeneric and inter-specific differences only.

Point (2) appears to be not true, and in any case can be criticized for the use of the term 'terminal I.Q.'.

The conclusion (3) follows neither in fact nor in logic. It is an inductive argument for which there is no other evidence in any other species. The biological extrapolation from species differences in infantile development to the slight but variable differences in intellectual development in man is completely unjustified. To escape the necessity of taking an inductive step from between-species to a within-species comparison, it would have to be argued that this general law is invariable, i.e., it has the status of a law of nature such as those around which physics and chemistry are built. If this were the case then it would have to apply to all individuals within all species, i.e., it would be a truly invariable law of nature. There are few (if any) biological rules of that status, and this is not one of them, as reference to De Beer's book *Embryos and Ancestors*, for instance, will make clear.

With regard to point (4) we have argued that this statement is too imprecise and indeed too inaccurate to have any meaning whatsoever, and the final implied conclusion (5) is therefore both illegitimate in terms of the argument used and, on the basis of the evidence offered, in fact incorrect. In the terms of the argument this final conclusion also implies an absolute or typological difference in intellectual abilities between the races which certainly does not exist, whatever may be the position with regard to possible mean values of I.Q. or scholastic achievement in these races.

This lengthy criticism of but one paragraph could be repeated over and again for Eysenck's book, for Shuey's book, and for parts of the present book. Genetic variation in man is universal and every individual is genetically unique. And yet time and again we refer to the effects of race crossing on this or that character as if one race is invariant for character A and one invariant for character B. Certainly the outcome of crosses between, for instance, very tall and very short people is of biological interest, but this is a marriage between individuals not races, and so it is for almost every other character which has been investigated in man.

Of course a great deal of the discussion and research on this topic of racial biology and racial differences is justified by the authors on the need to understand the differences and possible deficiencies of one race *vis-à-vis* another, in order that the truth might be known and justice might be done regarding the attainment of social equality. Eysenck (1971), for instance, says the following: 'A realistic policy [regarding compensatory education] needs not only compassion, but also knowledge. . . . The situation is too grave for name-calling; it requires serious discussion, and above all research of a quality and on a scale which has hitherto been sadly missing . . . only by proper research will the thesis of (partial) genetic determination of low Negro I.Q. be defeated.'

Dwight J. Ingle began a paper at the 1966 symposium of the American Association for the Advancement of Science with the following sentence: 'I am among those who prefer, when possible, to attend to individuality rather than racial origin'. He concludes with the following one, 'We seek knowledge, not in order to discriminate against the disadvantaged, but in order to aid intelligently their advancement'. The title of his paper is *The need to investigate average biological differences among racial groups*! (Ingle, 1968).

Arthur Jensen, in an article in 1970 which was largely a defence of his earlier lengthy paper, states that 'These estimates of heritability (of I.Q.) are based on tests administered to European and North American populations and cannot properly be generalized to other populations. I believe we need similar heritability studies in minority populations if we are to increase our understanding of what our tests measure in these populations and how these abilities can be most effectively used in the educational process'.

One could be forgiven for thinking from the above quotations that perhaps the range of I.Q. in Whites was from 100 to 150, and in Negroes from 50 to 100, rather than from about 50 to 150 in both races, and that in excess of 10 per cent of American Negroes score as high or higher than the average American White.

Certainly more money and research in education is desperately needed. But the educational problems of the White child with an I.Q. of 80 are just as different from a White child with an I.Q. of 120 as are those of a Negro child with an I.Q. of 80; and vice versa. The question of skin colour, or blood group, or stature, in other words, is immaterial to most educational problems. To link educational psychology and programmes of environmental supplementation to variables such as these is in this way self-evidently absurd, and yet this is what we do when we link education and race together in some sweeping generalizations. It seems as if in this area sociological concepts of racial identity and definition (*see* Chapter 1, p. 6) have intruded upon and largely replaced a genuine biological approach.

In Chapter 1 we decided that a race was a population which differed in some of its gene frequencies from other populations. In Chapter 2 we saw that genetic variation was universal and subject to continuing fluctuations as the result of natural selection, drift, recombination, and mutation. To imply or assume some uniformity and invariance by use of the terms Black or White, Negroid, or Mongoloid is biologically absurd.

It might indeed be the case that some races differ, on average, in the frequencies of some of the genes controlling intelligence. Until we can identify more precisely some of the 50 or 100 genes controlling I.Q. and intellectual achievement, we cannot possibly answer this question with any certainty. To concentrate on possible racial differences in intelligence (or other similar traits), even with the best of stated motives, seems largely to ignore what we already know of the genetics of the trait. It also seems to ignore the history of racial amalgamations which have taken place in the past and are still taking place today, and to ignore the genetic variability which exists for this trait within all races anyway. It is, in addition, the worst of all possible procedures with respect to the social consequences.

Although this book can be criticized on exactly these same grounds in parts, it is hoped that by exposing the biological reality of the situation, and the facts on which the arguments are based, people may be persuaded to adopt different approaches to racial integration in the future from those they have adopted in the past.

9.2. The Consequences to Whom?

Investigations of the biology of racial crosses often carry with them typological associations of racial attributes. A question more directly associated with this book is: why study racial crossing? In Chapter 4 we considered some long-term evolutionary reasons why, in general, racial hybridization is worthy of study. If we assert that this is the reason for studying racial crossing in man it must be because it is anticipated that the present racial distribution in man will be drastically changed in the future. So it might be argued that part of the justification for studying racial mixing is that we are concerned with the evolutionary consequences for the human species as a whole.

On the other hand many people are concerned about the consequences for one particular race if it mixes to any extent with another. Will undesirable traits be introduced? Will the average level of intelligence decline? Finally there is the interest of the hybrids themselves, particularly the first two or three generations after race crossing. Will they express any incompatibilities? Will they be disadvantaged in any biologically important way? We can consider each of these three aspects of integration separately.

So far as the evolutionary consequences are concerned we have considered in Chapter 8 some of the more recent examples of race crossing in man and their demographic effects. Can we assert that 5 per cent or 10 per cent of the world's population show disadvantageous effects or can it really be denied, if we consider a longer time span, that we are all racial hybrids? Since archaeology and history tell us that race mixing has continually occurred in the past, we must, in effect, look to the whole human population for its consequences. And it is not underpopulation or problems of extinction which are threatening us today.

We have seen in the present book that the genetic variety of combining races is not lost, merely rearranged. The more successful of these interracial offspring combine the advantageous features of both races. There are no examples of mixed races becoming extinct; there are many examples of mixed races enormously increasing in numbers and importance.

We turn then to the shorter term; to the point of view often expressed on behalf of one race that it would be disadvantageous for genes of another race to spread into it. As we pointed out in Chapter 6 this point of view assumes positions of inferiority or superiority regarding the races which are rarely justified. Furthermore it normally ignores the fact of enormous genetic variation in both races and the fact that marriages do not take place between statistical means but between individuals. Thus with regard to intelligence, for example, mating is still likely to be non-random with respect to this trait whatever might be the position with regard to skin colour. And despite predictions to the contrary, the numbers of gifted individuals who would be produced in a racially mixed population in the U.S.A., for instance, would probably be just as great as are now produced when the Whites and Negroes are partially isolated.

With respect to a few traits like the malarial protective haemoglobin and enzyme variants, it can be argued that if they are introduced into a race which lacked them in a non-malarial area then this could be considered a disadvantage in future generations. But note that this disadvantage could only manifest itself in the case of two heterozygous individuals marrying. Similarly with the Rhesus-negative gene, as we discussed in Chapter 5, its introduction into a population might result in some cases of erythroblastosis foetalis where previously none occurred. However this possibility can in fact now be prevented by appropriate medical action. To set against this slight disadvantage is the fact that the influx of new genes will decrease the incidence of homozygosity for the detrimental genes that the recipient race already possesses. As we saw in Chapters 5–7, in the long-term the break-up of isolates is favourable genetically.

This whole attitude though, of adding or subtracting benefits or disadvantages to one race or another, is one associated with the preservation of races as they are today. It is an argument in favour of racial isolation and evolutionary conservatism which is taken to its ultimate conclusion in South Africa. It is an argument which says that the present races of man can and should be preserved as they are and that introgression from another race will be harmful now and in the future; that the two races will preserve their relationship to one another for as long ahead as can be predicted. Going on past and present experience this will not be very long (*see*, in particular, Hulse, 1962, in conjunction with Chapters 3 and 8 of this book) and therefore this short-term argument would seem to be without any social or biological relevance. Note, however, that a great deal of the evidence concerning the consequences of racial amalgamation that has been presented in this book is based on the hidden assumption that this whole argument is a valid one, that races can and will be preserved and that we should consider man's future evolution in terms of his racial structure today. It is not merely enough, therefore, to show that racial amalgamation is not harmful, because that leaves open the possibility that with future discoveries it could be shown to be a very undesirable process. We have to show that this whole approach is an invalid one, resting on incorrect assumptions about man's past, incorrect conclusions about his future, and socially unacceptable beliefs about the status of the races of man today.

Finally we come to the possibilities that detrimental effects would show up in first or second generation offspring of racial crosses. There is amongst the wide range of biological phenomena examined in this book, almost no evidence that this is so. Indeed, since man is one species and the races are often closely related, one would hardly expect it to be otherwise. Again, though, false analogies and arguments have often been used in this context and have in places been quoted in this book just as they have, on occasion, been quoted by other commentators, enlightened and otherwise.

This possibility of disharmony in hybrid offspring is stated in an extreme form by Stockard, quoted in Montague (1964): 'Since prehistoric time, hybrid breedings of many kinds have occurred at random among the different races of human beings. Such race crossings may have tended to stimulate mutations and genic instability, thus bringing about freak reactions and functional disharmonies just as are found to occur among dogs'. As has often been pointed out, quite what the relevance is between dog crosses, where the participants have long been subject to extreme forces of inbreeding and artificial selection to give monstrous body forms, and the relatively slight human racial differences, is hard to see. Nevertheless because the assertion has been made it is dealt with at length in this and other books. The suspicion lingers, however, that although the evidence is today unfavourable to such arguments, tomorrow it might not be. In fact on the basis of virtually everything we know, we can agree with Montague that 'the whole notion of disharmony as a result of ethnic crossing is a pure myth'. It should hardly need a book this length to convince us of that.

Our final conclusion, then, can be that race crossing in man has no harmful or detrimental effects in the long term or the short term, to the races involved, or indeed to man in general. The long-term consequences are possibly of some significance, in that they result from the rearrangement and reassortment of genetic variability through the species which, on the basis of what happens in other species and what has happened in man's own past, is a stimulus to further evolution not a detraction from it.

The gist of this last chapter is that it hardly needed a whole book to arrive at that conclusion. Furthermore, most of the presentation of the evidence by which we have arrived at that conclusion has been in terms which have not taken full account of the enormous genetic variability universally present in man, and which have assumed that races are distinct entities which might be expected to show some harmful results on mixing.

The sexual isolation of the various races of man can be attributed to psychological and geographical causes, not biological. It seems likely that greater familiarity with the rest of mankind resulting from the growth of education and global population movements will serve to breakdown the barriers between the races. We need have no worries on the biological consequences of this racial mixing.

APPENDIX 1

A SUMMARY OF THE PRINCIPAL BLOOD GROUP SYSTEMS

Brief outlines of the genetics and anthropological usefulness of seven of the major blood groups are given here, together with different commonly used systems of nomenclature. For comprehensive information refer to Race and Sanger (1968) and Giblett (1969).

1. The ABO System

There are four major different blood types controlled by two codominant and one recessive allele at this locus

$$I^A = I^B > i$$

or

A and B codominant, O recessive

so that we have the following

Blood Type	*Possible Genotypes*
A	I^AI^A, I^Ai (or AA, AO)
B	I^BI^B, I^Bi (BB, OB)
AB	I^AI^B (AB)
O	ii (OO)

The blood type A is divisible into two subtypes A_1 and A_2 which is of some anthropological usefulness.

The most common group around the world is the O group and among Amerindians almost the whole population is O. B is virtually absent in Australia as well as America but becomes progressively more common from western to eastern Europe and reaches its highest frequency of over 30 per cent in northern India and southern Asia. It is also fairly common in Africa. A_2 is almost entirely confined to Europe. In most people the saliva and other mucous secretions contain water-soluble glycoproteins with A or B specificities. A minority of individuals do not exhibit A or B specificities in their saliva and these people are 'non-secretors'. This dimorphism is due to a pair of alleles, Se and se, not linked to the ABO locus. About 80 per cent of Europeans are secretors and about 20 per cent non-secretors.

2. The Rhesus System

This is the most complex of the blood groups associated with the red blood-cells and there is still considerable disagreement as to the genetic control of the system.

At the level of foetal incompatibility the system principally concerns two alleles showing simple dominance relationships, D and d. People who are DD and Dd are Rhesus positive, since they react to an antibody anti-Rh, whereas dd people, who are Rhesus negative, do not. This is responsible for haemolytic disease of the newborn in the following way: blood from the foetal circulation crosses the placenta into the mother and carries with it some of the paternal antigens. If there is an incompatibility between the parents, i.e., if the mother is Rh(—) and the father is Rh(+), the mother produces an antibody which in later pregnancies may cross the placenta from mother to foetus and be responsible for destroying foetal tissues.

It is apparent, therefore, that the d allele must, in the past, have been subject to some considerable degree of selection because of the incidence of largely fatal haemolytic disease of the newborn. The frequency of Rhesus-negative individuals in most parts of the world is, not surprisingly therefore, very low. Only in Europeans is there a significant frequency, averaging about 12 per cent (nearly 30 per cent in Basques), with lower frequencies among the extra-European Caucasoids of India, Pakistan, and Afghanistan.

However, a number of other Rhesus specificities have been discovered and these have been interpreted either as alleles at separate but very closely linked loci or further alleles at the one complex locus. The main Rhesus specificities are as follows:—

	Allelic Combinations							
	R_1	R_z	r	Ro	R″	R′	R_2	Ry
Antibodies	CDe	cDE	cde	cDe	cdE	Cde	CDE	CdE
Anti-Rho or D	+	+	—	+	—	—	+	—
Anti-rh″ or E	—	+	—	—	+	—	+	+
Anti-rh′ or C	+	—	—	—	—	+	+	+
Anti-hr′ or c	—	+	+	+	+	—	—	—
Anti-hr″ or e	+	—	+	+	—	+	—	—
Anti-d not found								

The CDE notation is the Fisher-Race notation the R_1R_z notation is that due to Alexander Weiner (Fisher, 1947; Weiner, 1948).

There are in fact more than 30 different Rhesus alleles known and these involve other alleles at the C, D, and E loci, designated D^u, e^s, C^w, etc. There are also a number of interaction phenotypes, deletion phenotypes, and some cases of additional alleles, i.e., those showing both C and c specificities.

The important reason for giving this brief outline of the complexities of the Rhesus system is that the particular allele combinations found in a given population must be a function of that population's evolutionary history. Thus, of the four commoner alleles, cDE(R_z), CDe(R_1), and cde(r) are each connected to cDe (Ro) by mutation with respect to a single antigenic specificity. This relationship, together with their global distribution, suggests that cDe(Ro) may have been the original allele of the system and that cDE(R_z), CDe(R_1), and cde(r) were derived by single independent mutations. Rarer alleles may then have been derived by crossing over within the appropriate genotypes (*see* Feldman, Nabholz, and Bodmer, 1969).

Some gene frequencies are:—

	Caucasoid	*Negroid*	*Mongoloid*
Ro, cDe	0·0186	0·7395	0·0409
R_1, CDe	0·4036	0·0256	0·7591
R_z, cDE	0·1670	0·0427	0·1951
R_2, CDE	0·0008	0	0·0049
r, cde	0·3820	0·1184	0
r′, Cde	0·0049	0·0707	0
r″, cdE	0·0029	0	0

3. The MNS-U System

The first alleles described for this system were M and N, which are co-dominant. For most of the world M and N are roughly equal and the genotypes MM, MN, and NN are all represented. Among Amerindians, however, M is very high, whereas throughout the Pacific region M is very low or absent.

The nomenclature of this blood group is sometimes set out:—

Ag^M Ag^M	M blood type
Ag^M Ag^N	MN blood type
Ag^N Ag^N	N blood type

An antibody known as anti-S was discovered in 1947. The S positives, agglutinated by anti-S, were not randomly distributed between the MN phenotypes. An anti-S antibody was later discovered and then, among Africans only, what appears to be a silent allele at the S locus. At the moment the system can be explained by two very closely linked loci, one with two alleles the other with three.

In Europe the gene frequency for S is 30–35 per cent and is more often associated with M. In Africa S is lower and S^u (no S or s produced) may reach 35 per cent. Although the MN frequencies in New Guinea and Australia are similar, S is rare or absent in Australia but reaches up to 12 per cent in Papua.

4. The Duffy System

This blood group was discovered in 1950 and comprises a Duffy-positive allele (Fy^a) and a Duffy-negative allele (Fy^b). Since Fy^a is dominant over Fy^b only two phenotypes can be recognized. However, when Negroes were examined a high proportion reacted neither with anti-Fy^a nor anti-Fy^b. It is possible that these individuals are homozygous for a third allele at this locus, although no antibody which agglutinates it has yet been found. This third allele is at the moment simply designated Fy. The frequencies of the Duffy-positive gene reach very high values in Australians and other Pacific peoples:—

Group Studied	*Gene Frequency* Fy^a
Cape York, Australia	1·00
Koreans	0·99
Chinese	0·91
Japanese	0·86
East Indians	0·73
Amerindians	0·5–1·00
English	0·41
American Whites	0·40
American Negroes	0·14

The rarity of Duffy-positive individuals in Africa and the relative high frequency of the dominant gene among Europeans makes the Duffy system one of the most sensitive for measuring the amount of European admixture in American Negroes.

5. The Diego System

The Diego blood group system involves two alleles Di^a and Di^b and with Di^a being dominant two phenotypes Diego-positive and Diego-negative can be recognized. Diego-positive individuals are restricted to Amerindians, particularly in South America where they may be over 40 per cent and although they are absent in Alaska to South Asiatics, Chinese, Japanese, and Koreans.

6. Other Useful Blood Groups

The P system has three alleles with apparently an order of dominance $P_1 > P_2 > p$. The P_2 allele usefully distinguishes Europeans, Africans, and Asians.

The Kidd system has two main alleles Jk^a and Jk^b with Jk as a rare silent allele. The allele frequencies distinguish Mongoloids from both Caucasoids and Negroids.

The Kell system, with KK and Kk being positive and kk negative, is not very useful anthropologically, nor is the Lutheran with two codominant alleles Lu^a and Lu^b.

APPENDIX 2

THE Gm POLYMORPHISM

THERE are a number of variants of the serum gamma globulin such that in addition to their antibody properties they have antigenic determinants. At first it seemed that this antigenic specificity could be determined in terms of two alleles at a single gene Gm^a and Gm^b which could give rise to three phenotypes Gm(a+b−), Gm(a+b+), and Gm(a−b+). Many other Gm factors are now known and they can be interpreted in terms of very closely linked genes. It is found that different racial groups exhibit different haplotypes (haploid genotypes), or allelic complexes inherited in coupling. For example in Caucasians Gm(a) and Gm(b) are allelic and there are almost no haplotypes giving rise to both Gm(a) and Gm(b), whereas among Africans virtually all are both Gm(a) and Gm(b) and therefore must be homozygous for a Gm(ab) haplotype.

It seems likely that crossing over is an important mechanism for the production of new Gm haplotypes and a few families have been found in which crossing over is clearly demonstrated.

In 1965 the W.H.O. proposed a numerical system of nomenclature and numbers were assigned to the previously recognized antigenic determinants broadly as below.

Original	*Current*	*Original*	*Current*
Gm a	1	Gm b^3	13 = 10
Gm x	2	Gm b^4	14
Gm $b^w = b^2$	3 = 4	Gm s	15
Gm f	4 = 3	Gm t	16
Gm $b = b^1$	5 = 12	Gm z	17
Gm c	6	Gm Rouen 2	18
Gm r	7	Gm Rouen 3	19
Gm e	8	Gm 20	20
Gm p	9	Gm g	21
Gm b^α	10 = 13	Gm y	22
Gm b^β	11	Gm n	23
Gm b^γ	12 = 5	Gm c^5	24

Although Gm^a and Gm^b were shown not to be homoalleles, i.e., they could in fact be inherited together and therefore are not absolute alternatives, various other possibilities for homoallelism have been investigated, but none have been definitely confirmed.

Steinberg (1969) and Steinberg and Kirk (1970) present the following data on Gm alleles commonly present in each of several races.

Caucasoid	$Gm^{1,2,17,21}$	$Gm^{1,17,21}$	$Gm^{3,5,13,14}$
Negroid	$Gm^{1,5,14,17}$	$Gm^{1,5,13,14,17}$ $Gm^{1,5,6,14,17}$	$Gm^{1,5,16,17}$
Mongoloid	$Gm^{1,2,17,21}$	$Gm^{1,17,21}$ $Gm^{1,3,5,13,14}$	$Gm^{1,13,17}$
Ainu	$Gm^{1,3,17}$	$Gm^{1,17,21}$ $Gm^{1,2,17,21}$	$Gm^{2,17,21}$
Bushmen	$Gm^{1,5,17}$	$Gm^{1,17,21}$ $Gm^{1,5,13,14,17}$	$Gm^{1,13,17}$
Pygmy	$Gm^{1,5,13,14,17}$	$Gm^{1,5,6,17}$	
Micronesian	$Gm^{1,3,5,13,14}$	$Gm^{1,17,21}$	
Melanesian			
New Guinea	$Gm^{1,2,17,21}$	$Gm^{1,17,21}$ $Gm^{1,5,13,14,17}$	$Gm^{1,3,5,13,14}$
Bougainville	$Gm^{1,2,17,21}$	$Gm^{1,17,21}$	$Gm^{1,3,5,13,14}$
Australian			
Northern Region	$Gm^{1,2,21}$	$Gm^{1,21}$	$Gm^{1,5,13,14}$
Central	$Gm^{1,2,21}$	$Gm^{1,21}$	

It is clear from this that the Gm alleles are potentially very useful in characterizing racial groups and in estimating degrees of racial admixture.

In particular the Gm antigens have been very useful in studying White admixture in U.S. Negroes, the amount of Bushmen admixture in several Negro tribes in southern Africa and the relationships and amalgamations of Australian Aborigines and New Guineans. From the table of Gm allele frequencies one can see, for instance, that Caucasoids have alleles Gm^1, $Gm^{1,2}$, and Gm^5 (in association with other determinants but not with each other), whereas $Gm^{1,5}$ characterizes Negroids in which race they never occur separately and never with Gm^2.

Reed (1969a) used this fact to calculate White admixture in U.S. Negroes on the assumption that the frequencies of these three Caucasian alleles should be proportional to the corresponding frequencies of the appropriate Caucasian 'parent' population (*see* Chapter 5, s. 5.*2*).

APPENDIX 3

SOME TESTS OF INTELLIGENCE, ATTAINMENT, AND ABILITY

1. The Stanford-Binet Scales

This test endeavours to measure intelligence where intelligence is regarded as general mental ability. It makes use of age standards, in that children of a given age are assessed against mean performance scores at different ages and it is particularly useful for testing children between about 2 and 16 years. The latest, 1960, revision of the Stanford-Binet test has improved its usefulness for intelligence testing of older children and adults. The early versions of the test, before the 1960 revision, were heavily biased in favour of those children with good understanding and use of language and weighted against those from poor cultural backgrounds or those with linguistic handicaps. This is shown in the accompanying table. The 1960 revision rectified this to some extent, although the test overall remains largely verbal in content. It is still rather a poor test for adults and is still imprecise in that it is endeavouring to measure intelligence overall and makes little attempt to measure creativity or imagination.

Alfred Binet, after whom the scale is named, was a Frenchman and his first scale dates back to 1905. After some revisions there came in 1916 the Stanford revision of the Binet-Simon Intelligence Scale, to give it its full name. The revision was undertaken at Stanford University by L. M. Terman and in 1937 and 1960 came further revisions by Terman and Merril. These latter two revisions are often referred to as Terman-Merril scales. In the 1937 version there were two forms of the test, the L and M form, which gave way to only one form in the last revision.

2. The Wechsler Scales

Since the Stanford-Binet scales are largely verbal in content there is an obvious need for other scales to examine deaf, illiterate, or non-English speaking children. The non-verbal tests developed with these purposes in mind used concrete objects such as formboards, cubes, mazes, geometric figures, jigsaw puzzles, etc. It was realized that a test which utilized both verbal and non-verbal scales was likely to be more reliable and to give more realistic results regarding overall intelligence than one using either of these singly, and the Wechsler scales were therefore developed with this in mind. They were furthermore designed particularly to provide a suitable test for adults in view

The Nature and Frequencies of Test Items used by the Stanford-Binet Scales at Different Ages (1937 Version)

	Ages				
	2–4½	5–9	10–14	Adult	Total
Practical ingenuity with formboards, blocks	6	3	1		10
Following simple instructions	3				3
Spatial: copying, drawing, reproducing designs	4	6	3	1	14
Identifying and naming objects or pictures	14				14
Vocabulary: defining concrete or abstract words; word fluency; differences between words	2	3	7	5	17
Word-relations: analogies, rhymes, similarities, sentence completion	1	4	3	6	14
Pictorial relations	3	1			4
Comprehension: seeing absurdities, interpreting stories or proverbs	2	4	4	4	14
Pictorial comprehension	1	2	3		6
Reasoning problems			2	3	5
Counting and number problems		3	2	3	8
Immediate memory for digits, words or sentences	4	5	5	4	18
Memory for objects or pictures	2				2
	42	31	30	26	129

of the Stanford-Binet Scale's deficiencies in this area. The Wechsler Scales differ from Stanford-Binet in being point scales not age scales.

There are up to eleven sub-tests comprising these scales for use in testing adults (WAIS), each of which cover the whole range of the different abilities tested. The whole test takes an hour to administer and so abbreviated versions are sometimes used. The I.Q. norms are usually five verbal and five non-verbal or performance tests from the eleven described below (Wechsler 1955).

1. *A General Information Test.* Twenty-five questions covering everyday rather than specialized knowledge, e.g., 'How many weeks are there in a year?'.

2. *A General Comprehension Test.* Ten questions involving commonsense or practical judgement, e.g., 'Why should people pay taxes?'. The subtest is one of general comprehension, but it also rests heavily on the ability to verbalize.

3. *A Test of Arithmetical Reasoning.* Ten problems to be done mentally in a short period of time. Designed to test mental alertness.

4. *A Similarities Test.* Questions are asked regarding twelve pairs of words, e.g., 'In what way are an orange and an apple similar?'. This test appears to sample general intelligence quite well.

5. *A Test of Immediate Memory Span.* The subject is required to repeat backwards and forwards a total of fourteen series of digits varying in length from two to nine which he has heard once. This test is probably of limited usefulness in assessing intelligence.

6. *A Vocabulary Test.* This comprises a total of forty words arranged in order of increasing difficulty. It is very useful for determining a person's general mental ability provided no developmental or environmental handicaps intervene; a proviso which, of course, can rarely be guaranteed.

7. *Picture Completion.* A total of fifteen pictures is given to the subject, in each of which a missing part has to be noted and named. The test is said to measure the individual's basic perceptual and conceptual abilities, although it is mainly useful at lower intelligence levels.

8. *Picture Arrangement.* Six series of from three to six pictures are given to the subject and he has to arrange each in an appropriate order to tell a story. The test measures a person's ability to comprehend and evaluate a total situation without the use of language.

9. *Object Assembly.* Three simple jigsaws are presented of familiar objects, for instance a disassembled hand, and the requirement is to reconstruct the whole. The test is said to measure the perception of parts and their reconstruction into a whole.

10. *Block Design.* A number of identical cubes are given to the subject, two of the sides being red, two white, and two diagonally divided between red and white. Using four, nine, or sixteen of these blocks, a number of designs presented to the subject have to be copied. This test measures visual analysis and synthesis.

11. *Digit-symbol Substitution.* A key shows nine simple symbols paired with the numbers 1 to 9. Below is a mixed list of numbers and the testee writes the corresponding symbols as quickly as possible. The test is measuring visual memory and speed and accuracy.

We have indicated, in briefly outlining each of these subtests, the functions of each, and the areas of intelligence they test. It is useful to summarize this information and also to indicate the cultural and other factors which might be influencing the performance of each one. It is in this area, after all, in which most of the objections to intelligence testing that are pertinent to this book have been raised.

The Wechsler Intelligence Scales for Children (WISC) are a downward extension of the previous scales, providing verbal performance and full-scale I.Q.s between the ages of 5 and 16 years. The main difference from the adult test is that *digit span* is usually omitted and a series of mazes used as a substitute. The test is well standardized and gives similar I.Q.s to Stanford-Binet over a more restricted range. The scale is weakest with younger, below average children, which is an immediate criticism for its use in interracial comparisons.

Subtest	*Functions*	*Influencing Factors*
Information	Long-range retention. Association and organization of experience	Cultural environment. Interests
Comprehension	Reasoning with abstractions.* Organization of knowledge. Concept formation	Cultural opportunities. Response to reality situations. Attention span
Arithmetic	Retention (of arithmetical processes)	Opportunity to acquire the fundamental arithmetical processes
Similarities	Analysis of relationships. Verbal concept formation	A minimum of cultural opportunities
Vocabulary	Language development. Concept formation	Cultural opportunities
Digit span	Immediate recall. Auditory imagery. Visual imagery at times	Attention span
Picture arrangement	Visual perception of relationships (visual insight). Synthesis of non-verbal material	A minimum of cultural opportunity. Visual acuity at times
Picture completion	Visual perception analysis. Visual imagery	Environmental experience. Visual acuity at times
Object assembly	Visual perception synthesis. Visual-motor integration	Rate of motor activity. Precision of motor activity
Block design	Perception of form. Visual perception analysis. Visual-motor integration	Rate of motor activity. Minimum of colour vision
Digit symbol	Immediate rote recall. Visual-motor integration. Visual imagery	Rate of motor activity

* 'Reasoning with abstractions' generally involves the processes of both analysis and synthesis, with the use of symbols—language and number. The testee must first analyse the relationships existing among the members, or parts, of the whole problem; then he must reorganize and interpret and, at times, create new wholes in order to reach the desired solution.

3. Goodenough Draw-a-Man Test

In this test the child is simply encouraged to draw the best picture of a man that he can. The presence or absence of up to 51 specific points is noted, varying from: Head present, legs present, at one end of the scale, to sleeves and trousers non-transparent, opposition of thumbs indicated, etc. at the other. Careful use of the scoring instructions is essential and the variability of even trained scorers, together with the variability of the child's performance on different occasions, means that the test is not altogether reliable.

4. Group Tests

The tests we have hitherto described are essentially individual tests; in other words, the concentration is upon the individual and the test requires trained personel for its administration and assessment. Other tests are designed to be used to test, for example, general intelligence levels in primary or secondary schools, or among entrants to the armed services, etc.

The California Test of Mental Maturity

This test provides scales for use throughout the primary and secondary education levels and for adults. At the primary level the tests are almost exclusively non-verbal and at higher levels increasingly difficult verbal and numerical materials are included. The numerical and verbal subtests at post-primary level are as follows:—

Inferences: Each item contains two premises, the logical conclusion following from them must be chosen from several possibilities provided.

Number series: There are two types of question. Firstly there are series of numbers that increase or decrease according to some rule; the testee has to select from each the number that does not obey the rule. Secondly the testee has to select from several alternatives the numbers that have been omitted from each series.

Numerical quantity: The testee has to indicate from several alternatives how many coins of each value are required to make up a specific sum of money. Secondly there are a series of increasingly difficult arithmetical problems.

Verbal concepts: A test of word similarities. A selection has to be made, from four possibilities, of a word that is synonymous with the key word.

The Lorge-Thorndike Intelligence Tests

The whole series of tests includes five levels from kindergarten level to grade 12. The first two levels use only non-verbal test items, the others use both verbal and non-verbal test items. The verbal series of tests includes familiar types such as word knowledge, sentence completion, verbal classification, verbal analogies, and arithmetical reasoning. The scales as a whole are meant to test abstract intelligence with the proviso that for young children, poor readers, and poorly educated individuals, non-verbal tests must be substituted for the verbal ones.

Tests specifically for Adults

Among the general tests available are: the Otis Self-Administering Test of Mental Ability: Higher Examination and the Army General Classification Test (AGCT) based upon the American Army Alpha Test which was used during the First World War. It is only recent developments of these which provide separate scores for both numerical and verbal test items. It is very hazardous, therefore, to make comparisons between different groups or different races on the basis of early results from tests such as these.

Special Aptitude Tests

An aptitude can be defined as a combination of characteristics indicative of an individual's capacity to acquire (with training) some specific knowledge, skill, or set of organized responses such as the ability to speak a language, to become a musician, or to do mechanical work (Freeman, 1962). Aptitude tests are different from general intelligence tests, from tests of skill or of skill achieved after training, and from educational achievement tests.

Standard tests in regular use include tests of visual acuity (monocular and biocular acuity, muscle balance, etc.), of colour vision by means of colour charts or cards or instruments, and auditory tests by means of audiometers.

Among the motor and manual tests used are:—

Strength of grip: Used principally for testing handedness and fatigue.

Reaction time: Testing the time interval between the onset of stimulus and the beginning of a person's overt intentional response.

Manual dexterity: A large number of tests involving gross movements of hand and arm or delicate finger dexterity and co-ordination are in common use. We mentioned in s. 6.*4.d*.ii the rotary pursuit apparatus in which the testee has to maintain contact with a metal disk rotating on a gramophone turntable.

Another category of tests are those for mechanical aptitude. This is not a single unitary function but involves a combination of sensory and motor capacities. Various tests include those for ability to match cut-outs on a number of boards (e.g., The Minnesota Spatial Relations Test) and tests of mechanical comprehension in which problems in picture form as opposed to a practical form are presented.

Finally there are a number of aptitude tests specifically designed for the fine arts and the professions. It is commonly believed that musical ability has a marked genetic component and a number of tests have been developed for use in this field. Among the most commonly used are the *Seashore Measures of Musical Talent* which measure the following six aspects of hearing given by means of records:—

Pitch discrimination;
Intensity or loudness discrimination;
Time discrimination;
Discrimination of timbre;
Judgement of rhythm;
Tonal memory.

In each case paired comparisons are asked for, such that the differences are at first readily discernible but become for each test very fine ones. A profile, not an overall score, representing the level of ability at each of the six auditory aspects is prepared.

There are a number of objections to the tests. For instance it is argued that each of the six tests should be separately validated against different kinds of musical activity. Furthermore the tests do not represent an actual musical experience and other tests have been developed to make good these deficiencies and to provide tests more appropriate for intending musicians or music teachers.

Bender-Gestalt Test

This is a clinical test of human perception in which a number of simple designs are presented individually on cards. The subject has to copy the designs with the pictures in front of him and the accuracy with which he does so can be assessed. It has been found that the tests reliably distinguish samples of normal psychotic and neurotic children. Gestalt psychology was originally a German theory of perception but is now more widely applied to other areas of psychology. It contends that mental processes and behaviour cannot be analysed into more elementary component units.

APPENDIX 4

CRANIAL AND DENTAL ANATOMY USEFUL IN ANTHROPOLOGY AND RACIAL STUDIES

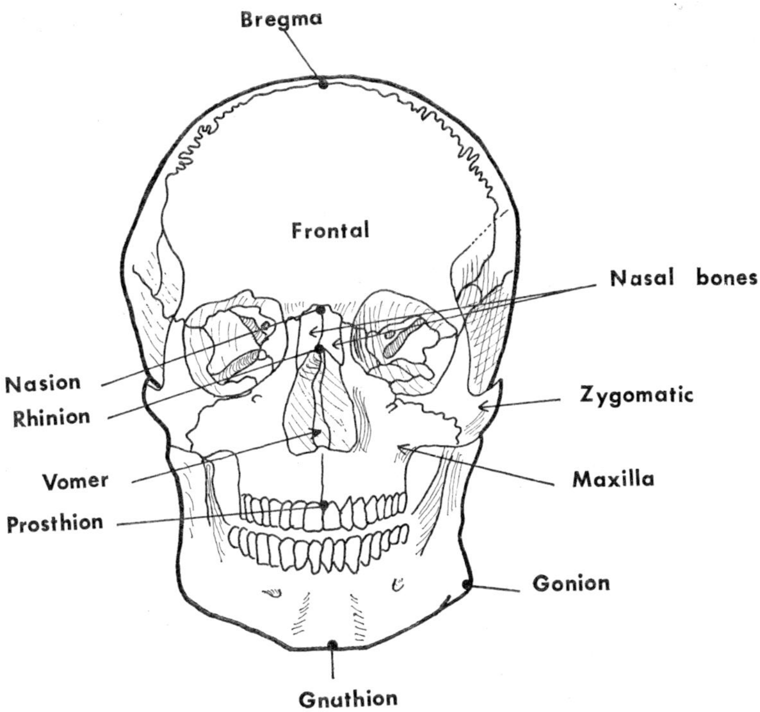

Fig. A4, 1.—Frontal view of the skull or *Norma frontalis* showing some of the landmarks or points and some of the bones visible from this view.

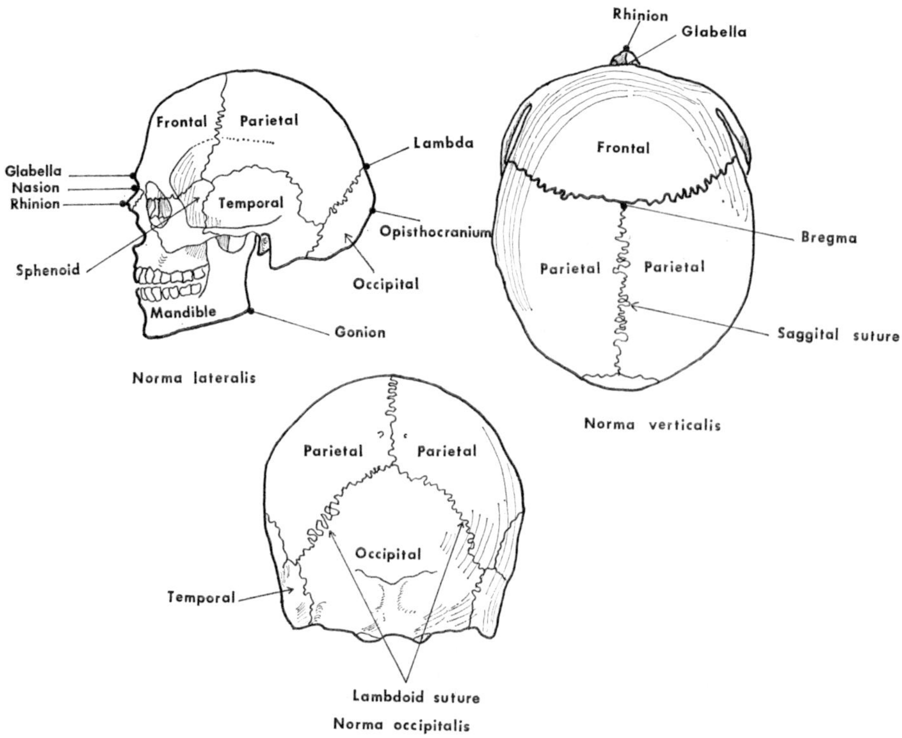

Fig. A4, 2.—*Norma lateralis, Norma occipitalis,* and *Norma verticalis* showing landmarks and bones visible in each position.

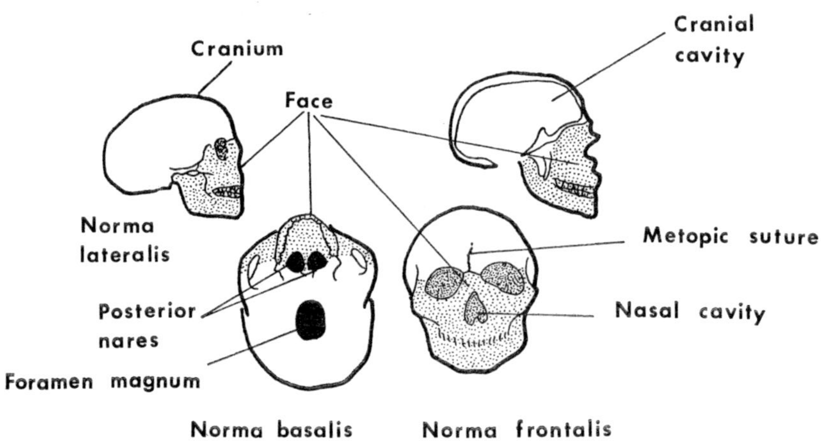

Fig. A4, 3.—Relationship of the face and cranium to the rest of the skull.

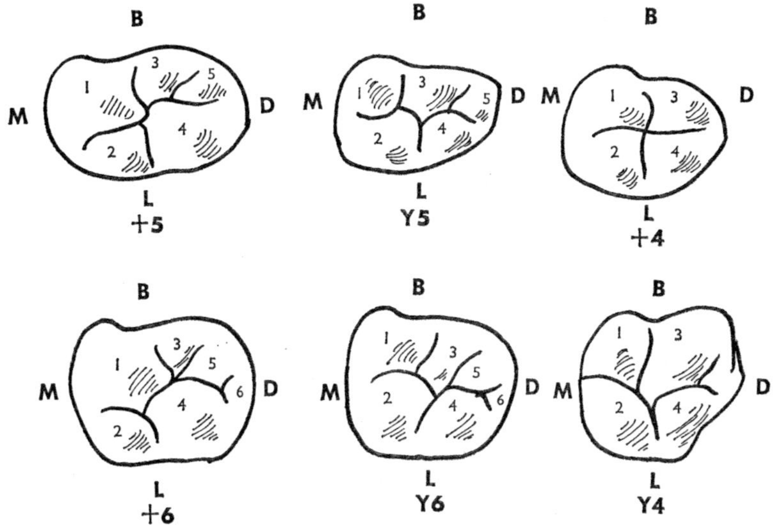

Fig. A4, 4.—Types of surface pattern in lower molars. (*From* Comas, 1960.) **B**, Buccal; **D**, Distal or posterior; **L**, Lingual; **M**, Mesial or anterior; **1**, Protoconid; **2**, Metaconid; **3**, Hypoconid; **4**, Entoconid; **5**, Hypoconulid.

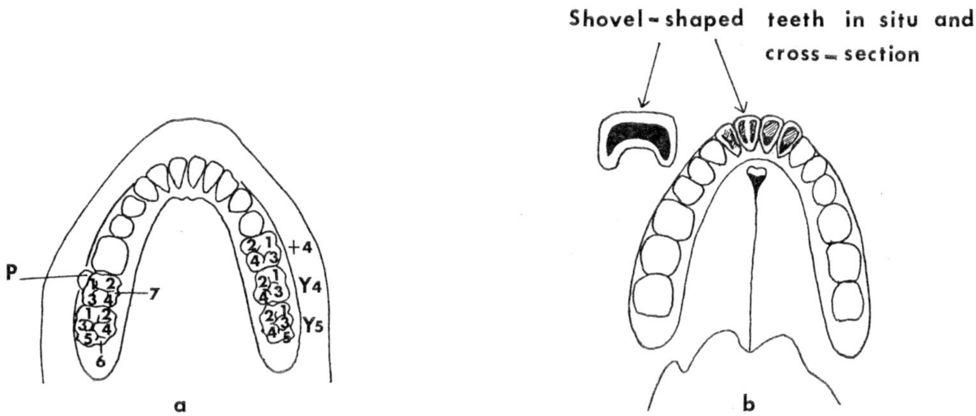

Fig. A4, 5.—**a**, Number and pattern of cusps of lower molars. The plus (+) pattern has no common boundary between cusps 2 and 3, whereas the Y pattern has cusps 2 and 3 adjacent. **P**, The protostylid lies adjacent to cusp 1. **b**, Shovel-shaped teeth of slightly different types together with a cross-section (or worn occlusal surface) of a shovel-shaped incisor. (*From* Lasker, 1950.)

GLOSSARY

The words in this glossary are only those which are not adequately defined within the body of the text. They are words which have a specific scientific usage and which are not likely to be universally understood by people referring to this book. And yet they are believed to be words an understanding of which is important to a full understanding of the subject-matter in the places where they appear. Italicized words appearing in a definition are themselves defined elsewhere.

It is intended to be a working glossary in that fairly full descriptions and examples are given. For more complete definitions and comprehensive coverages consult Rieger, Michaelis, and Green (1968) *A Glossary of Genetics and Cytogenetics* and King (1968) *Dictionary of Genetics.*

Adaptation

The process by which populations or organisms undergo changes that make them better able to survive and to reproduce in a given environment. In evolutionary terms it is a process which is the sum of all the various genetic changes which take place in the organisms concerned, with the exception of those which are eliminated through natural selection. The word may also be used to indicate physiological or sensory changes without any genetic component.

In addition the word 'adaptation' is used for those particular characteristics of an individual which improve its chances of survival and of leaving offspring, i.e., adaptations are the characteristics established in a population by natural selection.

Agenesis

The absence of an organ. The word is frequently used to designate such absence resulting from failure of appearance of the primordium of an organ during embryonic development.

Allele

An alternative form of a *gene* at a given position or *locus* on a *chromosome*. These alternative alleles may be designated *A* and *a*, particularly if one is *dominant* in expression to the other. In many cases where there is no manifestation of dominance the alleles may be designated as for instance I^A and I^B for the A and B blood groups and *i* for the O blood group, which is allelic with A and B but *recessive* to them.

Antibody

A specific and complex protein which is produced as a defence mechanism by the bodies of higher animals in response to the invasion by a particular *antigen*.

Antigen

A substance that is capable of stimulating the body to produce specific or neutralizing *antibodies*. It may be a foreign substance such as pollen, or bacteria, but any blood or tissue from an individual who is genetically different from the recipient will stimulate antigen production. In the case of the Rhesus incompatibility, for instance, it is the blood from the genetically distinct baby which stimulates antibody production in the mother.

Assortative Mating

In the case of *panmixia* any individual of one sex has an equal probability of mating with any individual of the opposite sex. Mating may depart from panmixia on the basis of genetic relationship. Inbreeding, for instance, would be positive genetic assortative mating whereas forced outbreeding would be negative genetic assortative mating. Alternatively mating may involve pairing on the basis of the similarities or dissimilarities of their *phenotypes*. Thus the mating of like with like more often than would be expected on the basis of chance alone is positive phenotypic assortative mating.

Autosomal

Chromosomes are autosomal or *genes* on such chromosomes are autosomal or inherited in an autosomal manner if they are not those concerned with the mechanism of sex determination. Man has 23 pairs of chromosomes of which one pair, the so-called X and Y chromosomes, are associated with sex determination, males and females having XY and XX chromosomes respectively. Genes on the X chromosome are said to be *sex-linked*. There are no known genes on the Y chromosome in man.

Balanced Polymorphism

A polymorphism which is maintained in a population usually as a result of the fact that the *heterozygotes* for the responsible *alleles* have a higher biological *fitness* than either of the *homozygotes*. The classic example in man is that of sickle-cell haemoglobin described in Chapter 2, s. 2.2.*b*.

Binomial Distribution

The probabilities of 0, 1, 2, . . . n successes in a series of n trials for an event of constant probability P, are the respective terms of the binomial expansion.

$$(q + p)^n = q^n + nq^{n-1}p + \ldots nqp^{n-1} + p^n$$

The probability distribution of the number of successes thus determined is called the binomial distribution. It finds important applications in genetics since with a probability of 0·5 for the segregation of each *allele* in a *heterozygote* to the *gametes* it can be used to analyse the probable outcome of various experimental or actual situations. As n increases the binomial approaches the *normal distribution*.

Brachycephalic, *see* Cephalic Index.

Canalization

This is the phenomenon of genes interacting successfully with one another through the developmental pathways of an individual to produce a standard *phenotype* without marked variation from the norm in spite of genetic or environmental disturbances. It is in other words a developmental buffering system serving to reduce potential variation. It is one aspect of the phenomenon of *homeostasis.*

Cephalic Index

The quotient of the maximum transverse diameter of the head (multiplied by 100) and its anteroposterior diameter. According to this index individuals are classified as *dolicocephalic* or longheaded (<76), *mesocephalic* (76 to 80·9), and *brachycephalic* or roundheaded ($>80{\cdot}9$).

Chi-square

A Chi-square test is essentially a test of whether the observed frequencies in a distribution differ significantly from the frequencies which might be expected according to some assumed hypothesis. Corresponding to each expected frequency on the basis of hypothesis will be an observed frequency, then Chi-square or $\chi^2 = \sum\frac{(O - E)^2}{E}$, where O and E are observed and expected frequencies respectively. Progressively higher values of χ^2 correspond to successively lower probabilities that the assumed hypothesis is correct.

Chromosome

The hereditary material, the *DNA*, is organized into distinct elongated structures which are found within the *nucleus* of the cell of higher organisms. In the body cells, or somatic cells, the chromosomes are found in pairs, each of these pairs comprising homologous chromosomes deriving from the maternal and paternal contributions. This is the *diploid* condition. In the mature sex cells there are only single representatives of each pair. This is the *haploid* condition. Man possesses 23 pairs of chromosomes, one pair of which the X and Y are concerned with sex determination the rest being *autosomal.* During the formation of the sex cells, i.e., in the passage from the diploid to the haploid state, the chromosome pairs segregate at random to the germ cells. Prior to segregation chromosomes may undergo *crossing-over.* By these two processes the genes are continually being sorted and recombined. The characters they control are therefore continually reappearing in new combinations. Genes which are close to one another on the chromosome are said to show *linkage.*

Cline

When a gradient exists within a continuous population in the frequency of different *genotypes* or *phenotypes* from one locality to another there exists a cline. Within any one population more than one cline may exist and these may vary in different directions. The amount of change with distance may be rapid or slow, when the cline is said to be steep or shallow. Examples in man are the progressive change in B blood group frequency from east to west across Europe and the general relationship of skin colour with latitude.

Coadaptation

The genes within any one organism have to co-operate and interact successfully to produce an individual. It is found in general that within any *Mendelian gene pool* most of the genes can interact with one another successfully. This phenomenon is coadaptation. It may be expressed at the level of the *gene*, the *chromosome*, or the whole *genome*. Genes introduced into such a coadapted gene pool from outside may give rise to biological upsets of greater or lesser severity.

CODOMINANCE, *see* DOMINANCE.

COEFFICIENT OF CORRELATION

Technically this is the ratio of the covariance of two variates to the geometrical mean of their variances. The correlation coefficient, denoted by r, cannot exceed $+1$ or be less than -1 in value. When the value is $+1$ there is a perfect functional relationship between two variables X and Y such that increasing X is associated with an increasing Y. The reverse situation holds when $r = -1$. If $r = 0$ there is no correlation at all between the variables.

COEFFICIENT OF VARIATION, *see* VARIANCE.

COMPLEMENTATION

This refers to the complementary activity or co-operation of homologous sets of genetic material involving the interaction of mutant genes or their products in double mutants. Those combinations which allow the development of a character or function which cannot be achieved by the individual action of single mutants are said to complement each other.

CONGENITAL

Any character or trait recognizable at birth. Congenital malformations are therefore those present at birth and they may be caused by genetic or environmental factors or both.

CONSANGUINITY

Relationship by descent from a common ancestor. Consanguineous marriages involve two such people. The relationship may be a very close one, such as brother/sister or first cousin. Alternatively it may be rather distant, provided the relationship is documented clearly.

CROSSING-OVER

The means by which genetic material is rearranged between homologous pairs of *chromosomes* is crossing-over. It is the reciprocal exchange of segments of genetic material at corresponding positions along pairs of homologous chromosomes accomplished in functional terms by symmetrical breakage and crosswise rejoining. The molecular processes involved in this process of recombination are not completely understood.

DERMATOGLYPHICS

This name refers both to the patterns of the ridged skin of palms, fingers, soles, and toes in primates and the study and classification of such patterns into different types.

DEUTERANOPIA

Defective colour vision of the dichromatic type, characterized by retention of the sensory mechanism for two hues only (blue and yellow) of the normal 4-primary quota and lacking that for red and green.

DIALLELE CROSS

The crossing of all inbred lines derived from the same base population in all possible combinations is called a 'diallele cross'. A diallele cross of *a* lines results in a^2 combinations, *a* of which are inbreds and $a(a - l)$ are crosses. When reciprocal combinations can be neglected $a(a - l)/2$ cross combinations are possible. The concept has been extended into human genetics by using races to represent different inbred lines. It is thus possible to analyse the situation in Hawaii, for instance, by means of the statistical techniques developed to analyse the diallele cross.

DIPLOID, *see* CHROMOSOME and MEIOSIS.

DIRECTIONAL SELECTION

In a biological sense selection is any non-random process which will lead to individuals of different *genotypes* being represented in different proportions in a later generation from that which they are in before selection is operative. This selection is said to be 'directional' with respect to one particular character if the mean expression of that character in a population is changed in a systematic way by the selection. In the evolution of the genus *Homo* for instance, the cranial capacity shows, at least until the time of Neanderthal Man, a continual increase. This can be ascribed to the effects of directional selection. It can be contrasted with stabilizing selection in which phenotypic extremes are in general selected against and with disruptive selection in which both phenotypic extremes may be selected for, and the mean value may be selected against to some extent.

DISCRIMINANT ANALYSIS

A statistical analysis designed to investigate and measure the degree of difference between two variables.

DISRUPTIVE SELECTION, *see* DIRECTIONAL SELECTION.

DIZYGOTIC, *see* MONOZYGOTIC.

DNA

Deoxyribonucleic acid is a polymer of deoxyribonucleotides and is the primary genetic material of all organisms cells bar a few aberrant viruses. The major components in all organisms are four nitrogenous bases adenine, guanine, thymine, and cytosine. DNA usually consists of two polynucleotide chains joined by hydrogen bonds between pairs of bases, with the two chains being coiled in an interlocking double helix. The structure of DNA is such that the molar amount of adenine (A) is equal to that for thymine (T), and that of guanine (G) is equal to that for cytosine (C). The overall composition of a particular DNA may therefore be described by the base ratio $(A + T)/(G + C)$.

The *chromosomes* of higher organisms contain a number of other compounds beside DNA but the details regarding the relationship between DNA and chromosome structure are not clearly understood.

DOLICOCEPHALIC, *see* CEPHALIC INDEX.

DOMINANCE

This is the relationship between two *alleles* such that the *heterozygote* containing both of these alleles resembles one of the *homozygous* parents and not the other (*see Fig.* 4.1). The homozygote for which the resemblance exists is said to carry the dominant allele. If the resemblance between the heterozygote and the homozygote is not complete then dominance is said to be 'partial'. If both alleles are expressed to an equal extent in the heterozygote they are said to be 'codominant'. If the *phenotype* of the heterozygote, measured on some arbitrary scale, exceeds that of either homozygote,the relationship between the alleles is said to be 'overdominant'.

ELECTROPHORESIS

This is the process of separation of molecules of different sizes and electric charges by means of sieving them through some dense medium such as starch, polyacrylamide, or acetate when an electric potential is applied across this medium. If a mixture of the different molecules, such as human blood-serum, is applied at one end of a strip of this material then the molecules travel at different speeds and take up different final positions dependent upon (1) the current applied, (2) the density of the gel, (3) the molecule itself. The technique is used to separate molecular species and, more importantly in the present context, to distinguish genetic variants of the same molecular species.

ENDOGAMY

Mating within a band, social group, or other defined breeding population is said to be endogamous. In a situation of caste endogamy, for instance, mating is permitted only with members of the same caste.

ENZYME

A protein molecule that catalyses a specific chemical reaction.

EPIGENETIC

This term refers in the broad sense to the interaction of genetic factors during the developmental process. It is used in particular in this book to refer to variation of a morphological nature which is caused by the operation of some threshold effect or switch mechanism during development, as described in detail in Chapter 2, section 2.1.

EUKARYOTES

Individuals whose cells have *nuclei* with nuclear envelopes, *chromosomes*, and *meiotic* and *mitotic* nuclear divisions during which the chromosomes are visible.

EXOGAMOUS

When mating occurs outside the bounds of the tribe, group, or normal mating population it is said to be exogamous.

FITNESS

The survival value and the reproductive capability, defined in terms of the proportionate contribution of offspring to the next generation, of a given *genotype*. If differences in fitness are in any way associated with the presence or absence of a particular *gene* in the individual's *genotype*, then selection is said to operate on that gene to cause a change of gene frequency.

GAMETE

A mature reproductive cell capable of fusing with a cell of similar origin but opposite sex to give a *zygote* or new whole organism of the next generation. The male gametes in man are sperm and the female gametes are eggs. (*See also heterogametic.*)

GAMMA GLOBULIN

The globulins in general are a group of proteins found in the blood-serum. In neutral or alkaline solutions the gamma globulins have the slowest *electrophoretic* mobility. Most *antibodies* in man are gamma globulins.

GENE

The gene is today defined operationally, in that it is a specific sequence of *nucleotides* which act as a functional unit and which determines thereby the sequence of amino-acids in a specific *polypeptide* or the sequence of nucleotides in a specific RNA molecule. The older ideas of the gene being the ultimate unit of activity are now superseded in that there are now recognized units of mutation, units of recombination, units of transcription, etc., which do not necessarily coincide and which are all contained within the gene conceived of as an operational unit.

GENOME, *see* GENOTYPE.

GENOTYPE

In the general sense this is the whole genetic constitution of the organism. It is essentially synonymous with genome. Alternatively it can be used to refer to one or a few specific genetic *loci*. As for instance, when indicating that a person who is blood group A may have a genotype which is either AA or AO.

GLACIATION

The last ice age, which occupied the whole of the Pleistocene era, was not a time of uninterrupted advance or occupation of the ice sheets. Up to four periods of advance can be recognized in the northern hemisphere, although the details differ from place to place. It is these periods which are termed 'glaciations'. In Europe, as shown in *Fig.* 3.8, the four glaciations are, in order, Gunz, Mindell, Riss, and Würm. The periods of maximum extent of the ice sheet are termed 'stadials' and there were often two or more stadials within any one glaciation, separated by interstadials. The periods between the glaciations are 'interglacials' and may well have been warmer and longer in extent than the period of time since the last retreat of the glaciers in the northern hemisphere.

GLUCOSE-6-PHOSPHATE DEHYDROGENASE, G6PD

This is an *enzyme* intermediate in carbohydrate metabolism common to a wide variety of animals. There is a genetically determined abnormality of this enzyme in man which results in a haemolytic anaemia on eating fava beans or being exposed to the drug primaquine. The distribution of this *gene* varies widely in different parts of the world and it is thought to play a part in protecting the carrier against malaria, although exactly how is not very well understood.

Group Specific Component

A protein found in human serum which migrates during *electrophoresis* with the rapidly migrating globulins but which is independent of them, the *haptoglobins*, and other serum groups.

Haemoglobinopathies

Any alteration in the genetically determined molecular structure of haemoglobin which gives rise to overt clinical symptoms is termed a 'haemoglobinapathy'. Most haemoglobinopathies are due to amino-acid substitutions alone one or other of the chains of which the molecule is constructed. Others, though, may be the result of differential rates of synthesis or faulty molecular structure.

Haploid, *see* Chromosome and Meiosis.

Haptoglobin

A serum α_2-globulin glycoprotein that serves to bind free haemoglobin. A number of polymorphic forms are known in man.

Hardy-Weinberg Law

In 1908 the English mathematician G. H. Hardy and the German physician W. Weinberg demonstrated that both *gene* frequencies and *genotype* frequencies will remain constant from generation to generation in a breeding population in the absence of those forces that change gene frequencies: selection, mutation, genetic drift, and abnormal processes of *meiosis*. With a single pair of alleles, *A* and *a*, genotype frequencies with be p^2 (*AA*): 2pq (*Aa*): q^2 (*aa*). If the genotypes in a population at any one *locus* are in these proportions then they are said to be in Hardy–Weinberg equilibrium.

Heritability

The proportion of the total *phenotypic variance* for a continuously varying character that is genetically determined is defined as the heritability of that character. It is often expressed as a percentage and decreases with an increasing environmental component of variance for the character under observation. It estimates the degree of resemblance between offspring and parent.

Heterogametic

In man, in common with many higher mammals, the sex-determining mechanism is based on possession by the two sexes of different *chromosomes*, the so-called 'X' or 'Y'. One of the sexes has two X chromosomes, the other has an X and Y, with the main function of the Y in man being not to carry normal *genes* but to stimulate the development of maleness. The heterogametic sex is that sex with both X and a Y chromosome since it can produce two distinct types of *gametes* depending on which sex chromosome they carry.

HETEROKARYON

Among some of the fungi, the ascomycetes or basidiomycetes, the cells, spores, or most commonly the mycelium, may possess genetically different *nuclei*, either two or more. This is in contrast to a homokaryon, where the nuclei are identical.

HETEROZYGOTE

An individual who possesses two different *alleles* at a particular genetic *locus* is said to be heterozygous at that locus and to be a heterozygote.

HETEROZYGOSITY

This is the average proportion of *genes* in any one individual of a species which are heterozygous, i.e., have two different *alleles* present at the one genetic *locus*. It is a function of the proportion of loci in that species which are polymorphic and the relative frequencies of the alleles at the varying genes. When an individual possesses two identical alleles at a genetic locus it is said to be *homozygous*. The proportion of such loci increases with inbreeding.

HIRSUITISM

Hairiness; particularly with respect to the body. Caucasoids generally show more hirsuitism than either Negroids or Mongoloids.

HOMEOSTASIS

In general terms this is the tendency of some biological systems to maintain a dynamic equilibrium and in the case of disturbance to restore the equilibrium by their own regulatory mechanisms. There are three important areas this concept extends into: (1) Developmental or *epigenetic* homeostasis in which a normal *phenotype* is produced despite developmental disturbances; (2) Physiological homeostasis in which phenotypic reactions buffer the organism against varying external conditions; and (3) Genetic homeostasis, which is the tendency of a population to maintain a genetic composition leading to an optimum balance in the environment in which the population lives. It encompasses the idea of *coadaptation* (*see* Chapter 4, s. 4.4).

HOMOKARYON, *see* HETEROKARYON.

HOMOZYGOSITY, *see* HETEROZYGOSITY.

INDEX

A mathematically expressed relationship between two anatomical measurements. (*See*, for example CEPHALIC INDEX.)

INTERGLACIAL, *see* GLACIATION.

INTERSTADIAL, *see* GLACIATION.

INTROGRESSION

The incorporation of *genes* of one species into the gene pool of another species by hybridization ('introgressive hybridization') and backcrossing.

INVERSION

A structural alteration of a *chromosome*, such that part of a chromosome and the *genes* it contains are reversed relative to the rest of the chromosome material. The usually accepted mode of origin is two breaks in a chromosome, 180° rotation of the chromosome segment between the breaks and then reunion at the site of breakage.

ISOVARIANT

A line on a map connecting populations with equal frequencies of some discontinuously or *epigenetically* varying character. Illustrated, for example, in *Fig.* 2.7 in which the world isovariant lines for *metopism* are shown.

LETHAL EQUIVALENT

A *gene* may be 100 per cent effective in killing the *zygote* at some stage of its development if it is present in the *homozygous* condition. Such a *gene* is a recessive lethal and is one lethal equivalent. On the other hand a homozygous gene may have only a 50 per cent probability of killing the zygote in which it is found. Two such genes in a population would constitute a lethal equivalent. Three homozygous genes with a killing probability of 33 per cent would also constitute a lethal equivalent and so on. One lethal equivalent may also be referred to as a 'lethon'.

LETHON, *see* LETHAL EQUIVALENT.

LEUKAEMIA

A disease of the blood in which there is a persistent increase in leucocytes, the white blood-cells, together with changes in the spleen, bone-marrow, or lymphoid system. It is a form of cancer.

LINKAGE

When *genes* are on the same *chromosome* they are said to be 'linked' and genes on homologous chromosomes are said to belong to the same 'linkage group'. Linkage is rarely complete since homologous chromosomes will exchange corresponding parts by *crossing-over* during the process of *meiosis*.

LOCUS

A genetic site on a *chromosome*. It is loosely equivalent to the word *gene*.

MACROGLOBULINAEMIA

Macroglobulin is a globulin of the blood with very high molecular weight. It is found in the blood in individuals suffering from one of a number of diseases and when showing marked increase in concentration it gives rise to the condition known as 'macroglobulinaemia'.

MACROMELANOPHORE

A large cell containing the pigment melanin. Found in the skin of fish, amphibians, and reptiles.

MALAR REGION

The malar bones are the cheek bones and therefore the malar region is the whole cheek area.

MALOCCLUSION, *see* OCCLUSION.

MANDIBLE

The lower jaw bone. A horeshoe-shaped body and two ascending rami which articulate with the skull. (*See* Appendix 4.)

MAXILLA

The irregularly shaped bone that, with others, forms the upper jaw. It also assists in the formation of the orbit, the nasal cavity, and the palate and holds the upper teeth. (*See* Appendix 4.)

MEIOSIS

Although the details of this process are incredibly varied in different higher organisms, the process of meiosis is essentially that which reduces the *zygotic* (*diploid*) *chromosome* number to the *gametic* (*haploid*) number. During meiosis homologous chromosomes pair, replicate once, and undergo assortment while the cell divides twice, so that each of the four meiotic products resulting from one meiosis receives one representative of each chromosome pair. Meiosis is the basis of sexuality of all higher plants and animals and serves the following main functions:

1. It provides for the random assortment of paired chromosomes (departures from randomness are termed 'meiotic drive'), the halving of the somatic chromosome number, and it marks the transition from the diplophase to the haplophase of the life cycle.
2. It determines the accurate segregation of *allelic* differences as well as the random *recombination* of unlinked *genes* and the non-random recombination of *linked* genes.

MELANIN

The principle dark pigment responsible for racial differences in skin colour is melanin. There are two common forms of the pigment found in mammals but each is constructed from the amino-acid tyrosine through an intermediary, dopa, by the action of the *enzyme* tyrosinase. Melanin is present in varying amounts in the skin, hair, and nerve tissues. Doses of ultra-violet light will lead to a marked increase in numbers of melanotic melanocytes at the junction of the dermis and epidermis in man and it is hypothesized that these have a protective effect. The detailed steps of melanin synthesis

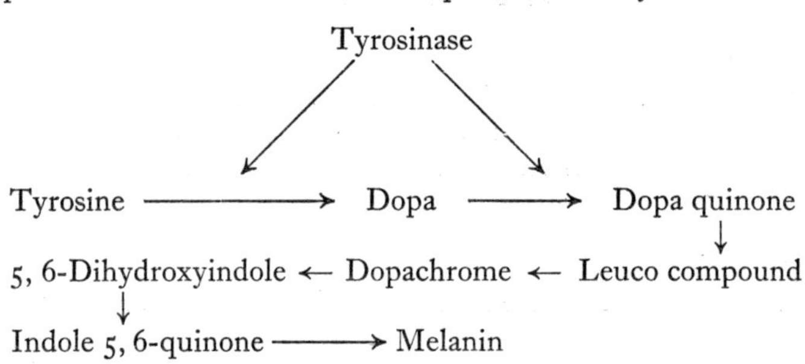

are illustrated here, indicating the number of points at which interruption of the synthetic cycle might occur as a result of mutations. Complete suppression of melanin production gives rise to albinism.

MENARCHE

The commencement of regular menstruation in women during puberty.

MENDELIAN POPULATION

An interbreeding group of individuals sharing a common *gene* pool is termed a 'Mendelian population'. Although it is the basic unit of study in population genetics there is usually no clear-cut boundary between two adjacent Mendelian populations owing to *gene* flow between them.

MESOCEPHALIC, *see* CEPHALIC INDEX.

MESOLITHIC, *see* NEOLITHIC.

MESTIZOS

A term, largely restricted to Latin America, for hybrids between Europeans and native Amerindians. It is often used with the adjective light or dark, indicative of greater or lesser European ancestry respectively. In most parts of Latin America this darkening of the skin would also probably indicate some Negro admixture since, as indicated in Chapter 7 particularly, various degrees of trihybrid mixtures are extremely common.

METOPION

The most prominent point in the midsagittal plane between the eyebrows. A standard anthropometric landmark. (*See* Appendix 4.)

METOPISM

The metopic suture is a suture between the two frontal bones of the skull which usually closes during infancy. If, however, the suture does not close the skull is said to exhibit 'metopism'. Different races have different characteristic frequencies of metopism.

MIMETIC

Phenotypes produced by different *genes* which are very similar in appearance to another, model phenotype, are said to be 'mimetic'. Mimicry may occur within genera or between closely related genera, usually in a polymorphic situation, when a mimetic species will model itself on one or more models which are distasteful or dangerous to predators. Otherwise mimicry may be for camouflage when the model may be unrelated or inanimate.

MITOCHONDRIA

The principal energy source of all *eukaryotic* cells are semi-autonomous cell organelles, structurally independent of other cytoplasmic components called mitochondria. They contain the *enzymes* of many important metabolic cycles and the mitochondria of many organisms have been shown to contain DNA which has important genetic functions.

MONOGAMY, *see* POLYGAMY.

MOUSTERIAN

A stone tool culture consisting of side scrapers, disk cores, and points widely represented in Europe, North Africa, and Western Asia; frequently associated with Neanderthal remains.

MULATTO

A term which has a wide meaning as any racial hybrid, but it is particularly used for people who have both Caucasoid and Negroid ancestry. It receives its widest usage in Latin America.

MULTIFACTORIAL INHERITANCE

Continuously varying characters whose expression in populations usually follow a *normal distribution* and which are controlled by several *genes* of small, usually additive effects, are said to show 'multifactorial inheritance'.

MONOZYGOTIC

Monozygotic twins are those arising from a single fertilized ovum which splits and develops into two embryos at some relatively early developmental stage. This is in contrast to dizygotic twins which are produced by two separate ova fertilized by separate sperm. Mono- and dizygotic twins are identical and fraternal twins respectively. Monozygotic twins must have identical *genotypes* whereas dizygotic twins are only as similar genetically as ordinary sibling pairs, although their environment, from conception on, may be more similar. The identity of genotype and similarity of environment make twins of both types important material for investigating genetic hypotheses of *heritability*, etc.

NEOLITHIC

Literally, the 'new stone age', this is a cultural division of man's history rather than a temporal one. As such, the time of the neolithic varies considerably in different parts of the world and it is often stated that such peoples as the South West African Bushmen, New Guineans, and Australian Aborigines are, or were at the time of European discovery, in a stone age period of cultural development. In Europe it corresponds to the earliest period of cultivation which is entirely post glacial, between perhaps 8000 and 3000 years B.P. In contrast the 'palaeolithic' is a much longer cultural period stretching back into the Pleistocene and divided up into a number of sub-cultures which can be roughly equated in Africa and Europe at least. It can be argued that the palaeolithic pre-dates the development of *H. sapiens*; in other words that *H. erectus* reached that stage of culture. (*See Figs*. 3.2, 3.8). 'Mesolithic' is a short cultural period occurring between these two.

NORMAL DISTRIBUTION

This is the familiar bell-shaped distribution curve. It is a curve such that 68 per cent of the observations lie within one *standard deviation* of the mean and nearly 95 per cent within 2 standard deviations of the mean. It is the limiting case of either the *binomial* or multinomial distribution and is the frequency distribution expected from a series of observations on a variable whose magnitude is affected by a large number of genes having small effect. For instance the distribution of intelligence in the population is normal with a mean of 100 and a standard deviation of 15.

NUCLEOTIDES

The molecules of *DNA* and *RNA* are composed of subunits called 'nucleotides'. These consist of three parts: a phosphate group, a sugar, and a complex basic group containing nitrogen. In the nucleotides which form DNA the sugar is deoxyribose and the base may be adenine (A), guanine (G), cytosine (C), or thymine (T). In the nucleotides which form RNA, the sugar is ribose and the bases are the same except that uracil (U) replaces thymine.

NUCLEUS

This structure is the most prominent structure found within most cells of all *eukaryotes*. It is the controlling centre of the cell and contains the *chromosomes*.

OCCLUSION

The closing and hence the meeting of the upper and lower sets of teeth. Successful occlusion depends on a suitable match between upper and lower jaws and the appropriate distribution of the cusps on the premolar and molar teeth. Malocclusion is therefore imperfect closing and meeting of the teeth.

OPISTHOCRANIUM

The back of the cranium; the whole rear area of the skull.

ORTHNOGRAPHIC

Possessing a chin which does not protrude relative to the rest of the face, the opposite of *prognathic*.

ORTHOGNATHOUS

Having jaws that do not protrude. The opposite of *prognathous* which means the possession of jaws which do protrude.

OVERDOMINANT, *see* DOMINANCE.

PALAEOLITHIC, *see* NEOLITHIC.

PANMICTIC

Mating is said to be 'panmictic' when it is completely random. In other words when each individual is equally likely to mate with any individual of the opposite sex. The situation is to be compared with *assortative mating*.

PANMIXIA, *see* PANMICTIC and ASSORTATIVE MATING.

PHENOTYPE

In contrast to *genotype* the phenotype is the observable character or characters of an individual. The term is used in two senses:

1. To classify individuals with respect to a single character—blue eyes or O blood group, for example;
2. To refer to the whole complex of characters possessed by an individual.

PHYLOGENETIC

A mode of biological classification which reflects as closely as possible the evolutionary relationship of the groups concerned.

POLYDACTYLY

A *congenital* abnormality which involves the possession of one or more extra digits on either fore or hind limbs. These extra digits may be variable in manifestation and well or poorly developed. There are a number of different *genes* known in mammals which may give rise to various types of polydactyly.

POLYGAMY

The state of multiple marriage. Usually the circumstance where a man has two or more wives simultaneously. A fairly common phenomenon in primitive tribes and a practice which tends to reduce the effective population size. To be compared with 'monogamy', marriage between one man and one woman.

POLYGENE, *see* GENE.

POLYPEPTIDE

A long chain of peptides which in turn are molecules composed of a series of amino-acids. They are the constituent parts of proteins and *enzymes* and the original epigram 'one gene—one enzyme' is now more accurately rendered 'one gene—one polypeptide chain'. This more accurately reflects the relationship between the genes and their products.

PROGNATHISM

The protruding of the jaws relative to the rest of the face.

PROTANOPY

Defective colour vision of the dichromatic type, characterized by retention of the sensory mechanism for two hues only (blue and yellow) of the normal 4-primary quota and lacking that for red and green. It differs from *deuteranopy* in that there is also a loss of dominance and a shift of brightness and hue curves toward the short-wave end of the spectrum as in twilight vision. It is due to a separate defective *gene*.

PTC

The chemical compound phenylthiocarbamide can be tasted as a very bitter substance by some people but it is completely tasteless to others, although the threshold concentrations at which PTC can still be tasted varies in individuals and between races. The proportion of non-tasters is highest among European, among whom it reaches 30 per cent or more, whereas among Mongoloids and Negroids non-tasters are usually less than 10 per cent of the population. It is an entirely artificial compound but similar compounds with similar tastes are found in various root crops.

QUASICONTINUOUS

Characters which are discontinuous in expression are usually controlled by one or a very small number of *genes*, whereas those which are continuously varying are usually controlled by several genes. The term 'quasicontinuous variation' was introduced to describe those cases where discontinuous characters are controlled by several genes of small effect, or *polygenes*. It usually involves the idea of a threshold in development and characteristic frequencies of expression in different populations as explained in Chapter 2, section *2.1*.

RECESSIVE, *see* DOMINANT.

RECOMBINATION, *see* CROSSING OVER.

SECRETOR

A person who secretes the ABO blood group substances into saliva and body fluids. The trait is due to a polymorphic *gene* quite distinct from that of the ABO *locus* itself.

SELECTIVE COEFFICIENT

Individuals with different *genotypes* will contribute different numbers of offspring to the next generation. When the proportions of these various genotypes are different in this succeeding generation from what they were in the parental then 'selection' can be said to have occurred. The selection coefficient is a quantitative measure of the intensity of selection. It is in fact the proportional reduction in the *gametic* contribution of a particular genotype compared with a standard genotype, usually the most favoured.

SERUM ALBUMIN

This is the chief protein of human blood-serum. It is among the fastest migrating fractions on *electrophoresis* and one which is relatively invariant throughout the human species.

SEX-LINKED

In man this refers to *genes* which are carried on the X-*chromosome* as opposed to genes on the other 22 pairs of chromosomes which are *autosomal.* Males which have an XY chromosomal constitution will always fully express such genes since there are no known genes on the Y-chromosome and they must pass them on to their daughters. In females, of course, X-linked genes may be dominant or recessive or in between and the genes may be passed to either sons or daughters.

SHOVEL TEETH

Incisor teeth that are scooped out on their lingual surfaces and having a variable degrees of inrolling of their lateral borders.

SIBLINGS

Progeny of the same parents, i.e., brothers and sisters. They may also be referred to as 'sibs'. Very closely related species may be termed 'sibling species'.

SOMATIC

Belonging to or associated with the body or body cells as opposed to the germ cells or sex cells. Not passed on to the next generation.

STABILIZING SELECTION, *see* DIRECTIONAL SELECTION.

STADIAL, *see* GLACIATION.

STANDARD DEVIATION, *see* VARIANCE.

STEATOPYGOUS

The Hottentots, particularly the females, have developed the system of depositing very large amounts of fat in their buttocks, possibly, it is suggested, as emergency reserves for harsh desert living. To exhibit this condition of having such enormous buttocks is to be 'steatopygous'.

SUTURE BONES

The series of small bones which lie between two major bones of the skull, i.e., within the suture between such bones; they are also called 'wormian bones'. They are variable in appearance and often vary in a fairly systematic way between the races.

SYNDACTYLY

A very variable *congenital* anomaly, particularly of the hand but also of the foot, in which the adjacent digits are to a greater or lesser extent joined together. The syndactyly may involve only the skin between the digits or it may mean the bones of the digits are wholly or partially fused.

THALASSAEMIA

A group of hereditary *haemoglobinopathies* giving rise to various degrees of haemolytic anaemias. Basically a malfunction of the rates of synthesis of different forms of haemoglobin. Thalassaemia major (Cooley's anaemia) is the most severe form and is due to the presence of a *homozygous gene*. When the gene is *heterozygous* thalassaemia minor results. The traits are common in people from the Mediterranean area.

TAURODONT

Teeth having enlarged pulp cavities, 'bull-toothed'.

TYPOLOGICAL

A system and philosophy of biological classification in which it is believed that each taxonomic category can be described in terms of an ideal 'type'. Variation around this type is regarded as irrelevant and to be ignored for taxonomic purposes. This way of thinking is now largely superseded by various systems of classification *phylogenetic* in approach. Nevertheless typological thinking has left a considerable heritage, particularly in thinking about racial classification.

VARIANCE

The mean square deviation of a variate from its mean. The square of the standard deviation. In a *normal distribution* more than 95 per cent of the values lie within the range of the mean $\pm$ 2 standard deviations and more than 99 per cent with ± 3 standard deviations. These facts can be used as the basis for deciding whether particular individuals derive from a given population, or, more particularly, whether two or more samples are really drawn from the same population or populations with the same mean and standard deviation. The coefficient of variation is the variance divided by the mean expressed as a percentage. This provides a useful comparative measure of relative variabilities.

VOMER

The unpaired flat bone that forms the inferior and posterior part of the nasal septum.

WORMIAN BONES, *see* SUTURE BONES.

ZYGOTE

A cell formed by the union of two *gametes*, the sperm and the egg, and the individual which develops from this fertilized egg.

REFERENCES

ALLISON, A. C. (1954a), 'Protection afforded by the sickle cell trait against subtertian malarial infection', *Br. med. J.*, **1**, 290.

ALLISON, A. C. (1954b), 'The distribution of the sickle cell trait in East Africa and elsewhere and its apparent relationship to the incidence of subtertian malaria', *Trans. R. Soc. trop. Med. Hyg.*, **48**, 312.

ALLISON, A. C. (1955), 'Aspects of polymorphism in man', *Cold Spring Harb. Symp. quant. Biol.*, **20**, 230.

ALLISON, A. C. (1964), 'Polymorphism and natural selection in human populations', *Cold Spring Harb. Symp. quant. Biol.*, **29**, 137.

AMADON, D. (1949), 'The seventy-five per cent rule for subspecies', *Condor*, **51**, 250.

ANASTASI, A. (1958), *Differential Psychology*, 3rd ed. New York: Macmillan.

ANGSEESING, J. P. A. (1972), 'Unquiet evolution: consideration of a previous synthesis', *J. biol. Educ.*, **6**, 41.

ARNHEIM, M., and TAYLOR, C. E. (1969), 'Non-Darwinian evolution: Consequences for neutral allelic variation', *Nature, Lond.*, **223**, 900.

ASHCROFT, M. T., HENEAGE, P., and LOVELL, H. G. (1966), 'Heights and weights of Jamaican schoolchildren of various ethnic groups', *Am. J. phys. Anthrop.*, **24**, 35.

ASHLEY, D. J. B. (1966), 'Observations on the epidemiology of prostatic hyperplasia in Wales', *Br. J. Urol.*, **38**, 567.

ASHLEY, D. J. B. (1967a), 'Diabetes in Wales', *J. med. Genet.*, **4**, 274.

ASHLEY, D. J. B. (1967b), 'Coronary artery disease in Wales', *J. med. Genet.*, **4**, 277.

ASHLEY, D. J. B. (1967c), 'Vascular disease of the central nervous system in Wales', *J. med. Genet.*, **4**, 280.

ASHLEY, D. J. B. (1969), 'Welshness and fertility', *J. med. Genet.*, **6**, 180.

ASHLEY, D. J. B., and DAVIES, H. D. (1966), 'The use of the surname as a genetic marker in Wales', *J. med. Genet.*, **3**, 203.

AYALA, F. J., POWELL, J. R., TRACEY, M. L., MOURNÃO, C. A., and PEREZ-SOLAS, S. (1972), 'Enzyme variability in the *Drosophila willistoni* group. 4. Genic variation in natural populations of *Drosophila willistoni*', *Genetics*, **70**, 113.

BACHE, R. M. (1895), 'Reaction time with reference to race', *Psych. Rev.*, **2**, 475.

BAJEMA, C. J. (1963), 'Estimation of the direction and intensity of natural selection in relation to human intelligence by means of the intrinsic rate of natural increase', *Eugen. Q.*, **10**, 175.

BAKER, C. M. A., and HANSON, H. C. (1965), 'Molecular genetics of avian proteins. 6. Evolutionary implications of blood proteins of eleven species of geese', *Comp. Biochem. Physiol.*, **17**, 997.

Baker, C. M. A., and Manwell, C. (1967), 'Molecular genetics of avian proteins. 8. Egg-white proteins of the migratory quail, *Coturnix coturnis*—New concepts of 'hybrid vigour', *Comp. Biochem. Physiol.*, **23**, 21.

Baker, C. M. A., Manwell, C., Labisky, R. F., and Harper, J. A. (1966), 'Molecular genetics of avian proteins. 5. Egg, blood and tissue proteins of the ring-necked pheasant, *Phasianus colchicus*', *Comp. Biochem. Physiol.*, **17**, 467.

Barnett, A. (1961), *The Human Species*. London: Penguin.

Barnicot, N. A. (1958), 'Reflectometry of the skin in Southern Nigerians and in some mulattoes', *Hum. Biol.*, **30**, 150.

Barrai, I., Cavalli-Sforza, L. L., and Mainardi, M. (1964), 'Testing a model of dominant inheritance for metric traits in man', *Heredity*, **19**, 651.

Bateman, A. J. (1950), 'Is gene dispersion normal?', *Heredity*, **4**, 353.

Bayley, N. (1943), 'Mental growth during the first three years', in *Child Behaviour and Development* (ed. Kounin and Wright). New York: McGraw Hill.

Bayley, N. (1965a), 'Research in child development: A longitudinal perspective', *Merril-Palmer Q. Behaviour Dev.*, **11**, 183.

Bayley, N. (1965b), 'Comparison of mental and motor test scores for ages one–fifteen months by sex, birth order, race, geographical location, and by education of parents', *Child Dev.*, **36**, 379.

Bayley, N. (1966), 'Learning in adulthood: The role of intelligence', in *Analyses of Concept Learning* (ed. Klausmeier and Harris). New York: Academic Press.

Beale, C. L. (1957), 'American triracial isolates, their status, and pertinence to genetic research', *Eugen. Q.*, **4**, 187.

Beardmore, J. A. (1970), 'Viral components in the genetic background', *Nature, Lond.*, **226**, 766.

Benoist, J. (1963), 'Les Martiniquais', *Bull. Soc. d'Anthropolo. Paris*, **11**, 260.

Berry, A. C., and Berry, R. J. (1967), 'Epigenetic variation in the human cranium', *J. Anat.*, **101**, 361.

Berry, R. J. (1963), 'Epigenetic polymorphism in wild populations of *Mus musculus*', *Genet. Res.*, **4**, 193.

Berry, R. J. (1967), 'Genetical changes in mice and men', *Eugen. Rev.*, **59**, 78.

Berry, R. J. (1968), 'The biology of non-metrical variation in mice and men', in *The Skeletal Biology of Earlier Human Populations* (ed. Brothwell). Oxford: Pergamon.

Berry, R. J. (1969a), 'Non-metrical skull variation in two Scottish colonies of the grey seal', *J. Zool.*, **157**, 11.

Berry, R. J. (1969b), 'Genetical factors in the aetiology of multiple sclerosis', *Acta neurol. scand.*, **45**, 459.

Berry, R. J., Berry, A. C., and Ucko, P. J. (1967), 'Genetical change in ancient Egypt', *Man*, **2**, 552.

Berry, R. J., and Crothers, J. H. (1968), 'Stabilizing selection in the dog-whelk *Nucella lapillus*', *J. Zool.*, **155**, 5.

Berry, R. J., and Searle, A. G. (1963), 'Epigenetic polymorphism of the rodent skeleton', *Proc. Zool. Soc. Lond.*, **140**, 577.

Bielicki, T., and Welon, Z. (1966), 'The operation of natural selection on human head form in an East European population', *Homo*, **15**, 22.

Bilsborough, A. (1969), 'Rates of evolutionary change in the hominid dentition', *Nature, Lond.*, **223**, 146.

Birdsell, J. B. (1950), 'Some implications of the genetical concept of race in terms of spatial analysis', *Cold Spring Harb. Symp. quant. Biol.*, **15**, 259.

Birdsell, J. B. (1963), Review of *The Origin of Races* (Coon), *Q. Rev. Biol.*, **38**, 178.

Birdsell, J. B. (1967), 'Preliminary data on the trihybrid origin of the Australian aborigines', *Archaeol. phys. Anthrop. Oceania*, **2**, 100.

Bishop, W. N., and Chapman, G. R. (1970), 'Early Pliocene sediments and fossils from the northern Kenya rift valley', *Nature, Lond.*, **226**, 914.

Black, J. A., and Dixon, G. H. (1968), 'Amino acid sequence of alpha chains of human haptoglobins', *Nature, Lond.*, **218**, 736.

Blake, N. M., and Kirk, R. L. (1969), 'A new genetic variant of 6-phosphogluconate dehydrogenase in Australian aborigines', *Nature, Lond.*, **221**, 278.

Blum, H. F. (1959), *Cancerogenesis by Ultra-violet Light.* Princeton: Princeton University Press.

Boas, F. (1895), 'Zur anthropologie der nordamerikanischen Indianer,' *Verh. berl. Ges. Anthrop.* (suppl. to *Z. ethn.*, **27**), 366.

Boas, F. (1948), *Race, Languages and Culture*, 2nd ed. New York: Macmillan.

Bodmer, J. G., and Bodmer, W. F. (1970), 'Studies on African pygmies. 4. A comparative study of the HL-A polymorphism in the Babinga Pygmies and other African and Caucasian populations', *Am. J. hum. Genet.*, **22**, 396.

Bodmer, W. F. (1968), 'Demographic approaches to the measurement of differential selection in human populations', *Proc. natn. Acad. Sci. U.S.A.*, **59**, 690.

Bodmer, W. F., and Cavalli-Sforza, L. L. (1970), 'Intelligence and race', *Scient. Am.*, **223**, 4, 19.

Bogue, D. J. (1959), *The Population of the United States.* New York: Free Press of Glencoe.

Böök, J. A. (1951), 'The incidence of congenital diseases and defects in a South Swedish population', *Acta Genet.*, **2**, 289.

Bowles, H. E. (1957), 'Maternal mortality in Hawaii 1950–1955', *Hawaii med. J.*, **16**, 622.

Boyce, A. J., Küchemann, C. F., and Harrison, G. A. (1967), 'Neighbourhood knowledge and the distribution of marriage distances', *Ann. hum. Gen.*, **30**, 335.

Boyd, W. C. (1950), *Genetics and the Races of Man.* Boston: Little Brown.

Boyd, W. C. (1964), 'Modern ideas on race in the light of our knowledge of blood groups and other characters with known modes of inheritance', in *Taxonomic Biochemistry and Serology* (ed. Leone). New York: Ronald Press.

Brace, C. L. (1967), 'Environment, tooth form, and size in the Pleistocene', *J. dent. Res.*, Suppl. to No. 5, 809.

Brace, C. L. (1969), 'A nonracial approach towards the understanding of human diversity', in *The Concept of Race* (ed. Montague). New York: Free Press of Glencoe.

Brandes, F. (1927), quoted by Tobias (1970).

Brereton, J. le G. (1960), 'An estimate of the assimilation rate of mixed bloods in New South Wales', in *Proceedings of Conference on Welfare Policies for Australian Aborigines*, Dept. of Adult Education, Univ. of New England, Armidale, N.S.W.

Bresler, J. B. (1962), 'The relationship between the fertility patterns of the F_1 generation and the number of countries of birth represented in the P_1 generation', *Am. J. Phys. Anthrop.*, **20**, 509.

Bresler, J. B. (1970), 'Outcrossings in Caucasians and foetal loss', *Social Biol.*, **17**, 17.

BRIDGES, E. (1948), quoted by PENROSE (1955).

BRNCIC, D. (1954), 'Heterosis and the integration of the genotype in geographic populations of *Drosophila pseudoobscura*', *Genetics*, **39**, 77.

BROTHWELL, D. R. (1958), 'The use of non-metrical characters of the skull in differentiating populations', *Dtsch. Ges. Anthrop.*, **6**, 103.

BROTHWELL, D. R. (1960), 'Upper Pleistocene human skull from the Niah cave', *Sarawak. Mus. J.*, **9**, 155.

BROTHWELL, D. R. (ed.) (1963), *Dental Anthropology*, Symp Soc. Stud. Human Biol., 5. Oxford: Pergamon.

BROTHWELL, D. R. (1965), 'Of mice and men. Epigenetic polymorphism in the skeleton', in *Homenaje a Juan Comas en su 65 Aniversario*, vol. 2. Mexico City.

BROTHWELL, D. R., and POWERS, R. (1968), 'Congenital malformations of the skeleton in earlier man', in *The Skeletal Biology of Earlier Human Populations* (ed. Brothwell). Oxford: Pergamon.

BRUES, A. M. (1954), 'Selection and polymorphism in the ABO blood groups', *Am. J. phys. Anthrop.*, **12**, 559.

BRUNER, F. G. (1908), 'The hearing of primitive peoples', *Archs Psychol.*, **11**, 1.

BULMER, R. N. H. (1969), 'Cultural Diversity and National Unity', Inaugural Lecture University of Papua and New Guinea, quoted by SIMONS (1971).

BURMA, J. H. (1963), 'Interethnic marriage in Los Angeles', in *Studies in Human Sexual Behaviour* (ed. Shiloh). Springfield, Ill.: Thomas.

BURNES, K. (1970), 'Patterns of WISC scores for children of two socio-economic classes and races', *Child Dev.*, **42**, 493.

BURT, C. (1966), 'The genetic determination of differences in intelligence: A study of monozygotic twins reared together and apart', *Br. J. Psychol.*, **57**, 137.

BURT, C., and HOWARD, M. (1956), 'The multifactorial theory of inheritance and its application to intelligence', *Br. J. statist. Psychol.*, **8**, 95.

CAMPBELL, B. G. (1963), 'Quantative taxonomy and human evolution', in *Classification and Human Evolution* (ed. Washburn). Chicago: Aldine Publishing Co.

CAMPBELL, D. T. (1964), 'Distinguishing differences of perception from failures of communication in cross-cultural studies', in *Cross-cultural Understanding: Epistemology in Anthropology* (ed. Northrop and Livingston). New York: Harper & Row.

CARSON, H. L. (1958), 'Increase in fitness in experimental populations resulting from heterosis', *Proc. natn. Acad. Sci. U.S.A.*, **44**, 1136.

CARTER, C. O. (1965), 'The inheritance of common congenital malformations', in *Progress in Medical Genetics* 4 (ed. Steinberg and Bearn). New York: Grune & Stratton.

CASTLE, W. E. (1930), 'Race mixture and physical disharmonies', *Science, N.Y.*, **71**, 603.

CAVALLI-SFORZA, L. L. (1957), 'Some notes on the breeding patterns of human populations', *Acta genet.*, **6**, 395.

CAVALLI-SFORZA, L. L. (1958), 'Some data on the genetic structure of human populations', *Proc. Tenth int. Congr. Genet.*, **1**, 389. Toronto: University of Toronto Press.

CAVALLI-SFORZA, L. L. (1966), 'Population structure and human evolution', *Proc. R. Soc. B.*, **164**, 362.

CAVALLI-SFORZA, L. L. (1969a), 'Human diversity', in *Proc. Twelfth int. Congr. Genet.*, **3**, 405. Tokyo: The Science Council of Japan.

CAVALLI-SFORZA, L. L. (1969b), 'Genetic drift in an Italian population', *Scient. Am.*, **221**, 2, 30.

CAVALLI-SFORZA, L. L., BARRAI, I., and EDWARDS, A. W. F. (1964), 'Analysis of human evolution under random genetic drift', *Cold Spring Harb. Symp. quant. Biol.*, **29**, 29.

CAVALLI-SFORZA, L. L., and BODMER, W. F. (1971), *The Genetics of Human Populations.* San Francisco: Freeman.

CAVALLI-SFORZA, L. L., and EDWARDS, A. W. F. (1963), 'Analysis of human evolution', in *Genetics Today* (ed. Geerts), Vol. 3 (*Proc. Eleventh int. Congr. Genet.*). New York: Pergamon Press.

CAVALLI-SFORZA, L. L., and EDWARDS, A. W. F. (1967), 'Phylogenetic analysis: models and estimation procedures', *Am. J. hum. Genet.*, **19**, 233.

CAVALLI-SFORZA, L. L., KIMURA, M., and BARRAI, I. (1966), 'The probability of consanguineous marriages', *Genetics*, **54**, 37.

CAVALLI-SFORZA, L. L., ZONTA, L. A., NUZZO, F., BERNINI, L., DEJONG, W. W., MEERA KHAN, P., RAY, A. K., WENT, L. N., SINISCALCO, M., NIJENHUIS, L. E., VANLOGHEM, E., and MODIANO, G. (1969), 'Studies on African pygmies. I. A pilot investigation of Babinga pygmies in the Central Africa Republic (with an analysis of genetic distances)', *Am. J. hum. Genet.*, **21**, 252.

CHOI, S., and TROTTER, M. (1970), 'A statistical study of the multivariate structure and race–sex difference of American White and Negro foetal skeletons', *Am. J. phys. Anthrop.*, **33**, 307.

CHRISTIE, A. (1949), 'Prevalence and distribution of ossification centres in the newborn infant', *Am. J. Dis. Child.*, **77**, 355.

CLARKE, B. (1970), 'Darwinian evolution of proteins', *Science, N.Y.*, **168**, 1009.

CLARKE, C. A., and SHEPPARD, P. M. (1960a), 'The evolution of dominance under disruptive selection', *Heredity*, **14**, 73.

CLARKE, C. A., and SHEPPARD, P. M. (1960b), 'The evolution of mimicry in the butterfly *Papilio dardanus*', *Heredity*, **14**, 163.

CLARKE, C. A., and SHEPPARD, P. M. (1963), 'The interaction between major genes and polygenes in the determination of the mimetic pattern of *Papilio dardanus*', *Evolution*, **17**, 404.

CLARKE, C. A., and SHEPPARD, P. M. (1971), 'Further studies on the genetics of the mimetic butterfly *Papilio memnon*', *Phil. Trans. R. Soc. B.*, **263**, 35.

CLARKE, C. A., SHEPPARD, P. M., and THORNTON, I. W. B. (1968), 'The genetics of the mimetic butterfly *Papilio memnon*', *Phil. Trans. R. Soc. B.*, **254**, 37.

CODWELL, J. E. (1949), 'Motor function and the hybridity of the American Negro', *J. Negro Educ.*, **18**, 452.

COMAS, J. (1960), *Manual of Physical Anthropology*. Springfield, Ill.: Thomas.

COON, C. S. (1963), *The Origin of Races*. London: Cape.

COON, C. S. (1966), *The Living Races of Man*. London: Cape.

COON, C. S., GARN, S. M., and BIRDSELL, J. B. (1950), *A Study of the Problems of Race Formation in Man*. Springfield, Ill.: Thomas.

CORDEIRO, A. R., SALZANO, F. M., and MARQUES, V. B. (1960), 'An inter-racial hybridization experiment in natural populations of *Drosophila willistoni*', *Heredity*, **15**, 35.

CROW, J. F. (1958a), 'Some possibilities for measuring selection intensities in man', in *Natural Selection in Man* (ed. Spuhler). Detroit: Wayne Univ. Press.

CROW, J. F. (1958b), 'Some possibilities for measuring selection intensities in man', *Hum. Biol.*, **30**, 1.

CROW, J. F. (1961), 'Population genetics', *Am. J. hum. Genet.*, **13**, 137.

CROW, J. F., and KIMURA, M. (1970), *An Introduction to Population Genetics Theory.* New York: Harper & Row.

CUMMINS, H., and MIDLO, C. (1943), *Fingerprints, Palms, and Soles.* Philadelphia: Blakiston.

CURTAIN, C., VAN LOGHEM, E., BAUMGARTEN, A., GOLAB, T., GORMAN, J., RUTGERS, C., and KIDSON, C. (1971), 'The ethnological significance of the gamma-globulin (Gm) factors in Melanesia', *Am. J. phys. Anthrop.*, **34**, 257.

CURTI, M. W. (1960), 'Intelligence tests of white and coloured school children in Grand Cayman', *J. Psychol.*, **49**, 13.

DA COSTA MARTINS, D. (1971), 'Height, weight, and chest circumference of children of different ethnic groups in Lourenço Marques, Moçambique, in 1965 with a note on the secular trend', *Hum. Biol.*, **43**, 253.

DAMON, A. (1971), 'Race, ethnic group, and disease', in *The Biological and Social Meaning of Race* (ed. Osborne). San Francisco: Freeman.

DANKS, D. M., ALLAN, J., and ANDERSON, C. M. (1965), 'A genetic study of fibrocystic disease of the pancreas', *Ann. hum. Genet.*, **28**, 323.

DARLINGTON, C. D. (1966), *Genetics and Man.* London: Penguin.

DAVENPORT, C. B. (1913a), *Fecundity of Hybrids.* Publ. Carnegie Inst. Wash., 122. Washington, D.C.

DAVENPORT, C. B. (1913b), *Heredity of Skin Colour in Negro–White Crosses.* Publ. Carnegie Inst. Wash., 188. Washington, D.C.

DAVENPORT, C. B., and STEGGERDA, M. (1929), *Race Crossing in Jamaica.* Publ. Carnegie Inst. Wash., 395. Washington, D.C.

DAVIVONGS, V. (1963), 'The pelvic girdle of the Australian aborigine, sex differences and sex determination', *Am. J. phys. Anthrop.*, **21**, 443–455.

DE BEER, G. R. (1951), *Embryos and Ancestors.* Oxford: Clarendon Press.

DEOL, M. S., GRUNEBERG, H., SEARLE, A. G., and TRUSLOVE, G. M. (1957), 'Genetical differentiation involving morphological characters in an inbred strain of mice', *J. Morph.*, **100**, 345.

DESÁ, F. R., BENEVIDES, E., and SALZANO, F. M. (1969), 'Fingerprints of whites and negroes from Southern Brazil', *Am. J. phys. Anthrop.*, **31**, 59.

DESSAUER, H. C., and NOVO, F. (1969), 'Geographic variation of blood and liver proteins in cricket frogs', *Biochem. Genet.*, **3**, 171.

DEUTSCH, M. (1969), 'Happenings on the way back to the forum, social science, I.Q., and race differences revisited', *Harvard educ. Rev.*, **39**, 523.

DEWEY, W. J., BARRAI, I., MORTON, N. E., and MI, M. P. (1965), 'Recessive genes in severe mental defect', *Am. J. hum. Genet.*, **17**, 237.

DOBZHANSKY, TH. (1951), *Genetics and the Origin of Species*, 3rd ed. New York: Columbia University Press.

DOBZHANSKY, TH. (1962), *Mankind Evolving.* New Haven: Yale University Press.

DOBZHANSKY, TH. (1964), 'How do the genetic loads affect the fitness of their carriers in Drosophila populations', *Am. Nat.*, **98**, 151.

DOBZHANSKY, TH. (1970), *Genetics of the Evolutionary Process.* New York: Columbia University Press.

DREGER, R. M., and MILLER, K. S. (1960), 'Comparative psychological studies of negroes and whites in the United States', *Psychol. Bull.*, **57**, 361.

DREGER, R. M., and MILLER, K. S. (1968), 'Comparative psychological studies of negroes and whites in the United States', *Psychol. Bull.*, Monogr. Suppl., **70**, 1.

DUNCAN, O. D., FEATHERMAN, D. L., and DUNCAN, B. (1968), *Socio-economic Background and Occupation Achievement: Extensions of a Basic Model.* Final Report, project No. 5.0074 (EO -191) U.S. Department of Health Education and Welfare Office of Education, Bureau of Research, May, 1968.

DUNN, L. C., and TOZZER, A. M. (1928), 'An anthropometric study of Hawaiians of pure and mixed blood', Papers of the *Peabody Mus. Am. Archaeol. Ethnol.*, Harvard University, **11**, 90.

DUNSDON, M. I., and HUNTLEY, R. M. C. (1965), cited by HUNTLEY (1966).

DYER, K. F. (1968a), 'Evolution observed: some examples of evolution in historical times', *J. biol. Educ.*, **2**, 317.

DYER, K. F. (1968b), 'The evolving story of evolution', *New Scient.*, **39**, 375.

DYER, K. F. (1968c), 'Scrutinizing human genes', *New Scient.*, **40**, 237.

DYER, K. F. (1969a), 'Biological assessment of racial integration', *J. biol. Educ.*, **3**, 321.

DYER, K. F. (1969b), 'Hidden variability in man', *New Scient.*, **44**, 72.

DYER, K. F. (1969c), 'The effect of radiation on small competing populations of *Drosophila melanogaster*. 1. The accumulation of genetic damage', *Genetics*, **61**, 227.

DYER, K. F. (1969d), 'The effect of radiation on small competing populations of *Drosophila melanogaster*. 2. The changing frequencies of autosomal lethals', *Genetics*, **61**, 245.

DYER, K. F. (1969e), 'The effect of radiation on small competing populations of *Drosophila melanogaster*. 3. Changes in competitive ability', *Genetics*, **61**, 275.

DYER, K. F. (1969f), 'Sunburn, mutation, and cancer', *New Scient.*, **43**, 228.

DYER, K. F. (1970a), 'How man's genes have evolved', *Science J.*, **6**, 7, 23.

DYER, K. F. (1970b), 'Thesis, antithesis, and synthesis in the development of modern biology', *J. biol. Educ.*, **4**, 279.

DYER, K. F. (1971a), 'Effects of radiation on populations of *Drosophila melanogaster* with different genetic structures. 1. Changes in fitness and productivity', *Aust. J. biol. Sci.*, **24**, 565.

DYER, K. F. (1971b), 'Effects of radiation on populations of *Drosophila melanogaster* with different genetic structures. 3. Interaction with population size', *Aust. J. biol. Sci.*, **24**, 585.

DYER, K. F. (1971c), 'The quiet revolution: a new synthesis of biological knowledge', *J. biol. Educ.*, **5**, 15.

ECKARD, E. W. (1947), 'How many Negroes "pass"?', *Am. J. Sociol.*, **52**, 498.

EDWARDS, A. W. F., and CAVALLI-SFORZA, L. L. (1964), 'Reconstruction of evolutionary trees', *Systematics Ass. Publ.*, **6**, 67.

ELLIOTT, R. B., MAXWELL, G. M., and VAWSER, N. (1967), 'Lactose maldigestion in Australian aboriginal children', *Med. J. Aust.*, **1**, 46.

EMERSON, S. (1948), 'A physiological basis for some suppressor mutations and possibly for one gene heterosis', *Proc. natn. Acad. Sci. U.S.A.*, **34**, 72.

EMERY, A. E. H. (1971), *Elements of Medical Genetics*, 2nd ed. Edinburgh: Livingstone.

EMSWELLER, S. L., ASEN, S., and UHRING, J. (1962), 'Tumor formation in interspecific hybrids of Lilium', *Science, N.Y.*, **136**, 266.

ERLENMEYER-KIMLING, L., and JARVIK, L. F. (1963), 'Genetics and intelligence: a review', *Science, N.Y.*, **142**, 1471.

EYSENCK, H. J. (1971), *Race, Intelligence, and Education.* Melbourne: Sun Books.

FAIRBRIDGE, R. W. (1960), 'The changing level of the sea', *Scient. Am.*, **202**, 70.

FALCONER, D. S. (1960), *Introduction to Quantitative Genetics.* Edinburgh: Oliver & Boyd.

FALCONER, D. S. (1965), 'The inheritance of liability to certain diseases, estimated from the incidence among relatives', *Ann. hum. Genet.*, **29**, 51.

FELDMAN, M. W., NABHOLZ, M., and BODMER, W. F. (1969), 'Evolution of the Rh polymorphism: a model for the interaction of incompatibility, reproductive compensation and heterozygote advantage', *Am. J. hum. Genet.*, **21**, 171.

FIRSCHEIN, I. L. (1961), 'Population dynamics of the sickle-cell trait in the Black Caribs of British Honduras, Central America', *Am. J. hum. Genet.*, **13**, 233.

FISCHER, E. (1913), *Die Rehobother Bastaards.* Jena: Gustav Fischer Verlag.

FISCHER, E. (1939), 'Versucheiner Phänogenetik der normalen korperlichen Eigenschaften des Menschen', *Z. induct. Abstamm. Vererb.*, **76**, 47.

FISHER, R. A. (1930), *The Genetical Theory of Natural Selection.* Oxford: Clarendon Press.

FISHER, R. A. (1947), 'The rhesus factor', *Am. Sci.*, **35**, 95.

FISHER, R. A., FORD, E. B., and HUXLEY, J. (1939), 'Taste-testing the anthropoid apes', *Nature, Lond.*, **144**, 750.

FITCH, W. M., and NEEL, J. V. (1969), 'Phylogeny of American tribes', *Am. J. hum. Genet.*, **21**, 384.

FLEMING, R. M. (1939), 'Physical heredity in human hybrids', *Ann. Eugen.*, **9**, 55.

FORD, E. B. (1940), 'Polymorphism and taxonomy', in *The New Systematics* (ed. Huxley). Oxford: Clarendon Press.

FORD, E. B. (1954), 'Problems in the evolution of geographical races', in *Evolution as a Process* (ed. Huxley, Hardy, and Ford). London: Allen & Unwin.

FORD, E. B. (1971), *Ecological Genetics*, 3rd ed. London: Chapman & Hall.

FRACCARO, M. (1956), 'A contribution to the study of birth weight based on an Italian sample', *Ann. hum. Genet.*, **20**, 282.

FRANK, R. (1971), 'Cave sediments as palaeoenvironmental indicators, and the sedimentary sequence in Koonalda cave', in *Aboriginal Man and Environment in Australia* (ed. Mulvaney and Golson). Canberra: Australian National University Press.

FRASER, F. C. (1970), 'The genetics of cleft lip and cleft palate', *Am. J. hum. Genet.*, **336**, 352.

FREEMAN, F. S. (1962), *Theory and Practice of Psychological Testing.* New York: Holt, Rinehart, & Winston.

FREIRE-MAIA, A. (1971), 'Inbreeding load in Whites and Negroes', *Acta genet. med. Gemellol.*, **1**, 59.

FREIRE-MAIA, A., and AZEVEDO, E. (1971), 'Brazilian inbreeding load: sib and cousin controls', *Am. J. hum. Genet.*, **23**, 1.

FREIRE-MAIA, N. (1963), 'The load of lethal mutations in White and Negro Brazilian populations. 2. Second survey', *Hum. Hered.*, **13**, 199.

FREIRE-MAIA, N., and FREIRE-MAIA, A. (1964), 'Estimate of the genetic load disclosed by inbreeding', *Genetics*, **50**, 527.

FRIEDLANDER, J. S., and STEINBERG, A. G. (1970), 'Anthropological significance of gamma globulin (Gm and Inv) antigens in Bougainville Island, Melanesia', *Nature, Lond.*, **228**, 59.
FROGGATT, P. (1960), 'Albinism in Northern Ireland', *Ann. hum. Genet.*, **24**, 213.
GALTON, F. (1894), *Natural Inheritance.* New York: Macmillan.
GARN, S. M. (1961), *Human Races*, 1st ed. Springfield, Ill.: Thomas.
GARN, S. M. (1962), 'The genetics of normal human growth', in *De Genetica Medica* (ed. Gedda). Rome: Inst. G. Mendel.
GARN, S. M. (1971), *Human Races*, 3rd ed. Springfield, Ill.: Thomas.
GARN, S. M., LEWIS, A. B., and VICINUS, J. H. (1963), 'Third molar polymorphism and its significance to dental genetics', *J. dent. Res.*, **42**, 1344.
GATES, R. R. (1928), 'A pedigree study of Amerindian crosses in Canada', *J. roy. anthrop. Inst.*, **58**, 511.
GATES, R. R. (1941), 'Tasmanoid tribes in North Queensland', *Rec. S. Aust. Mus.*, **7**, 1.
GATES, R. R. (1953), 'Studies of interracial crossing', *Int. Anthrop. Linguist Rev.*, **1**, 15.
GATES, R. R. (1956), 'Studies in race crossing', *Z. Morph. Anthrop.*, **47**, 233.
GATES, R. R. (1957), 'Forms of hair in South African races', *Man.*, **57**, 98.
GATES, R. R. (1958a), 'Studies of interracial crossing: Zulus and East Indians in Natal', *Z. Morph. Anthrop.*, **49**, 22.
GATES, R. R. (1958b), 'Studies in race crossing: The Japanese war children', *Z. Morph. Anthrop.*, **49**, 129.
GATES, R. R. (1960a), 'Racial elements in the aborigines of Queensland', *Z. Morph. Anthrop.*, **50**, 150.
GATES, R. R. (1960b), 'The genetics of the Australian aborigines', *Acta genet. med. Gemell.*, **9**, 7.
GATES, R. R. (1962), 'Race crossing', in *De Genetica Medica* (ed. Gedda). Rome: Inst. G. Mendel.
GEBER, M., and DEAN, R. F. A. (1957), 'The state of development of newborn African children', *Lancet*, 1216.
GEDDA, L., SERIO, A., and MERCURI, A. (1961), quoted by GATES (1962).
GIBLETT, E. R. (1969), *Genetic Markers in Human Blood.* Oxford: Blackwell.
GIBSON, J. B., and LIGHT, P. (1967), 'Intelligence among university scientists', *Nature, Lond.*, **213**, 441.
GLASS, B. (1950), 'The action of selection on the principal Rh alleles', *Am. J. hum. Genet.*, **2**, 269.
GLASS, B. (1954), 'Genetic changes in human populations, especially those due to gene flow and genetic drift', *Adv. Genet.*, **6**, 95.
GLASS, B. (1955), 'On the unlikelihood of significant admixture of genes from the North American Indians in the present composition of the Negroes of the United States', *Am. J. hum. Genet.*, **7**, 368.
GLASS, B. (1956), 'On the evidence of random genetic drift in human populations', *Am. J. phys. Anthrop.*, **14**, 541.
GLASS, B., and LI, C. C. (1953), 'The dynamics of racial intermixture—an analysis based on the American Negro', *Am. J. hum. Genet.*, **1**, 1.
GLICK, C. E. (1970), 'Interracial marriage and admixture in Hawaii', *Soc. Biol.*, **17**, 278.

GLICK, P. L. (1970), 'Intermarriage among ethnic groups in the United States', *Soc. Biol.*, **17**, 292.

GOFFENEY, B., HENDERSON, N., and BUTLER, B. (1971), 'Negro–white, male–female, eight-month developmental scores compared with seven year WISC and Bender test scores', *Child Dev.*, **42**, 595.

GOLDEN, B., BJRNS, B., BRIDGER, W., and MOSS, A. (1971), 'Social class differentiation in cognitive development among black pre-school children', *Child Dev.*, **42**, 37.

GOLDSTEIN, M. S. (1969), 'Human palaeopathology and some diseases in living primitive societies: a review of the recent literature', *Am. J. phys. Anthrop.*, **31**, 285.

GOLSON, J. (1968), 'Archaeological prospects for Melanesia', in *Prehistoric Culture ni Oceania: a Symposium* (ed. Yawata and Sinoto). Honolulu: Bishop Museum Press.

GORDON, H., and GORDON, M. (1957), 'Maintenance of polymorphism by potentially injurious genes in eight natural populations of the platyfish *Xiphophorus maculatus*', *J. Genet.*, **55**, 1.

GORDON, M. (1947), 'Speciation in fishes. Distribution in time and space of seven dominant multiple alleles in *Platypoecilus maculatus*', *Adv. Genet.*, **1**, 95.

GOWEN, J. W. (1964), *Heterosis*. New York: Hafner.

GRAHNEN, H. (1962), 'Hereditary factors in relation to dental caries and congenitally missing teeth', in *Genetics and Dental Health* (ed. Witkop). New York: McGraw-Hill.

GREWAL, M. S. (1962), 'The rate of genetic divergence of sublines in the C57BL strain of mice', *Genet. Res.*, **3**, 226.

GRINDER, R. E., SPOTTS, W. S., and CURTI, M. W. (1964), 'Relationships between Goodenough Draw-A-Man Test performance and skin colour among preadolescent Jamaican children', *J. soc. Psychol.*, **62**, 181.

GRUNEBERG, H. (1950), 'Genetical studies on the skeleton of the mouse. 1. Minor variation of the vertebral column', *J. Genet.*, **50**, 112.

GRUNEBERG, H. (1951), 'The genetics of a tooth defect in the mouse', *Proc. R. Soc. B.*, **138**, 437.

GRUNEBERG, H. (1952), 'Genetical studies on the skeleton of the mouse. 4. Quasi-continuous variation', *J. Genet.*, **51**, 239.

GRUNEBERG, H. (1970), 'Is there a viral component in the genetic background?', *Nature, Lond.*, **225**, 39.

GUSINDE, M. (1942), 'Pygmäen Neger-Bastarde in östlichen Kongogebiet', *Z. Morph. Anthrop.*, **40**, 92.

HAAG, W. G. (1962), 'The Bering Strait land bridge', *Scient. Am.*, **206**, 112.

HAJNAL, J. (1963), 'Random mating and the frequency of consanguineous marriages', *Proc. R. Soc. B.*, **159**, 125.

HALDANE, J. B. S. (1922), 'Sex ratio and unisexual sterility in hybrid animals', *J. Genet.*, **12**, 101.

HALDANE, J. B. S. (1949), 'Suggestions as to quantitative measurements of rates of evolution', *Evolution*, **3**, 51.

HALDANE, J. B. S. (1956), 'The argument from animals to men: an examination of its validity for anthropology', *J. roy. anthrop. Inst.*, **36**, 1.

HALLER, J. S., jun. (1970), 'The physician versus the Negro', *Bull. Hist. Med.*, **44**, 154.

HANIHARA, K. (1963), 'Deciduous dentition of the Japanese-American hybrids', in *Dental Anthropology* (ed. Brothwell). Oxford: Pergamon.

HARRIS, H. (1966), 'Enzyme polymorphisms in man', *Proc. R. Soc. B.*, **164**, 298.

HARRIS, H. (1969), 'Enzyme and protein polymorphism', *Br. med. Bull.*, **25**, 5.

HARRIS, H. (1970), *The Principles of Human Biochemical Genetics.* Amsterdam: North Holland.

HARRISON, G. A., and OWEN, J. J. T. (1956), 'The application of spectrophotometry to the study of skin colour inheritance', *Acta genet.*, **6**, 481.

HARRISON, G. A., and OWEN, J. J. T. (1964), 'Studies on the inheritance of human skin colour', *Ann. hum. Genet.*, **28**, 27.

HARRISON, G. A., OWEN, J., DA ROCHA, F., and SALZANO, F. M. (1967), 'Skin colour in South Brazilian populations', *Hum. Biol.*, **39**, 21.

HARRISON, G. A., WEINER, J. S., TANNER, J. M., and BARNICOT, N. (1964), *Human Biology. An Introduction to Human Evolution, Variation, and Growth.* London: Oxford University Press.

HEDGES, S. R. (1969), 'Epigenetic polymorphism in populations of *Apodemus sylvaticus* and *Apodemus flavicolis* Rodentia, Muridae', *J. Zool.*, **159**, 425.

HEER, D. M. (1967), 'Intermarriage and racial amalgamation in the United States', *Eugen. Q.*, **14**, 112.

HENRY, L. (1961), 'Some data on natural fertility', *Eugen. Q.*, **8**, 81.

HERSKOVITZ, M. J. (1928), *The American Negro: A Study in Racial Crossing.* New York: Knopf.

HERSKOVITZ, M. J. (1930), *The Anthropometry of the American Negro.* New York: Columbia University Press.

HIERNAUX, J., and HEINTZ, N. (1967), 'Croissance biométrique des Franco-Vietnamiens', *Bull. Soc. d'Anthrop. Paris*, **11**, 55.

HIGGINSON, J., and OETTLÉ, A. G. (1960). 'Cancer in the S. African Bantu', *J. natn. Cancer Inst.*, **24**, 608.

HIRAIZUMI, Y. (1964), 'Prezygotic selection as a factor in the maintenance of variability', *Cold Spring Harb. Symp. quant. Biol.*, **29**, 51.

HOLLINGSWORTH, M. J. (1965), 'Observations on the birthweights and survival of African babies: single births', *Ann. hum. Genet.*, **28**, 291.

HOLT, S. B. (1952), 'Genetics of dermal ridges: inheritance of total finger ridge count', *Ann. Eugen.*, **17**, 140.

HOLT, S. B. (1968), *The Genetics of Dermal Ridges.* Springfield, Ill.: Thomas.

HONZIK, M. P., MACFARLANE, J. W., and ALLEN, L. (1948), 'The stability of mental test performance between two and eighteen years', *J. expl. Educ.*, **17**, 309.

HOOTON, E. A. (1947), *Up from the ape.* London: MacMillan.

HOPKINSON, D. A., and HARRIS, H. (1971), 'Recent work on isozymes in man', *Ann. Rev. Genet.*, **5**, 5.

HOWELL, F. C. (1957), 'The evolutionary significance of variation and varieties of "Neanderthal" man', *Q. Rev. Biol.*, **32**, 330.

HOWELL, F. C. (1969), 'Temporal relations of the Pliocene/Pleistocene Hominidae recovered from the Lower Omo Basin', *Nature, Lond.*, **223**, 1234.

HOWELLS, W. W. (1966), '*Homo erectus*', *Scient. Am.*, Nov., 1966, reprinted in *Human Variation and Origins* (ed. Laughlin and Osborne). San Francisco: Freeman.

HRDLICKA, A. (1925), *The Old Americans.* Baltimore: Williams & Wilkins.

HUBBY, J. L., and LEWONTIN, R. C. (1966), 'A molecular approach to the study of genic heterozygosity in natural populations. 1. The number of alleles at different loci in *Drosophila pseudoobscura*', *Genetics*, **54**, 577.

HUBBY, J. L., and THROCKMORTON, L. H. (1965), 'Protein differences in Drosophila. 2. Comparative species genetics and evolutionary problems', *Genetics*, **52**, 203.

HUBBY, J. L., and THROCKMORTON, L. H. (1968), 'Protein differences in Drosophila. 4. A study of sibling species', *Am. Nat.*, **102**, 193.

HUDSON, L. (1960), 'Degree class and attainment in scientific research', *Br. J. Psychol.*, **51**, 67.

HULSE, F. S. (1957), 'Exogamie et heterosis', *Archs. Suisses Anthrop. Gen.*, **22**, 103.

HULSE, F. S. (1962), 'Race as an evolutionary episode', *Am. Anthrop.*, **64**, 929.

HUNTLEY, R. M. C. (1966), 'Heritability of Intelligence', in *Genetic and Environmental Factors in Human Ability* (ed. Meade and Parkes). Edinburgh: Oliver & Boyd.

INGLE, D. J. (1968), 'The need to investigate average biological differences among racial groups', in *Science and the Concept of Race* (ed. Mead, Dobzhansky, Tobach, and Light). New York: Columbia University Press.

IVANHOE, F. (1970), 'Was Virchow right about Neanderthal?', *Nature, Lond.*, **227**, 577.

JAYANT, K. (1964), 'Birth weight and some other factors in relation to infant survival. A study of an Indian sample', *Ann. hum. Genet.*, **27**, 261.

JAYANT, K. (1966), 'Birth weight and survival: A hospital survey repeated after 15 years', *Ann. hum. Genet.*, **29**, 367.

JENKS, A. E. (1916), 'Indian–White amalgamation. An anthropometric study', *Univ. Minn. Stud. soc. Sci.*, **6**, 1.

JENSEN, A. R. (1967), 'Estimation of the limits of heritability of traits by comparison of monozygotic and dizygotic twins', *Proc. natn. Acad. Sci. U.S.A.*, **58**, 149.

JENSEN, A. R. (1969), 'How much can we boost I.Q. and scholastic achievement?', *Harvard educ. Rev.*, **39**, 1.

JENSEN, A. R. (1970), 'Race and genetics of intelligence. A reply to Lewontin', *Bull. atom. Sci.*, **26**, (5) 17.

JINKS, J. L., and FULKER, D. W. (1970), 'Comparison of the biometrical, MAVA and classical approaches to the analysis of human behaviour', *Psychol. Bull.*, **73**, 311.

JOHNSON, L. C., and CORAH, N. L. (1963), 'Racial differences in skin resistance', *Science, N.Y.*, **139**, 766.

JOSE, D. G., and WELCH, J. S. (1970), 'The syndrome of protein calorie malnutrition in Australian aboriginal children', *Med. J. Aust.*, **1**, 349.

JUEL-NIELSEN, N. (1965), 'Individual and environment: a psychiatric-psychological investigation of monozygous twins reared apart', *Acta psychiat. neurol. scand.* (Monogr. Suppl. 183).

KALLA, A. K. (1969), 'Affinities in skin pigmentation of some Indian populations', *Hum. Hered.*, **19**, 499.

KARN, M. N., and PENROSE, L. S. (1951), 'Birth weight and gestation time in relation to maternal age, parity and infant survival', *Ann. Eugen.*, **15**, 206.

KARPINOS, B. D. (1960), 'Racial differences in visual acuity', *Publ. Hlth Rep.*, **75**, 1045.

KATZENMEYER, W. G. (1962), quoted by SHUEY (1966).

KEITH, A. (1925), *The Antiquity of Man.* London: Williams & Norgate.

KELLOCK, W. L., and PARSONS, P. A. (1970), 'A comparison of the incidence of minor nonmetrical cranial variants in Australian aborigines with those of Melanesia and Polynesia', *Am. J. phys. Anthrop.*, **33**, 235.

KELLY, H. J., and REYNOLDS, L. (1947), 'Appearance and growth of ossification centres and increase in the body dimensions of White and Negro infants', *Am. J. Roentgenol.*, **57**, 417.

KENNEDY, W. A., VAN DE RIET, V., and WHITE, J. C. (1963), 'A normative sample of intelligence and achievement of Negro elementary school children in the South-eastern United States', *Monogr. Soc. Res. Child Dev.*, 28.

KING, J. L., and JUKES, T. H. (1969), 'Non-Darwinian evolution', *Science, N.Y.*, **164**, 788.

KING, R. C. (1968), *A Dictionary of Genetics.* New York: Oxford University Press.

KINZEY, W. G. (1970), 'Rates of evolutionary change in the hominid canine teeth', *Nature, Lond.*, **225**, 296.

KIRCHNER, R. (1952), quoted by GATES (1962).

KIRK, R. L. (1968), *The Haptoglobin Groups in Man.* Basel: Karger.

KIRK, R. L. (1971), 'Genetic evidence and its implications for Aboriginal prehistory', in *Aboriginal Man and Environment in Australia* (ed. Mulvaney and Golson). Canberra: Australian National University Press.

KIRK, R. L., BLAKE, N. M., MOODIE, P. M., and TIBBS, G. J. (1971), 'Population genetic studies in Australian aborigines', *Hum. Biol. Oceania*, **1**, 1.

KISER, C. V. (1968), 'Assortative mating by educational attainment in relation to fertility', *Eugen. Q.*, **15**, 98.

KLEIN, J. W., and DEFRIES, J. L. (1969), 'Similar polymorphism of taste sensitivity to PTC in mice and men', *Nature, Lond.*, **225**, 555.

KLINEBERG, O. (1935), *Race Differences.* New York: Harper.

KNUDSON, A. G., WAYNE, L., and HALLETT, W. (1967), 'On the selective advantage of cystic fibrosis heterozygotes', *Am. J. hum. Genet.*, **19**, 388.

KRIEGER, H. (1970), 'Racial admixture effects in North East Brazil', *Ann. hum. Genet.*, **33**, 423.

KRIEGER, H., MORTON, N. E., AZEVEDO, E., FREIRE-MAIA, A., and YASUDA, N. (1965), 'Racial admixture in North East Brazil', *Ann. hum. Genet.*, **29**, 113.

KÜCHEMANN, C. F., BOYCE, A. J., and HARRISON, G. A. (1967), 'A demographic and genetic study of a group of Oxfordshire villages', *Hum. Biol.*, **39**, 251.

KUGELMAN, J. P., and VAN SCOTT, E. J. (1961), 'Tyrosinase activity in melanocytes of human albinos', *J. invest. Dermat.*, **37**, 73.

LABERGE, C. (1969), 'Hereditary tyrosinaemia in a French Canadian isolate', *Am. J. hum. Genet.*, **21**, 36.

LAKOVAARA, S., and SAURA, A. (1971), 'Genetic variation in natural populations of *Drosophila obscura*', *Genetics*, **69**, 377.

LANCASTER JONES, F. (1970), *The Structure and Growth of Australia's Aboriginal Population.* Canberra: Australian National University Press.

LARNACH, S. L., and MACINTOSH, N. W. G. (1970), *The Craniology of Aborigines of Queensland.* The Oceania Monographs, No. 15.

LASKA-MIERZEJEWSKA, T. (1966), 'Menarche in Cuban girls', *Przeglad Antrop.*, **32**, 29.

LASKA-MIERZEJEWSKA, T. (1967), 'Desarrollo y maduracion de los ninos y jovenes Habaneros', *Fuera de Serie Rev. Polaca*, **74**, 9.

LASKER, G. W. (1950), 'Genetic analysis of racial traits of the teeth', *Cold Spring Harb. Symp. quant. Biol.*, **25**, 191.
LASKER, G. W. (1952), 'Mixture and genetic drift in ongoing human evolution', *Am. Anthrop.*, **54**, 433.
LASKER, G. W. (1969), 'Human biological adaptability', *Science, N.Y.*, **166**, 1480.
LEAKEY, L. S. B. (1962), 'A new lower Pliocene fossil primate from Kenya', *A. Mag. nat. Hist.*, Ser. 13, **4**, 689.
LEAKEY, L. S. B. (1967), 'An early Miocene member of Hominidae', *Nature, Lond.*, **213**, 155.
LEAKEY, L. S. B., TOBIAS, P. V., and NAPIER, J. R. (1964), 'A new species of the genus *Homo* from Olduvai Gorge', *Nature, Lond.*, **202**, 7.
LEAKEY, R. E. F. (1970), 'New hominid remains and early artefacts from Northern Kenya: Artefacts from a new Plio-Pleistocene locality near Lake Rudolf in Kenya', *Nature, Lond.*, **226**, 223.
LEVINSON, B. M. (1964), quoted by SHUEY (1966).
LEVINTHAL, C., FAN, D. P., HIGA, A., and ZIMMERMAN, S. (1963), 'The decay and protection of messenger RNA in bacteria', *Cold Spring Harb. Symp. quant. Biol.*, **28**, 183.
LEWIS, G. E. (1934), 'Preliminary notice of the new man-like apes from India', *Am. J. Sci.*, **27**, 161.
LEWONTIN, R. C. (1967), 'An estimate of average heterozygosity in man', *Am. J. hum. Genet.*, **19**, 681.
LEWONTIN, R. C. (1970a), 'Race and intelligence', *Bull. atom. Sci.*, **26** (3), 2.
LEWONTIN, R. C. (1970b), 'Further remarks on race and the genetics of intelligence', *Bull. atom. Sci.*, **26**, (5), 23.
LEWONTIN, R. C., and HUBBY, J. L. (1966), 'A molecular approach to the study of genetic heterozygosity in natural populations. 2. Amount of variation and degree of heterozygosity in natural populations of *Drosophila pseudoobscura*', *Genetics*, **54**, 595.
LI, C. C. (1955), *Population Genetics*. Chicago: Chicago University Press.
LILIENFELD, A. M. (1970), 'Population differences in frequency of malformations at birth', in *Congenital Malformations* (ed. Fraser and McKusick). Amsterdam: Excerpta Medica.
LINTS, F. A., and LINTS, C. (1968), 'Respiration in Drosophila. 2. Respiration in relation to age by wild, inbred and hybrid *Drosophila melanogaster* imagos', *Expl Geront.*, **3**, 341.
LITTLE, K. L. (1943), 'Some anthropological characteristics of Anglo-Negro children', *J. roy. anthrop. Inst.*, **73**, 57.
LIVESEY, T. M., and LOUTTIT, C. M. (1930), 'Reaction time experiments with certain racial groups', *J. app. Psych.*, **14**, 557.
LIVINGSTONE, F. B. (1958), 'Anthropological implications of sickle cell gene distribution in West Africa', *Am. Anthrop.*, **60**, 533.
LIVINGSTONE, F. B. (1964), 'Aspects of the population dynamics of the abnormal haemoglobin and glucose-6-phosphate dehydrogenase deficiency genes', *Am. J. hum. Genet.*, **16**, 435.
LIVINGSTONE, F. B. (1965), 'The distribution of the abnormal haemoglobin genes and their significance for human evolution', *Evolution*, **18**, 685.

LIVINGSTONE, F. B. (1967), *Abnormal Haemoglobins in Human Populations.* Chicago: Aldine.

LIVINGSTONE, F. B. (1969a), 'Evolution of quantitative characteristics, which are determined by several additive loci', *Am. J. phys. Anthrop.*, **31**, 355.

LIVINGSTONE, F. B. (1969b), 'Gene frequency clines of the β-haemoglobin locus in various human populations and their simulation by models involving differential selection', *Hum. Biol.*, **41**, 224.

LIVINGSTONE, F. B. (1969c), in *The Concept of Race* (ed. Montague). London: Collier-MacMillan.

LIVINGSTONE, F. B. (1969d), 'Polygenic models for the evolution of human skin colour differences', *Hum. Biol.*, **41**, 480.

LIVINGSTONE, F. B. (1971), 'Malaria and human polymorphism', *A. Rev. Genet.*, **5**, 33.

LOTSY, J., and GODDIJN, W. A. (1928), 'Voyages of exploration to judge of the bearing of hybridization upon evolution. 1. South Africa', *Genetica* 10, 2 vols.

LUSH, I. E. (1970), 'The extent of biochemical variation in mammalian populations', in *Variation in Mammalian Populations* (ed. Berry and Southern). London: Academic Press.

MCKEOWN, T., and RECORD, R. G. (1960), 'Malformations in a population observed for five years after birth', in *Ciba Foundation Symposium on Congenital Malformations.* London: Churchill.

MCKUSICK, V. A. (1967), 'The ethnic distribution of disease in the United States', *J. chron. Dis.*, **20**, 115.

MCLAREN, A., and WALKER, P. M. B. (1965), 'Genetic discrimination by means of DNA/DNA binding', *Genet. Res.*, **6**, 230.

MCLAREN, A., and WALKER, P. M. B. (1968), 'The comparison of closely related groups of rodents by DNA/DNA anealing', *Genet. Res.*, **12**, 117.

MCLAREN, A., and WALKER, P. M. B. (1970), 'Rodent DNA: Comparisons between species', *Evolution*, **24**, 199.

MCLEISH, A., and THOMA, P. (1970), 'Welsh musicality', *Nature, Lond.*, **230**, 337.

MACNEISH, R. S. (1971), 'Early man in the Andes', *Scient. Am.*, **224**, 4, 36.

MCNEMAR, Q. (1933), 'Twin resemblances in motor skills, and the effect of practice thereon', *J. genet. Psychol.*, **42**, 70.

MANGE, A. P. (1964), 'Growth and inbreeding of a human isolate', *Hum. Biol.*, **36**, 104.

MANN, I. (1966), *Culture, Race, Climate and Eye Disease.* Springfield, Ill.: Thomas.

MANWELL, C., and BAKER, C. M. A. (1968), 'Genetic variation of isocitrate, malate, and 6-phosphogluconate dehydrogenase in snails of the genus Cepaea—introgression, hybridization, polymorphism, and pollution', *Comp. Biochem. Physiol.*, **26**, 195.

MANWELL, C., and BAKER, C. M. A. (1970), *Molecular Biology and the Origin of Species.* London: Sidgwick and Jackson.

MANWELL, C., BAKER, C. M. A., and CHILDERS, W. (1963), 'The genetics of haemoglobin in hybrids—1. A molecular basis for hybrid vigour', *Comp. Biochem. Physiol.*, **10**, 103.

MANWELL, C., and KERST, K. V. (1966), 'Possibilities of biochemical taxonomy of bats using haemoglobin, lactate dehydrogenase, esterases and other proteins', *Comp. Biochem. Physiol.*, **17**, 741.

MATSUNAGA, E. (1955), 'Intra-uterine selection by the ABO incompatibility of mother and foetus', *Am. J. hum. Genet.*, **7**, 66.

MAXWELL, J., and PILLINER, A. E. G. (1960), 'The intellectual resemblance between sibs', *Ann. hum. Genet.*, **24**, 23.
MAYNARD SMITH, J. (1970), 'The causes of polymorphism', *Symp. zool. Soc. Lond.*, **26**, 371.
MAYR, E. (1932), 'Notes on thickheads (*Pachycephala*) from Polynesia', *Am. Mus. Novitates*, **531**, 1.
MAYR, E. (1963), *Animal Species and Evolution*. London: Oxford University Press.
MAYR, E., and GILLIARD, E. T. (1952), 'Altitudinal hybridization in New Guinea honeyeaters', *Condor*, **54**, 325.
MAYR, E., and ROSEN, C. B. (1956), 'Geographic variation and hybridization in populations of Bahama snails (*Cerion*)', *Am. Mus. Novitates*, **1806**, 1.
MEREDITH, H. V. (1970), 'Body weight at birth of viable human infants. A world wide comparative treatise', *Hum. Biol.*, **42**, 217.
MEREDITH, H. V. (1971), 'World wide somatic comparisons among contemporary human groups of adult females', *Am. J. phys. Anthrop.*, **34**, 89.
MERRILL, G. G., and COOK, E. E. (1957), 'The electroencephalogram in the Negro. A comparison of electrical activity of the brain in White and Negro patients', *Electroenceph. clin. Neurophysiol.*, **9**, 531.
METTLER, F. A. (1955), 'Culture and the structural evolution of the neural system', James Arthur lecture on the evolution of the human brain. New York: American Museum of Natural History.
MILLER, A. H. (1941), *Speciation in the Avian Genus Junco*, Univ. Calif. Pub. in Zoology, 44, 173.
MILLS, J. R. E. (1963), 'Occlusion and malocclusion of the teeth of primates', in *Dental Anthropology* (ed. Brothwell). Oxford: Pergamon.
MOLL, R. H., LONNQUIST, J. H., VELEZ-FORTUNO, J., and JOHNSON, E. C. (1966), 'The relationship of heterosis and genetic divergence in maize', *Genetics*, **52**, 139.
MOLL, R. H., SALHVANA, W. S., and ROBINSON, H. F. (1962), 'Heterosis and genetic diversity in variety crosses of maize', *Crop Sci.*, **2**, 197.
MONTAGUE, A. (1960), *Physical Anthropology*, 3rd ed. Springfield, Ill.: Thomas.
MONTAGUE, A. (1964), *Man's Most Dangerous Myth: The Fallacy of Race*. Cleveland: World Publishing Co.
MONTAGUE, A. (ed.) (1969), *The Concept of Race*. London: Collier-Macmillan.
MOORE, J. A. (1950), 'Further studies on *Rana pipiens* racial hybrids', *Am. Nat.*, **84**, 247.
MOORE, J. A. (1963), 'Diploid and haploid interracial hybrids in *Rana pipiens*', *Proc. 11th int. Congr. Genet.* New York: Pergamon Press.
MORTON, N. E. (1955), 'The inheritance of human birthweight', *Ann. Eugen.*, **20**, 125.
MORTON, N. E. (1958), 'Empirical risks in consanguineous marriages. Birth weight, gestation time, and measurements of infants', *Am. J. hum. Genet.*, **10**, 344.
MORTON, N. E. (1963), 'The components of genetic variability', in *The Genetics of Migrant and Isolate Populations* (ed. Goldschmidt). Baltimore: Williams & Wilkins.
MORTON, N. E. (1964), 'Genetic studies of north-eastern Brazil', *Cold Spring Harb. Symp. quant. Biol.*, **29**, 69.
MORTON, N. E. (1970), 'Birth defects in racial crosses', in *Congenital Malformations* (ed. Fraser and McKusick). Amsterdam: Excerpta Medica.

MORTON, N. E., CHUNG, C. S., and MI, M. P. (1967), *Genetics of Interracial Crosses in Hawaii.* New York: Karger.

MOURANT, A. E. (1954), *The distribution of the Human Blood Groups.* Oxford: Blackwell.

MUKAI, T. (1964), 'The genetic structure of natural populations in *Drosophila melanogaster.* 1. Spontaneous mutation rate of polygenes controlling viability', *Genetics,* **50**, 1.

MUKAI, T. (1969a), 'The genetical structure of natural populations of *Drosophila melanogaster.* 6. Further studies on the optimum heterozygosity hypothesis', *Genetics,* **61**, 479.

MUKAI, T. (1969b), 'The genetical structure of natural populations of *Drosophila melanogaster.* 7. Synergistic interaction of spontaneous mutant polygenes controlling viability', *Genetics,* **61**, 749.

MULLER, H. J. (1936), 'On the variability of mixed races', *Am. Nat.,* **70**, 409.

MULVANEY, D. J. (1971), in *Aboriginal Man and Environment in Australia* (ed. Mulvaney and Golson). Canberra: Australian National University Press.

MYERS, C. S. (1901). *Report of the Cambridge Anthropological Expedition to Torres Straits,* vol. 2. London: Cambridge University Press.

NEEL, J. V. (1958), 'A study of major congenital defects in Japanese infants', *Am. J. hum. Genet.,* **10**, 398.

NEEL, J. V. (1970), 'Lessons from a "primitive" people', *Science, N.Y.,* **170**, 815.

NEEL, J. V., and POST, R. (1963), 'Transitory "positive" selection for colour blindness?', *Eugen. Q.,* **10**, 33.

NEEL, J. V., and SALZANO, F. M. (1966), 'A prospectus for genetic studies on the American Indians', in *The Biology of Human Adaptability* (ed. Baker and Weiner). Oxford: Clarendon Press.

NEEL, J. V., SHAW, M., and SCHULL, W. J. (eds.) (1965), *Genetics and the Epidemiology of Chronic Diseases.* Washington, D.C.: U.S. Dept. Health Education and Welfare.

NEEL, J. V., and VALENTINE, W. N. (1947), 'Further studies on the genetics of thalassemia', *Genetics,* **32**, 38.

NEEL, J. V., and WARD, R. H. (1970), 'Village and tribal genetic distances among American Indians and the possible implications for human evolution', *Proc. natu. Acad. Sci. U.S.A.,* **65**, 322.

NEI, M. (1965), 'Variation and covariation of gene frequencies in subdivided populations', *Evolution,* **19**, 256.

NEI, M., and IMAIZUMI, Y. (1966a), 'Genetic structure of human populations. 1. Local differentiation of blood group gene frequencies in Japan', *Heredity,* **21**, 9.

NEI, M., and IMAIZUMI, Y. (1966b), 'Genetic structure of human populations. 2. Differentiation of blood group gene frequencies among isolated populations', *Heredity,* **21**, 183.

NEWMAN, M. T. (1953), 'The application of ecological rules to the racial anthropology of the New World', *Am. Anthrop.,* **55**, 311.

NEWMAN, M. T. (1961), 'Biological adaptation of man to his environment: heat, cold, altitude and nutrition', *Ann. N.Y. Acad. Sci.,* **91**, 617.

NEWMAN, H. H., FREEMAN, F. N., and HOLZINGER, K. J. (1937), *Twins: A Study of Heredity and Environment.* Chicago: University of Chicago Press.

NOBLE, C. E. (1968), 'The learning of psychomotor skills', *A. Rev. Psych.,* **19**, 203.

NOBLE, C. E. (1969), 'Race and experimental psychology', *Perspect. Biol. Med.,* **13**, 10.

NOBLE, C. E. (1972), 'Psychomotor learning', *Encyclopaedia Britannica.*

NORDSKOG, A. W., and GHOSTLEY, F. J. (1964), 'Heterosis in poultry', *Poultry Sci.*, **33**, 704.

O'BRIEN, S. J., and MACINTYRE, R. F. (1969), 'Gene-enzyme variability in *Drosophila*', *Am. Nat.*, **103**, 97.

OSBORNE, R. T. (1965), unpublished data quoted by SHUEY (1966).

PARSONS, P. A. (1965), 'Growth in a sample of Australian Aborigines from one to twenty-three months', *Growth*, **29**, 207.

PATERNIANI, E., and LONNQUIST, J. H. (1963), 'Heterosis in interracial crosses of corn. (*Zea mays* L.)', *Crop Sci.*, **3**, 504.

PENROSE, L. S. (1955), 'Evidence of heterosis in man', *Proc. R. Soc. B*, **144**, 203.

PENROSE, L. S. (1957), 'Genetics of anencephaly', *J. ment. Defic. Res.*, **1**, 4.

PETTIGREW, T. F., and NUTTALL, R. L. (1963), 'Negro American perception of the irradiation illusion', *Percept. Motor Skills*, **17**, 98.

PILBEAM, D. (1970), *The Evolution of Man.* London: Thames and Hudson.

POLLITZER, W. S. (1969), 'Ancestral traits, parental population and hybrids', *Am. J. phys. Anthrop.*, **30**, 415.

POLLITZER, W. S., and BROWN, W. (1969), 'Isolate of hybrid origin in process of dissolution', *Hum. Biol.*, **41**, 388.

POLLITZER, W. S., MENEGAZ-BOCK, R. M., and HERION, J. C. (1966), 'Factors in the microevolution of a triracial isolate', *Am. J. hum. Genet.*, **18**, 26.

POST, R. H. (1962), 'Population differences in vision acuity', *Eugen. Q.*, **9**, 189.

POST, R. H. (1963), 'Colour blindness distribution in Britain, France, China, and Japan: a review', *Eugen. Q.*, **10**, 110.

POST, R. H. (1964), 'Hearing acuity variation among Negroes and Whites', *Eugen. Q.*, **11**, 65.

POST, R. H. (1966), 'Deformed nasal septa and relaxed selection', *Eugen. Q.*, **13**, 101.

POST, R. H. (1969a), 'Deformed nasal septa and relaxed selection II', *Social Biol.*, **16**, 179.

POST, R. H. (1969b), 'Population differences in tear duct size: implications of relaxed selection', *Social Biol.*, **16**, 257.

POST, R. H. (1971), 'Possible cases of relaxed selection in civilized populations', *Humangenetik*, **13**, 253.

POTTER, J. (1965), 'The growth of population in America 1700–1860', in *Population in History: Essays in Historical Demography* (ed. Glass and Eversley). Chicago: University of Chicago Press.

PRAKASH, S. (1969), 'Genic variation in a natural population of *Drosophila persimilis*', *Proc. natn. Acad. Sci. U.S.A.*, **62**, 778.

PRAKASH, S., LEWONTIN, R. C., and HUBBY, J. L. (1969), 'A molecular approach to the study of genic heterozygosity in natural populations. 4. Patterns of genic variation in central, marginal and isolated populations of *Drosophila pseudoobscura*', *Genetics*, **61**, 841.

PRICE, J. (1967), 'Human polymorphism', *J. med. Genet.*, **4**, 44.

PROPERT, D. N., EDMUNDS, R., and PARSONS, P. A. (1968), 'Birth weights and growth rates up to one year of mixed blood and full blood Australian aborigines', *Aust. paediat. J.*, **4**, 134.

RACE, R. R., and SANGER, R. (1968), *Blood Groups in Man*, 5th ed. Oxford: Blackwell.

RADCLIFFE-BROWN, A. R. (1930), 'Former numbers and distribution of the Australian Aborigines', in *Official Year Book of the Commonwealth of Australia*, No. 23. Melbourne: Government Printer.

RAMOS, A. (1961), *Introdução a Antropologia Brasiliera*, 3rd ed., vol. 3. Livraria editora da casa do estudante do Brasil.

REED, T. E. (1968), 'Research on blood groups and selection from the child health and development studies Oakland, California. III. Couple mating type and reproductive performance', *Am. J. hum. Genet.*, **20**, 129.

REED, T. E. (1969a), 'Critical tests of hypotheses for race mixture using Gm data on American Caucasians and Negroes', *Am. J. hum. Genet.*, **21**, 71.

REED, T. E. (1969b), 'Caucasian genes in American Negroes', *Science, N.Y.*, **165**, 762.

RENSCH, B. (1929), *Das Prinzip geographischer Rassenkreise und das problem der Artbildung*. Berlin: Boentraeger.

RICHMOND, R. C. (1972), 'Enzyme variability in the *Drosophila Willistoni* group. III. Amounts of variability in the superspecies *Drosophila paulistorum*', *Genetics*, **70**, 87.

RIEGER, R., MICHAELIS, A., and GREEN, M. M. (1968), *A Glossary of Genetics and Cytogenetics*. New York: Springer-Verlag.

RIFE, D. C. (1954a), 'Distributions of skin pigmentation, dermatoglyphics, tasting ability and ABO blood groups within mixed Negro-white populations', *Acta. genet. med. Gemell.*, **3**, 259.

RIFE, D. C. (1954b), 'Populations of hybrid origin as a source material for the detection of linkage', *Am. J. hum. Genet.*, **6**, 26.

RIVERS, W. H. R. (1901), *Report of the Cambridge Anthropological Expedition to Torres Straits*, vol. 2. London: Cambridge University Press.

RIVERS, W. H. R. (1905), 'Observations on the senses of the Todas', *Br. J. Psych.*, **1**, 321.

ROBERTS, D. F. (1953), 'Body weight, race and climate', *Am. J. phys. Anthrop.*, **11**, 533.

ROBERTS, D. F. (1955), 'Dynamics of racial intermixture in the American Negro—some anthropological considerations', *Am. J. hum. Genet.*, **7**, 361.

ROBERTS, D. F., and HIORNS, R. W. (1962), 'The dynamics of racial intermixture', *Am. J. hum. Genet.*, **14**, 261.

ROBERTS, J. A. F. (1940), 'Studies on a child population. 5. The resemblance in intelligence between sibs', *Ann. Eugen.*, **10**, 293.

ROBSON, M. K., and PARSONS, P. A. (1967), 'Fingerprint studies on four central Australian aboriginal tribes', *Archaeol. phys. Anthrop. Oceania*, 69.

RUCKNAGEL, D. L., and LAROS, R. K. (1969), 'Haemoglobinopathies: genetics and implications for studies of human reproduction', *Clin. Obstet. Gynec.*, **12**, 49.

RUDDLE, F. H., RODERICK, T. H., SHOWS, T. B., WEIGH, P. G., CHIPMAN, R. K., and ANDERSON, P. K. (1969), 'Measurement of genetic heterogenicity by means of enzyme polymorphism in wild populations of the mouse', *J. Hered.*, **60**, 321.

SALDANHA, P. H. (1957), 'Gene flow from White into Negro populations in Brazil', *Am. J. hum. Genet.*, **9**, 299.

SALDANHA, P. H. (1960), 'Race mixture among northeastern Brazilian populations', *Am. Anthrop.*, **64**, 751.

SALDANHA, P. H. (1961), 'The effect of immigration on the incidence of some congenital abnormalities and on the distribution of stature in a rural community of Sao Paulo, Brazil', *Proc. 2nd int. Congr. Hum. genet.*, **73**. Rome: Excerpta Medica.

SALDANHA, P. H. (1962a), 'Frequency of congenital malformations in a mixed population from southern Brazil', *Proc. 2nd int. Congr. Congen. Malformations*, 323. New York: International Medical Congress.

SALDANHA, P. H. (1962b), 'The genetic effects of immigration in a rural community of Sao Paulo, Brazil', *Acta Genet. Med. Gemel.*, **11**, 158.

SALZANO, F. M., DA ROCHA, F. J., and TONDO, C. V. (1968), 'Haemoglobin types and gene flow in Pôrto Alegre, Brazil', *Acta Genet.*, **18**, 449.

SALZANO, F. M., and FREIRE-MAIA, N. (1970), *Problems in Human Biology. A Study of Brazilian Populations.* Detroit: Wayne State University Press.

SALZANO, F. M., and HIRSCHFELD, H. (1965), 'The dynamics of the Gc polymorphism in a Brazilian population', *Acta Genet.*, **15**, 116.

SANGHVI, L. D., KIRK, R. L., and BALAKRISHNAN, V. (1971), 'A study of genetic distance among some populations of Australian aborigines', *Hum. Biol.*, **43**, 445.

SARGENT, C. W., WESTFALL, C. H. P., and ADAMS, F. M. (1958), 'The obstetric risk of the Japanese woman with a Caucasian husband', *Am. J. Obstet. Gynec.*, **76**, 137.

SARKISSIAN, I. V., and MCDANIEL, R. G. (1967), 'Mitochondrial polymorphism in maize. 1. Putative evidence for de novo origin of hybrid-specific mitochondria', *Proc. natn. Acad. Sci. U.S.A.*, **57**, 1262.

SCHLAGER, G., and DICKIE, M. M. (1971), 'Natural mutation rates in the house mouse. Estimates for five specific loci and dominant mutations', *Mutation Res.*, **11**, 89.

SCHREIDER, E. (1967), 'Body height and inbreeding in France', *Am. J. phys. Anthrop.*, **12**, 1.

SCHULL, W. J., and NEEL, J. V. (1965), *The Effects of Inbreeding on Japanese Children.* New York: Harper & Row.

SCHULTZE, L. (1928), 'Zur Kenntnis des Körpers der Hottentotten und Buschmänner', *Denkschr. med-naturw. Ges. Jena*, **17**, 145.

SCHWARTZ, D. (1960), 'Genetic studies on mutant enzymes in maize: synthesis of hybrid enzymes by heterozygotes', *Proc. natn. Acad. Sci. U.S.A.*, **46**, 1210.

SCHWARTZ, D. (1964), 'A second hybrid enzyme in maize', *Proc. natn. Acad. Sci. U.S.A.*, **51**, 602.

SEARLE, A. G. (1959), 'The incidence of anencephaly in a polytypic population', *Ann. hum. Genet.*, **23**, 279.

SEARLE, A. G. (1964), 'Effects of low-level irradiation on fitness and skeletal variation in an inbred mouse strain', *Genetics*, **50**, 1160.

SEARLE, A. G. (1968), *The Comparative Genetics of Coat Colour in Mammals.* London: Logos.

SELANDER, R. K., HUNT, W. G., and YANG, S. Y. (1969), 'Protein polymorphism and genic heterozygosity in two European subspecies of the house mouse', *Evolution*, **23**, 379.

SELANDER, R. K., and YANG, S. Y. (1969), 'Protein polymorphism and genic heterozygosity in a wild population of the house mouse (*Mus musculus*)', *Genetics*, **63**, 653.

SELANDER, R. K., YANG, S. Y., LEWONTIN, R. C., and JOHNSON, W. E. (1970), 'Genetic variation in the horseshoe crab (*Limulus polyphemus*), a phylogenetic relic', *Evolution*, **24**, 462.

SEMLER, I. J., and ISCOE, I. (1963), quoted by SHUEY (1966).
SHAPIRO, H. L. (1929), 'Descendants of the mutineers of the Bounty', *Mem. Bishop Mus.*, **9**, 1. Honolulu.
SHAPIRO, H. L. (1930), 'The physical characters of the Society Islanders', *Mem. Bishop Mus.*, **11**, 273. Honolulu.
SHAW, R. F., and SMITH, A. P. (1969), 'Is Tay-Sachs disease increasing?', *Nature, Lond.*, **224**, 1214.
SHEPPARD, P. M. (1961), 'Some contributions to genetics from the study of the Lepidoptera', *Adv. Genet.*, **10**, 165.
SHIELDS, J. (1962), *Monozygotic Twins brought up Apart and brought up Together.* London: Oxford University Press.
SHUEY, A. M. (1966), *The Testing of Negro Intelligence*, 2nd ed. New York: Social Science Press.
SICK, K., FRYDENBERG, O., and NIELSEN, J. T. (1963), 'Haemoglobin patterns of plaice, flounder and their natural and artificial hybrids', *Nature, Lond.*, **198**, 411.
SIEG, R. (1955), quoted by GATES (1962).
SIMMONS, R. T., GRAYDON, J. J., and TINDALE, N. B. (1964), 'Further blood group genetical studies on Australian aborigines of Bentinck, Mornington and Forsyth Islands and the mainland Gulf of Carpentaria, together with frequencies for natives of the Western Desert, Western Australia', *Oceania*, **35**, 66.
SIMMONS, R. T., GRAYDON, J. J., ZIGAS, V., BAKER, L. L., and GAJDUSEK, H. (1961), 'Studies on Kuru V. A blood group genetical survey of the Kuru region and other parts of Papua—New Guinea'. *Am. J. trop. Med. Hyg.*, **10**, 639.
SIMONS, E. L. (1963), 'A critical reappraisal of tertiary primates', in *Evolutionary and Genetic Biology of Primates* (ed. Buettner-Janusch), vol. 1. New York: Academic Press.
SIMONS, M. J. (1971), 'ANZ—ASIA: Proposals for human population studies', *Search*, **2**, 369.
SIMPSON, G. G. (1953), *The Major Features of Evolution.* New York: Columbia University Press.
SPUHLER, J. N. (1961), quoted by DOBZHANSKY (1962).
SPUHLER, J. N. (ed.) (1967), *Genetic Diversity and Human Behaviour.* New York: Aldine.
SPUHLER, J. N. (1968), 'Assortative mating with respect to physical characteristics', *Eugen. Q.*, **15**, 128.
SPUHLER, J. N., and LINDZEY, G. (1967), 'Racial differences in behaviour', in *Behaviour Genetic Analysis* (ed. Hirsch). New York: McGraw-Hill.
STAINBROOK, E., and SIEGEL, P. S. (1944), quoted by SHUEY (1966).
STEGGERDA, M. (1932a), 'Physical measurements of Dutch men and women', *Am. J. phys. Anthrop.*, **16**, 309.
STEGGERDA, M. (1932b), 'Anthropometry of adult Maya Indian. A study of their physical and physiological characteristics', *Publ. Carnegie Instn Wash.*, 434. Washington, D.C.
STEINBERG, A. G. (1969), 'Globulin polymorphisms in man', *A. Rev. Genet.*, **3**, 25.
STEINBERG, A. G., and KIRK, R. L. (1970), 'Gm and Inv types of aborigines in the Northern Territory of Australia', *Archaeol. phys. Anthrop. Oceania*, **5**, 163.
STERN, C. (1953), 'Model estimates of the frequency of white and near-white segregants in the American Negro', *Acta. genet.*, **4**, 281.

STERN, C. (1960), *Principles of Human Genetics*, 2nd ed. San Francisco: Freeman.

STERN, C. (1970), 'Model estimates of the number of gene pairs involved in pigmentation variability of the Negro-American', *Hum. Hered.*, **20**, 165.

STEVENSON, A. C., JOHNSTON, H. A., GOLDING, D. R., and STEWART, M. I. P. (1966), *Comparative Study of Congenital Malformations: Basic Tabulations in respect of Consecutive Post 28-week Births recorded in the Cooperating Centres*. Oxford: Medical Research Council.

STEVENSON, A. C., JOHNSTON, H. A., STEWART, M. I. P., and GOLDING, D. R. (1966), 'Congenital Malformations', *Bull. W.H.O.*, Suppl. to vol. 34.

STINCHCOMBE, A. L. (1969), 'Environment: The cumulation of effects is yet to be understood', *Harvard. educ. Rev.*, **39**, 511.

STONE, W. S., KOJIMA, K., and JOHNSON, F. M. (1968), 'Enzyme polymorphism in animal populations', *Proc. 12th int. Congr. Genet.*, **2**, 153. Tokyo: The Science Council of Japan.

STRICKBERGER, M. W. (1968), *Genetics*. New York: MacMillan.

STROUHAL, E. (1971), 'Anthropometric and functional evidence of heterosis from Egyptian Nubia', *Hum. Biol.*, **43**, 271.

STUCKERT, R. P. (1964), 'Race mixture: The African ancestry of White Americans', in *Physical Anthropology and Archaeology, Selected Readings* (ed. Hammond). New York: MacMillan.

SULLIVAN, L. R. (1920), 'Anthropometry of the Siouan tribes', *Anthrop. Papers, Am. Mus. nat. Hist.*, **23**, 81.

SUTTER, J., and TRAN-NGOC-TOAN (1957), 'The problem of the structure of isolates and their evolution among human populations', *Cold Spring Harb. Symp. quant. Biol.*, **22**, 379.

TANNER, J. M. (1962), *Growth at Adolescence*, 2nd ed. Oxford: Blackwell.

TANNER, J. M. (1964), 'Human growth and constitution', in *Human Biology* (Harrison, Weiner, Tanner, and Barnicot). London: Oxford University Press.

TERMAN, L. M., and MERRIL, M. (1937), *Measuring Intelligence: A Guide to the Administration of the New Revised Stanford-Binet Tests*. Boston: Houghton Mifflin.

THORNE, A. G. (1971), 'The racial affinities and origins of the Australian aborigines', in *Aboriginal Man and Environment in Australia* (ed. Mulvaney and Golson). Canberra: Australian National University Press.

THORNE, A. G., and MACUMBER, P. G. (1972), 'Discoveries of Pleistocene man at Kow Swamp Australia', *Nature, Lond.*, **238**, 316.

TIWARI, S. C. (1963), 'Studies of crossing between Indians and Europeans', *Ann. hum. Genet.*, **26**, 219.

TOBIAS, P. V. (1955), 'On a Bushman-European hybrid family', *Man*, **54**, 179.

TOBIAS, P. V. (1959), 'Studies on the occipital bone in Africa. 5. The occipital curvature in fossil man and the light it throws on the morphogenesis of the Bushman', *Am. J. phys. Anthrop.*, **17**, 1.

TOBIAS, P. V. (1961), 'Studies on skin reflectance in Bushman-European hybrids', *Proc. 2nd int. Congr. hum. Genet.*, 461. Rome: Istituto G. Mendel.

TOBIAS, P. V. (1966), 'The peoples of Africa south of the Sahara', in *The Biology of Human Adaptability* (ed. Baker and Weiner). London: Oxford University Press.

TOBIAS, P. V. (1967), *Olduvai Gorge*. Vol. 2. *The Cranium and Maxillary Dentition of Australopithecus (Zinjanthropus) boisei*. London: Cambridge University Press.

TOBIAS, P. V. (1970), 'Brain-size, grey matter and race—fact or fiction?', *Am. J. phys. Anthrop.*, **32**, 3.

TOBIAS, P. V. (1971), 'Remains from Cave of Hearths, Makapansgat', *Am. J. phys. Anthrop.*, **34**, 335.

TOBIAS, P. V., and VON KOENIGSWALD, G. H. R. (1964), 'Comparison between the Olduvai hominines and those of Java and some implications for hominid phylogeny', *Nature, Lond.*, **204**, 515.

TONDO, C. V., and SALZANO, F. M. (1962), 'Abnormal haemoglobins in a Brazilian Negro population', *Am. J. hum. Genet.*, **14**, 401.

TREVOR, J. C. (1953), *Race Crossing in Man*, Galton Laboratory Memoirs 36. London: Cambridge University Press.

TROTTER, M., BROMAN, G. E., and PETERSON, R. R. (1960), 'Densities of bones of White and Negro skeletons', *J. Bone Jt Surg.*, **42**, 50.

TROTTER, M., and PETERSON, R. R. (1970), 'Weight of skeleton during postnatal development', *Am. J. phys. Anthrop.*, **33**, 313.

TURNER, C. G. (1969), 'Microevolutionary interpretations from the dentition', *Am. J. phys. Anthrop.*, **30**, 421.

TURNER, J. R. G. (1971), 'Two thousand generations of hybridisation in a *Heliconius* butterfly', *Evolution*, **25**, 471.

VAN VALEN, L., and MELLIN, G. W. (1967), 'Selection in natural populations. 7, New York babies', *Ann. hum. Genet.*, **31**, 109.

VERNON, P. E. (1966), 'Development of current ideas about intelligence tests', in *Genetic and Environmental Factors in Human Ability* (ed. Meade and Parkes). Edinburgh: Oliver & Boyd.

VERNON, P. E. (1969), *Intelligence and Cultural Environment*. London: Methuen.

VETUKHIV, M. (1956), 'Fecundity of hybrids between geographical populations of *Drosophila pseudoobscura*', *Evolution*, **10**, 139.

VINT, F. W. (1934), 'The brain of the Kenya native', *J. Anat.*, **68**, 216.

VIRCHOW (1872), quoted by IVANHOE (1970).

VOGEL, F., and CHAKRAVARTTI, M. R. (1966), 'ABO bloodgroups and smallpox in a rural population of West Bengal and Bihar (India)', *Humangenetik*, **3**, 166.

VON ECONOMO, C. (1929), *The Cyto-architectonics of the Human Cerebral Cortex*. London: Oxford University Press.

WAARDENBERG, P. J. (1961), in *Genetics and Ophthalmology* (ed. Waardenberg, Franschetti, and Klein). Springfield, Ill.: Thomas.

WADDINGTON, C. H. (1957), *The Strategy of the Genes*. London: Allen & Unwin.

WALKER, A. C. (1969), 'Albinism in a full blood Aboriginal child', *Med. J. Aust.*, **2**, 1105.

WALLACE, B. (1954), 'Inter-population hybrids in *Drosophila melanogaster*', *Evolution*, **9**, 302.

WALLACE, B. (1966), 'On the dispersal of *Drosophila*', *Am. Nat.*, **100**, 551.

WALLACE, B. (1968), *Topics in Population Genetics*. New York: Norton.

WALLACE, B., and VETUKHIV, M. (1955), 'Adaptive organisation of the gene pools of *Drosophila* populations', *Cold Spring Harb. Symp. quant. Biol.*, **20**, 303.

WALLIS, W. D. (1938), 'Variability in race hybrids', *Am. Anthrop.*, **40**, 680.

WALSH, R. J. (1971), 'Pigment of the skin in New Guinea indigenes', *Ann. hum. Genet.*, **34**, 379.

WALSH, R. J., and PRICE, A. V. G. (1963), 'Studies in melanin pigmentation of the skin of racial crosses in Port Moresby', *Oceania*, **33**, 287.

WASHBURN, S. L. (1963), 'The study of race', *Am. Anthrop.*, **45**, 521.

WASHBURN, S. L. (ed.) (1964), *Classification and Human Evolution.* London: Methuen.

WATKIN, I. M. (1956), 'ABO blood groups and racial characteristics in rural Wales', *Heredity*, **10**, 161.

WATKIN, I. M. (1960), 'A Viking settlement in Little England beyond Wales: ABO bloodgroup evidence', *Man*, **60**, 1.

WATKIN, I. M. (1965), 'ABO Blood-groups, human history and language in Herefordshire with special reference to the low B frequency in Europe', *Heredity*, **20**, 83.

WECHSLER, D. (1955), *Manual for the Wechsler Adult Intelligence Scale.* New York: The Psychological Corporation.

WEIDENREICH, F. (1945), 'Giant early man from Java and South China', *Anthrop. Pap. Am. Mus. nat. Hist.*, **43**, 205.

WEINER, A. S. (1948), 'A classification of the ABO and Rh blood systems', *Am. J. phys. Anthrop.*, **6**, 236.

WEINER, J. S. (1951), 'A spectrophotometer for measurement of skin colour', *Man*, **51**, 253.

WEINER, J. S., HARRISON, G. A., SINGER, R., HARRIS, R., and JOPP, W. (1964), 'Skin colour in S. Africa', *Hum. Biol.*, **36**, 294.

WEINER, J. S., SEBAG-MONTEFIORE, M., and PETERSON, J. N. (1963), 'A note on the skin colour of Aguarana Indians of Peru', *Hum. Biol.*, **35**, 470.

WELLHAUSEN, E. J., ROBERTS, L. M., HERNANDEZ, E. X., and MANGELSDORF, E. (1952), *Races of maize in Mexico.* Cambridge, Mass.: The Bussey Inst., Harvard Univ.

WILLERMAN, L., NAYLOR, A. F., and MYRIANTHOPOULOS, M. C. (1970), 'Intellectual development of children from interracial matings', *Science, N.Y.*, **170**, 1329.

WILLIAMS, G. D. (1931), 'Maya-Spanish crosses in Yucatan', *Pap. Peabody Mus.*, 13.

WISSLER, C. (1930), 'Growth of children in Hawaii: based on observations by Louis R. Sullivan', *Mem. Bishop Mus.*, **11**, 105–257.

WITKOP, C. J. (1954), 'A study of tri-racial isolates in Eastern United States', *Acta genet.*, **6**, 410.

WITKOP, C. J. (1971), 'Albinism' in *Advances in Human Genetics*, vol. 2 (ed. Harris and Hirschorn). New York-London: Plenum Press.

WITTY, P. A., and JENKINS, M. D. (1936), 'Intra-race testing and Negro intelligence', *J. Psychol.*, **1**, 179.

WOLANSKI, N., JAROSZ, E., and PYZUK, M. (1970), 'Heterosis in man: growth in offspring and distance between parents birthplaces', *Social Biol.*, **17**, 1.

WOLPOFF, M. H. (1969), 'Cranial capacity and taxonomy of Olduvai Hominid 7. *Nature, Lond.*, **223**, 182.

WOODS, R. C., and MARTIN, C. R. (1943), 'Testing in musical education', *Educ. psychol. Measmt*, **3**, 29.

WOODWORTH, R. S. (1910), 'Racial differences in mental traits', *Science, N.Y.*, **31**, 171.

WOOLF, C. M., and DUKEPOO, F. C. (1969), 'Hopi Indians, inbreeding and albinism', *Science, N.Y.*, 164.

WORKMAN, P. L. (1968), 'Gene flow and the search for natural selection in man', *Hum. Biol.*, **40**, 260.

WORKMAN, P. L., BLUMBERG, B. S., and COOPER, A. J. (1963), 'Selection, gene migration and polymorphic stability in a U.S. White and Negro population', *Am. J. hum. Genet.*, **15**, 429.

WURM, S. A. (1964), 'Australian New Guinea highland languages and the distribution of their typological features', *Am. Anthrop.*, **66**, 77.

WURM, S. A. (1969), 'Classification of Australian languages including Tasmania', in *Current Trends in Linguistics*. 8, *Linguistics in Oceania*. Leiden: Mouton et Cie.

YAMAGUCHI, B. (1967), 'A comparative osteological study of the Ainu and the Australian Aborigine', Occ. Pap. No. 10 Australian Institute of Aboriginal Studies, Canberra.

YASUDA, N., and MORTON, N. E. (1967), 'Studies on human population structure', *Proc. 3rd int. Congr. hum. Genet.*, 249. Baltimore: Johns Hopkins Press.

YOUNG, J. Z. (1971), *An Introduction to the Study of Man*. London: Oxford University Press.

AUTHOR INDEX

The names of all authors are given in this index although only the first one of a citation may appear in the text.

SUBJECT INDEX